Integral Equation Methods in Scattering Theory

Books in the Classics in Applied Mathematics series are monographs and textbooks declared out of print by their original publishers, though they are of continued importance and interest to the mathematical community. SIAM publishes this series to ensure that the information presented in these texts is not lost to today's students and researchers.

Classics in Applied Mathematics

C. C. Lin and L. A. Segel, *Mathematics Applied to Deterministic Problems in the Natural Sciences*

Johan G. F. Belinfante and Bernard Kolman, *A Survey of Lie Groups and Lie Algebras with Applications and Computational Methods*

James M. Ortega, *Numerical Analysis: A Second Course*

Anthony V. Fiacco and Garth P. McCormick, *Nonlinear Programming: Sequential Unconstrained Minimization Techniques*

F. H. Clarke, *Optimization and Nonsmooth Analysis*

George F. Carrier and Carl E. Pearson, *Ordinary Differential Equations*

Leo Breiman, *Probability*

R. Bellman and G. M. Wing, *An Introduction to Invariant Imbedding*

Abraham Berman and Robert J. Plemmons, *Nonnegative Matrices in the Mathematical Sciences*

Olvi L. Mangasarian, *Nonlinear Programming*

*Carl Friedrich Gauss, *Theory of the Combination of Observations Least Subject to Errors: Part One, Part Two, Supplement*. Translated by G. W. Stewart

U. M. Ascher, R. M. M. Mattheij, and R. D. Russell, *Numerical Solution of Boundary Value Problems for Ordinary Differential Equations*

K. E. Brenan, S. L. Campbell, and L. R. Petzold, *Numerical Solution of Initial-Value Problems in Differential-Algebraic Equations*

Charles L. Lawson and Richard J. Hanson, *Solving Least Squares Problems*

J. E. Dennis, Jr. and Robert B. Schnabel, *Numerical Methods for Unconstrained Optimization and Nonlinear Equations*

Richard E. Barlow and Frank Proschan, *Mathematical Theory of Reliability*

Cornelius Lanczos, *Linear Differential Operators*

Richard Bellman, *Introduction to Matrix Analysis, Second Edition*

Beresford N. Parlett, *The Symmetric Eigenvalue Problem*

Richard Haberman, *Mathematical Models: Mechanical Vibrations, Population Dynamics, and Traffic Flow*

Peter W. M. John, *Statistical Design and Analysis of Experiments*

Tamer Başar and Geert Jan Olsder, *Dynamic Noncooperative Game Theory, Second Edition*

Emanuel Parzen, *Stochastic Processes*

Petar Kokotović, Hassan K. Khalil, and John O'Reilly, *Singular Perturbation Methods in Control: Analysis and Design*

*First time in print.

Classics in Applied Mathematics (continued)

Jean Dickinson Gibbons, Ingram Olkin, and Milton Sobel, *Selecting and Ordering Populations: A New Statistical Methodology*

James A. Murdock, *Perturbations: Theory and Methods*

Ivar Ekeland and Roger Témam, *Convex Analysis and Variational Problems*

Ivar Stakgold, *Boundary Value Problems of Mathematical Physics, Volumes I and II*

J. M. Ortega and W. C. Rheinboldt, *Iterative Solution of Nonlinear Equations in Several Variables*

David Kinderlehrer and Guido Stampacchia, *An Introduction to Variational Inequalities and Their Applications*

F. Natterer, *The Mathematics of Computerized Tomography*

Avinash C. Kak and Malcolm Slaney, *Principles of Computerized Tomographic Imaging*

R. Wong, *Asymptotic Approximations of Integrals*

O. Axelsson and V. A. Barker, *Finite Element Solution of Boundary Value Problems: Theory and Computation*

David R. Brillinger, *Time Series: Data Analysis and Theory*

Joel N. Franklin, *Methods of Mathematical Economics: Linear and Nonlinear Programming, Fixed-Point Theorems*

Philip Hartman, *Ordinary Differential Equations, Second Edition*

Michael D. Intriligator, *Mathematical Optimization and Economic Theory*

Philippe G. Ciarlet, *The Finite Element Method for Elliptic Problems*

Jane K. Cullum and Ralph A. Willoughby, *Lanczos Algorithms for Large Symmetric Eigenvalue Computations, Vol. I: Theory*

M. Vidyasagar, *Nonlinear Systems Analysis, Second Edition*

Robert Mattheij and Jaap Molenaar, *Ordinary Differential Equations in Theory and Practice*

Shanti S. Gupta and S. Panchapakesan, *Multiple Decision Procedures: Theory and Methodology of Selecting and Ranking Populations*

Eugene L. Allgower and Kurt Georg, *Introduction to Numerical Continuation Methods*

Leah Edelstein-Keshet, *Mathematical Models in Biology*

Heinz-Otto Kreiss and Jens Lorenz, *Initial-Boundary Value Problems and the Navier-Stokes Equations*

J. L. Hodges, Jr. and E. L. Lehmann, *Basic Concepts of Probability and Statistics, Second Edition*

George F. Carrier, Max Krook, and Carl E. Pearson, *Functions of a Complex Variable: Theory and Technique*

Friedrich Pukelsheim, *Optimal Design of Experiments*

Israel Gohberg, Peter Lancaster, and Leiba Rodman, *Invariant Subspaces of Matrices with Applications*

Lee A. Segel with G. H. Handelman, *Mathematics Applied to Continuum Mechanics*

Rajendra Bhatia, *Perturbation Bounds for Matrix Eigenvalues*

Barry C. Arnold, N. Balakrishnan, and H. N. Nagaraja, *A First Course in Order Statistics*

Charles A. Desoer and M. Vidyasagar, *Feedback Systems: Input-Output Properties*

Stephen L. Campbell and Carl D. Meyer, *Generalized Inverses of Linear Transformations*

Alexander Morgan, *Solving Polynomial Systems Using Continuation for Engineering and Scientific Problems*

I. Gohberg, P. Lancaster, and L. Rodman, *Matrix Polynomials*

Galen R. Shorack and Jon A. Wellner, *Empirical Processes with Applications to Statistics*

Richard W. Cottle, Jong-Shi Pang, and Richard E. Stone, *The Linear Complementarity Problem*

Rabi N. Bhattacharya and Edward C. Waymire, *Stochastic Processes with Applications*

Robert J. Adler, *The Geometry of Random Fields*

Mordecai Avriel, Walter E. Diewert, Siegfried Schaible, and Israel Zang, *Generalized Concavity*

Rabi N. Bhattacharya and R. Ranga Rao, *Normal Approximation and Asymptotic Expansions*

Françoise Chatelin, *Spectral Approximation of Linear Operators*

(continued)

Classics in Applied Mathematics (continued)

Yousef Saad, *Numerical Methods for Large Eigenvalue Problems, Revised Edition*

Achi Brandt and Oren E. Livne, *Multigrid Techniques: 1984 Guide with Applications to Fluid Dynamics, Revised Edition*

Bernd Fischer, *Polynomial Based Iteration Methods for Symmetric Linear Systems*

Pierre Grisvard, *Elliptic Problems in Nonsmooth Domains*

E. J. Hannan and Manfred Deistler, *The Statistical Theory of Linear Systems*

Françoise Chatelin, *Eigenvalues of Matrices, Revised Edition*

David Colton and Rainer Kress, *Integral Equation Methods in Scattering Theory*

Integral Equation Methods in Scattering Theory

David Colton
University of Delaware
Newark, Delaware

Rainer Kress
Georg-August-Universität Göttingen
Göttingen, Germany

siam.
Society for Industrial and Applied Mathematics
Philadelphia

This SIAM edition is a republication of the work first published by John Wiley & Sons, Inc., in 1983.

This book was published in 1992 by Krieger Publishing Company as a reprint of the original 1983 book published by John Wiley and Sons, Inc.

10 9 8 7 6 5 4 3 2 1

Library of Congress Cataloging-in-Publication Data

Colton, David L., 1943- author.

Integral equation methods in scattering theory / David Colton, University of Delaware, Newark, Delaware, Rainer Kress, Georg-August-Universität Göttingen, Göttingen, Germany.

pages cm

Reprint of: New York : Wiley, ©1983.

Includes bibliographical references and index.

ISBN 978-1-611973-15-0

1. Integral equations. 2. Boundary value problems. 3. Scattering (Mathematics) I. Kress, Rainer, 1941- author. II. Title.

QA431.C59 2013

515'.43--dc23

2013030732

Contents

Preface to the Classics Edition

We are pleased and honored that SIAM has decided to reissue our book *Integral Equation Methods in Scattering Theory* in their Classics in Applied Mathematics series. On this occasion some explanation seems appropriate as to why this work could be considered a "classic" and why interest should remain in a book that has been out of print for a number of years.

When we wrote this book in 1983 the application of integral equations to problems in acoustic and electromagnetic scattering theory was mainly confined to the electrical engineering community and to a rather small group of mathematicians. At that time the primary mathematical focus in scattering theory was based on a rather abstract point of view as exemplified by the seminal monograph *Scattering Theory* by Lax and Phillips. Meanwhile, the field of inverse scattering was almost entirely dominated by physicists interested in quantum scattering and the methods of Gelfand–Levitan and Marchenko. In view of this situation, we thought it was appropriate to attempt to write a book that presented a mathematically rigorous development of the use of integral equation techniques to solve acoustic and electromagnetic scattering problems that would be appealing to classical analysts as well as mathematically sophisticated engineers. In addition we wanted to suggest a direction in inverse scattering theory that emphasized the nonlinear and ill-posed nature of multi-dimensional inverse scattering problems. The result of these considerations was the original edition of this book.

Our book quickly became a standard reference in the field and was soon translated into Russian. This recognition was mainly due to our treatment of the direct scattering problem where the Riesz–Fredholm theory in dual systems was systematically presented together with a careful and rigorous derivation of the mapping properties of surface potentials in Hölder spaces. These results were then applied to an in-depth examination of the direct scattering problem for acoustic and electromagnetic waves. The lasting value of this book lies in the fact that these basic results have not changed in the past thirty years, although nowadays the theory is more often presented in a Sobolev space setting. However, as discussed in our second book, *Inverse Acoustic and Electromagnetic Scattering Theory*, the mapping properties of surface potentials in Sobolev spaces can be derived from the Hölder space results by using a functional analytic tool due to Lax. As far as inverse scattering problems are concerned, the material in our first book is by and large outdated. However, since 1983 the themes of nonlinearity and ill-posedness have dominated the field of inverse scattering theory, and the fruitfulness of these ideas can be seen in our above-mentioned second book on scattering theory, in particular the third edition, which appeared in 2013.

In closing, we would like to alert the reader to several inaccuracies in our presentation on inverse scattering and optimal control problems.

- Our proof of Theorem 6.10 suffers from the same deficiency as the original proof in the monograph by Lax and Phillips and is correct only for convex scattering objects. A correct proof for arbitrary shaped scattering objects is given in our second book on scattering theory.

- Due to regularity issues, Schiffer's proof does not carry over to other boundary conditions, and therefore Theorem 6.11 requires another proof which can again be found in our second book.

- The proofs of Theorems 9.11 and 9.13 are incorrect (as they are in the research paper in which these theorems were originally presented). A correct proof of Theorem 9.11 can be found in the book *Optimization Methods in Electromagnetic Radiation* by Angell and Kirsch, whereas at this time it is an open question as to whether or not Theorem 9.13 is valid.

DAVID COLTON
RAINER KRESS

Newark, Delaware
Göttingen, Germany

Preface

Integral equation methods play a central role in the study of boundary-value problems associated with the scattering of acoustic or electromagnetic waves by bounded obstacles. This is primarily due to the fact that the mathematical formulation of such problems leads to equations defined over unbounded domains, and hence their reformulation in terms of boundary integral equations not only reduces the dimensionality of the problem, but also allows one to replace a problem over an unbounded domain by one over a bounded domain. From a numerical point of view, both of these advantages are extremely important, and in the past four decades a consistent and concentrated effort has been exerted by engineers, physicists, and mathematicians to develop and utilize boundary integral equation methods in scattering theory. In recent years the development of integral equation methods for the direct scattering problem seems to be nearing completion, whereas the use of such an approach to study the inverse scattering problem has progressed to an extent that a "state of the art" survey appears highly desirable. These considerations, combined with the continued scientific interaction between the Department of Mathematical Sciences at the University of Delaware and the Institut für Numerische und Angewandte Mathematik at the University of Göttingen, have motivated us to attempt to present a rigorous and reasonably self-contained treatise on the use of integral equation methods in scattering theory.

In view of the overwhelming amount of literature in the field, we found it necessary at the beginning to set clear and well-defined goals concerning the content of the present monograph. This choice of subject matter has obviously been influenced by our own research interests. Hence, in addition to the basic properties of solutions to the Helmholtz and Maxwell equations, we have decided to emphasize the following themes: (1) the regularity properties of acoustic and electromagnetic potentials; (2) the close relationship between Maxwell's equations, the vector Helmholtz equation, and the scalar Helmholtz equation; (3) the reformulation of the boundary-value problems of scattering theory as integral equations that are uniquely solvable for all values of the wave number; (4) the low frequency behavior of solutions to the boundary-value problems of scattering theory; (5) the use of function theoretic methods to study the inverse scattering problem; (6) the role of compactness in stabilizing the inverse scattering problem; and (7) the use of integral equation methods to reformulate the inverse scattering problem as a problem in constrained optimization, as well as the consideration of various related optimization problems in scattering theory. Although references are made throughout the text to papers concerned with the numerical implementation of our methods, we have decided not to include such material in our presentation. Furthermore, since integral equation methods are basically restricted to scattering problems for low or intermediate values of the wave number, we have chosen not to attempt any treatment of the wealth of material concerned with high frequency methods in scattering theory.

Having formulated the above goals, a major problem arose as to what demands to place on the reader as far as mathematical background was concerned. In order for our book to be accessible to engineers and physicists as well as mathematicians, we have decided to assume only a minimal background in functional analysis and analytic function theory, and to include sections in the text covering the more advanced aspects of these areas that are needed for an understanding of the material on scattering theory. Hence we have included a discussion of such topics as the Riesz–Fredholm theory of compact operators, entire functions of a complex variable, univalent function theory, improperly posed problems, and weak compactness. Hopefully this will make our book digestible to a wider audience than would otherwise have been possible.

The first author would like to gratefully acknowledge financial support from the Air Force Office of Scientific Research under grant AFOSR 81-0103. This book was completed while the second author was on sabbatical leave at the University of Delaware, and both authors would like to thank the University of Delaware and the University of Göttingen for making this visit possible. A particular note of thanks is given to Alison Chandler for her careful typing of the manuscript.

DAVID COLTON
RAINER KRESS

Newark, Delaware
Göttingen, Germany

Symbols

Sets

∂G	Boundary of set G
$\overline{G}$	Closure of set G
G^{ext}	Set of extreme points of set G
$\varnothing$	Empty set
$\mathbb{N}$	Set of natural numbers

Scalars

$\overline{\lim}$	Limit superior
$\underline{\lim}$	Limit inferior
$\bar{z}$	Complex conjugate of z

Linear Spaces

$\mathbb{R}^n$	Euclidean n space
$\mathbb{C}$	Space of complex variables
$(\cdot,\cdot)$	Scalar product
$[\cdot,\cdot]$	Vector product
$(\cdot,\cdot,\cdot)$	Triple product
$\lvert\cdot\rvert$	Euclidean norm of a vector in $\mathbb{R}^n$
$\hat{x}$	Unit vector in $\mathbb{R}^n$

Function Spaces

$C(G)$	Normed space of real or complex-valued continuous functions defined on G
$C^{0,\alpha}(G)$	Normed space of real or complex-valued uniformly Hölder continuous functions defined on G
$L^p(G)$	Set of functions whose pth power is integrable over G in the sense of Lebesgue

$\|\cdot\|$	Norm
$\|\cdot\|_\infty$	Maximum norm
$\langle\cdot,\cdot\rangle$	Bilinear form
	See also the index under **Function Spaces**

Functions

exp	Exponential
log	Logarithm to the base e
χ_G	Characteristic function of set G
$\mu(G)$	Lebesgue measure of set G
O, o	Landau symbols

Operators

Δ	Laplace operator
arg	Argument
Re	Real part
Im	Imaginary part
grad	Gradient
div	Divergence
curl	Curl
Grad	Surface gradient
Div	Surface divergence
	See also the index under ***Operators***

1

THE RIESZ–FREDHOLM THEORY FOR COMPACT OPERATORS

The basic tool used in this book for the investigation of both the direct and the inverse scattering problems for acoustic and electromagnetic waves is the method of boundary integral equations. The integral equations that will arise in the course of our investigations are of the Fredholm type with either a weakly singular or a strongly singular kernel. We will show that in both cases, perhaps after the application of a suitable regularizing operator, the resulting integral equation can be reduced to one of the form

$$(\mathbf{I}-\mathbf{A})\phi = f$$

where $\mathbf{A}$ is a compact integral operator and f is an element of an appropriately chosen Banach space. The questions of existence and uniqueness of solutions to operator equations of this form are answered by the Riesz–Fredholm theory and hence is the subject matter of this chapter. In order to present this theory, it is necessary for us to assume that the reader is familiar with the elementary properties of normed spaces and bounded linear operators. However, aside from this prerequisite we shall try and make the analysis as self-contained as possible.

Our plan of this chapter is as follows. We first introduce the concept of compact operators and study their basic properties. We then present the Riesz theory for linear operator equations of the second kind and follow this by the Fredholm theory in dual systems. A new aspect of our presentation is that we do not assume that the bilinear form associated with the dual system is bounded. After this discussion of the Riesz–Fredholm theory, we shall consider operator equations depending on a parameter as well as singular perturbation problems associated with such equations. These results will be applied in Chapter 5 to study the low frequency behavior of solutions to acoustic and electromagnetic scattering problems. We conclude the chapter by an elementary discussion of spectral theory and the method of successive approximations for solving operator equations of the second kind.

1.1 COMPACT OPERATORS

We begin by introducing the concept of compact operators and the study of their basic properties.

Definition 1.1. A linear operator $\mathbf{A}: X \to Y$ from a normed space X into a normed space Y is called compact if it maps any bounded set in X into a relatively compact set in Y.

We recall that a set U in a normed space X is called relatively compact if its closure is compact, that is, if each sequence in U contains a subsequence converging in X. Therefore we have the following equivalent condition for an operator to be compact.

Theorem 1.2. A linear operator $\mathbf{A}: X \to Y$ is compact if and only if for each bounded sequence (ϕ_n) in X the sequence $(\mathbf{A}\phi_n)$ contains a convergent subsequence.

Theorem 1.3. Compact linear operators are bounded.

Proof. Assume the compact linear operator $\mathbf{A}: X \to Y$ is not bounded. Then there exists a sequence (ϕ_n) in X such that $\|\phi_n\| = 1$ and $\|\mathbf{A}\phi_n\| \geqslant n$ for all $n \in \mathbb{N}$. Since $\mathbf{A}$ is compact, there exists a subsequence such that $\mathbf{A}\phi_{n(k)} \to \psi \in Y$, $k \to \infty$. In particular, $\|\mathbf{A}\phi_{n(k)}\| \to \|\psi\|$, $k \to \infty$, which is a contradiction to $\|\mathbf{A}\phi_{n(k)}\| \geqslant n(k) \to \infty$, $k \to \infty$.

Theorem 1.4. Any linear combination of compact linear operators is compact.

Proof. Let $\mathbf{A}, \mathbf{B}: X \to Y$ be compact linear operators and let $\alpha, \beta \in \mathbb{C}$. Then each bounded sequence (ϕ_n) in X contains a subsequence $(\phi_{n(k)})$ such that $(\mathbf{A}\phi_{n(k)})$ and $(\mathbf{B}\phi_{n(k)})$ converge. Hence $(\alpha\mathbf{A} + \beta\mathbf{B})\phi_{n(k)} = \alpha\mathbf{A}\phi_{n(k)} + \beta\mathbf{B}\phi_{n(k)}$ converges and therefore $\alpha\mathbf{A} + \beta\mathbf{B}$ is compact.

Theorem 1.5. Let X, Y, Z be normed spaces and let $\mathbf{A}: X \to Y$ and $\mathbf{B}: Y \to Z$ be bounded linear operators. Then the product $\mathbf{BA}: X \to Z$ is compact if one of the two operators $\mathbf{A}$ or $\mathbf{B}$ is compact.

Proof. Let (ϕ_n) be a bounded sequence in X: $\|\phi_n\| \leqslant C$ for all $n \in \mathbb{N}$. If $\mathbf{A}$ is compact, then there exists a subsequence $(\phi_{n(k)})$ such that $\mathbf{A}\phi_{n(k)} \to \psi \in Y$, $k \to \infty$. Since $\mathbf{B}$ is bounded and therefore continuous, we have $(\mathbf{BA})\phi_{n(k)} = \mathbf{B}(\mathbf{A}\phi_{n(k)}) \to \mathbf{B}\psi \in Z$, $k \to \infty$. Hence $\mathbf{BA}$ is compact.

If $\mathbf{A}$ is bounded and $\mathbf{B}$ compact, the sequence $(\mathbf{A}\phi_n)$ in Y is bounded since $\|\mathbf{A}\phi_n\| \leqslant \|\mathbf{A}\|\|\phi_n\| \leqslant \|\mathbf{A}\|C$ for all $n \in \mathbb{N}$. Therefore, there exists a subsequence

$(\phi_{n(k)})$ such that $(\mathbf{BA})\phi_{n(k)} = \mathbf{B}(\mathbf{A}\phi_{n(k)}) \to \chi \in Z$, $k \to \infty$. Hence again $\mathbf{BA}$ is compact.

Theorem 1.6. Let X be a normed space and Y be a Banach space. Let the sequence $\mathbf{A}_n \colon X \to Y$ of compact linear operators converge uniformly to a linear operator $\mathbf{A} \colon X \to Y$, that is, $\|\mathbf{A}_n - \mathbf{A}\| \to 0$, $n \to \infty$. Then $\mathbf{A}$ is compact.

Proof. Let (ϕ_m) be a bounded sequence in $X \colon \|\phi_m\| \leqslant C$ for all $m \in \mathbb{N}$. Since the $\mathbf{A}_n$ are compact, we can choose, by the standard diagonalization procedure, a subsequence $(\phi_{m(k)})$ such that $(\mathbf{A}_n\phi_{m(k)})$ converges for every fixed n as $k \to \infty$. More precisely, since $\mathbf{A}_1$ is compact, we can choose a subsequence $(\phi_{m_1(k)})$ such that $(\mathbf{A}_1\phi_{m_1(k)})$ converges as $k \to \infty$. The sequence $(\phi_{m_1(k)})$ again is bounded and, since $\mathbf{A}_2$ is compact, we can choose a subsequence $(\phi_{m_2(k)})$ of $(\phi_{m_1(k)})$ such that $(\mathbf{A}_2\phi_{m_2(k)})$ converges as $k \to \infty$. Repeating this process of selecting subsequences, we arrive at a double array $(\phi_{m_n(k)})$ such that each row $(\phi_{m_n(k)})$ is a subsequence of the previous row $(\phi_{m_{n-1}(k)})$ and each sequence $(\mathbf{A}_n\phi_{m_n(k)})$ converges as $k \to \infty$. For the diagonal sequence $\phi_{m(k)} := \phi_{m_k(k)}$ we have $(\mathbf{A}_n\phi_{m(k)})$ converges as $k \to \infty$ for all $n \in \mathbb{N}$.

Let $\varepsilon > 0$ be arbitrary. Since $\|\mathbf{A}_n - \mathbf{A}\| \to 0$, $n \to \infty$, there exists $n_0 \in \mathbb{N}$ such that $\|\mathbf{A}_{n_0} - \mathbf{A}\| < \varepsilon/3C$. Because $(\mathbf{A}_{n_0}\phi_{m(k)})$ converges there exists $N(\varepsilon) \in \mathbb{N}$ such that

$$\|\mathbf{A}_{n_0}\phi_{m(k)} - \mathbf{A}_{n_0}\phi_{m(l)}\| < \frac{\varepsilon}{3}, \qquad k, l \geqslant N(\varepsilon).$$

But then for all $k, l \geqslant N(\varepsilon)$ we have

$$\|\mathbf{A}\phi_{m(k)} - \mathbf{A}\phi_{m(l)}\| \leqslant \|\mathbf{A}\phi_{m(k)} - \mathbf{A}_{n_0}\phi_{m(k)}\|$$

$$+ \|\mathbf{A}_{n_0}\phi_{m(k)} - \mathbf{A}_{n_0}\phi_{m(l)}\| + \|\mathbf{A}_{n_0}\phi_{m(l)} - \mathbf{A}\phi_{m(l)}\| < \varepsilon.$$

Thus $(\mathbf{A}\phi_{m(k)})$ is a Cauchy sequence and therefore convergent in the Banach space Y.

Theorem 1.7. Let $\mathbf{A} \colon X \to Y$ be a bounded linear operator with finite dimensional range $\mathbf{A}(X)$. Then $\mathbf{A}$ is compact.

Proof. Let (ϕ_n) be a bounded sequence in $X \colon \|\phi_n\| \leqslant C$ for all $n \in \mathbb{N}$. Then, since $\|\mathbf{A}\phi_n\| \leqslant \|\mathbf{A}\|\|\phi_n\| \leqslant \|\mathbf{A}\|C$ the sequence $(\mathbf{A}\phi_n)$ is bounded in the finite dimensional subspace $\mathbf{A}(X)$. By the Bolzano–Weierstrass theorem any bounded sequence in a finite dimensional normed space contains a convergent subsequence. Therefore $\mathbf{A}$ is compact.

Lemma 1.8 (Riesz). Let X be a normed space, $U \subsetneqq X$ a closed subspace, and $\alpha \in (0,1)$. Then there exists an element $\psi \in X$ with $\|\psi\| = 1$ such that

$$\|\psi - \phi\| \geqslant \alpha$$

for all $\phi \in U$.

Proof. Since $U \neq X$, there exists an element $f \in X$ with $f \notin U$ and because U is closed we have

$$\beta := \inf_{\phi \in U} \|f - \phi\| > 0.$$

We can choose $g \in U$ such that

$$\beta \leqslant \|f - g\| \leqslant \frac{\beta}{\alpha}.$$

Now we define

$$\psi := \frac{f - g}{\|f - g\|}.$$

Then $\|\psi\| = 1$ and for all $\phi \in U$ we have

$$\|\psi - \phi\| = \frac{1}{\|f - g\|}\|f - (g + \|f - g\|\phi)\| \geqslant \frac{\beta}{\|f - g\|} \geqslant \alpha$$

since $g + \|f - g\|\phi \in U$.

Theorem 1.9. The identity operator $\mathbf{I}: X \to X$ is compact if and only if X has finite dimension.

Proof. Assume $\mathbf{I}$ is compact and X is not finite dimensional. Choose an arbitrary $\phi_1 \in X$ with $\|\phi_1\| = 1$. Then $U_1 := \text{span}\,(\phi_1)$ is a finite dimensional and therefore closed subspace of X. By Lemma 1.8 there exists $\phi_2 \in X$ with $\|\phi_2\| = 1$ and $\|\phi_2 - \phi_1\| \geqslant \frac{1}{2}$. Now consider $U_2 := \text{span}\,(\phi_1, \phi_2)$. Again by Lemma 1.8 there exists $\phi_3 \in X$ with $\|\phi_3\| = 1$ and $\|\phi_3 - \phi_1\| \geqslant \frac{1}{2}$, $\|\phi_3 - \phi_2\| \geqslant \frac{1}{2}$. Repeating this procedure, we obtain a sequence (ϕ_n) with the properties $\|\phi_n\| = 1$ and

$$\|\phi_n - \phi_m\| \geqslant \tfrac{1}{2}, \qquad n \neq m.$$

This implies that the bounded sequence (ϕ_n) does not contain a convergent subsequence which contradicts the compactness of the identity operator. Therefore, if the identity operator is compact, X has finite dimension.

The converse is an immediate consequence of Theorem 1.7.

This theorem, in particular, implies that the converse of Theorem 1.3 is false.

Let $G \subset \mathbb{R}^2$ be a Jordan-measurable (with nonzero measure) and compact set and let $C(G)$ be the Banach space of complex-valued continuous functions defined in G equipped with the maximum norm

$$\|\phi\|_\infty := \max_{x \in G} |\phi(x)|.$$

Consider the integral operator $\mathbf{A}: C(G) \to C(G)$ defined by

$$(\mathbf{A}\phi)(x) := \int_G K(x, y)\phi(y)\,dy, \qquad x \in G, \tag{1.1}$$

where $K: G \times G \to \mathbf{C}$ is a continuous kernel.

Theorem 1.10. The integral operator $\mathbf{A}$ with continuous kernel is a compact operator on $C(G)$.

Proof. Choose a sequence of subdivisions

$$G = \bigcup_{i=1}^{n} \overline{\Delta_{i,n}}$$

such that the measurable open sets $\Delta_{i,n}$ are disjoint, $\Delta_{i,n} \cap \Delta_{j,n} = \varnothing$, $i \neq j$, and the diameters satisfy

$$\max_{1 \leqslant i \leqslant n} (\operatorname{diam} \Delta_{i,n}) \to 0, \qquad n \to \infty. \tag{1.2}$$

Select a point $y_{i,n}$ from each $\Delta_{i,n}$ and consider the sequence of operators $\mathbf{A}_n: C(G) \to C(G)$ defined by

$$(\mathbf{A}_n\phi)(x) := \sum_{i=1}^{n} K(x, y_{i,n}) \int_{\Delta_{i,n}} \phi(y)\,dy.$$

The $\mathbf{A}_n$ are bounded

$$\|\mathbf{A}_n\|_\infty \leqslant \max_{x, y \in G} |K(x, y)| \int_G dy$$

and have finite dimensional range

$$\mathbf{A}_n(C(G)) = \operatorname{span}(K(\cdot, y_{i,n}), \qquad i = 1, \ldots, n).$$

Thus by Theorem 1.7 the operators $\mathbf{A}_n$ are compact.

Define $K_n: G \times G \to \mathbf{C}$ by

$$K_n(x, y) := K(x, y_{i,n}), \qquad x \in G, \qquad y \in \Delta_{i,n}.$$

Then we can rewrite the definition of $\mathbf{A}_n$ in the form

$$(\mathbf{A}_n\phi)(x) = \int_G K_n(x, y)\phi(y)\,dy.$$

Since K is uniformly continuous on the compact set $G\times G$, given $\varepsilon>0$, it follows from (1.2) that there exists $N(\varepsilon)\in\mathbb{N}$ such that

$$|K(x,y)-K_n(x,y)|<\frac{\varepsilon}{\int_G dy},\qquad x,y\in G,\qquad n\geqslant N(\varepsilon).$$

Then

$$|(\mathbf{A}\phi)(x)-(\mathbf{A}_n\phi)(x)|\leqslant\varepsilon\|\phi\|_\infty,\qquad x\in G$$

and therefore $\|\mathbf{A}-\mathbf{A}_n\|_\infty\leqslant\varepsilon$ for all $n\geqslant N(\varepsilon)$. Hence, by Theorem 1.6 the operator $\mathbf{A}$ is compact.

Now we consider the integral operator $\mathbf{A}$ defined by (1.1) where K is a weakly singular kernel, that is, K is defined and continuous for all $x,y\in G$, $x\neq y$, and there exist positive constants M and $\alpha\in(0,2]$ such that for all $x,y\in G$, $x\neq y$, we have

$$|K(x,y)|\leqslant M|x-y|^{\alpha-2}. \tag{1.3}$$

By $|x|$ we denote the euclidean norm of a point $x\in\mathbb{R}^2$.

Theorem 1.11. The integral operator $\mathbf{A}$ with weakly singular kernel is a compact operator on $C(G)$.

Proof. The integral in (1.1) defining the operator $\mathbf{A}$ exists as an improper integral since

$$|K(x,y)\phi(y)|\leqslant M\|\phi\|_\infty|x-y|^{\alpha-2}$$

and

$$\int_G|x-y|^{\alpha-2}\,dy\leqslant 2\pi\int_0^d\rho^{\alpha-2}\rho\,d\rho=\frac{2\pi}{\alpha}d^\alpha$$

where we have introduced polar coordinates with origin at x and d is the diameter of G.

Now we introduce piecewise linear continuous functions $k_n\colon[0,\infty)\to\mathbb{R}$, $n\in\mathbb{N}$, by

$$k_n(t):=\begin{cases}0 & 0\leqslant t\leqslant\dfrac{1}{2n}\\[2mm] 2nt-1, & \dfrac{1}{2n}\leqslant t\leqslant\dfrac{1}{n}\\[2mm] 1, & \dfrac{1}{n}\leqslant t<\infty\end{cases}$$

and define continuous kernels $K_n: G\times G\to\mathbf{C}$ by

$$K_n(x,y):=\begin{cases} k_n(|x-y|)K(x,y), & x\neq y,\\ 0, & x=y.\end{cases}$$

The corresponding integral operators $\mathbf{A}_n: C(G)\to C(G)$ are compact by Theorem 1.10. We have the estimate

$$\begin{aligned}|(\mathbf{A}\phi)(x)-(\mathbf{A}_n\phi)(x)| &= \left|\int_G [K(x,y)-K_n(x,y)]\phi(y)\,dy\right|\\ &\leqslant \int_{G_{x,1/n}} |K(x,y)|\,\|\phi\|_\infty\,dy\\ &\leqslant M\|\phi\|_\infty 2\pi\int_0^{1/n}\rho^{\alpha-2}\rho\,d\rho\\ &= M\|\phi\|_\infty\frac{2\pi}{\alpha}\left(\frac{1}{n}\right)^\alpha, \qquad x\in G,\end{aligned}$$

where $G_{x,1/n}:=\{y\in G \mid |y-x|\leqslant 1/n\}$. From this we observe that $\mathbf{A}_n\phi\to\mathbf{A}\phi$, $n\to\infty$, uniformly and therefore $\mathbf{A}\phi\in C(G)$. Furthermore, it follows that

$$\|\mathbf{A}-\mathbf{A}_n\|_\infty\leqslant M\frac{2\pi}{\alpha}\left(\frac{1}{n}\right)^\alpha\to 0, \qquad n\to\infty,$$

and thus $\mathbf{A}$ is compact by Theorem 1.6.

Theorems 1.10 and 1.11 obviously can be extended to Euclidean spaces of arbitrary dimension $\mathbb{R}^s$, where the condition (1.3) is replaced by $|K(x,y)|\leqslant M|x-y|^{\alpha-s}$.

The compactness of integral operators with continuous kernels also can be derived from the following theorem.

Theorem 1.12 (Arzelá–Ascoli). Let $G\subset\mathbb{R}^s$ be a compact set. A set $K\subset C(G)$ is relatively compact (with respect to the maximum norm on $C(G)$) if and only if it is bounded and equicontinuous, that is, there exists a constant C such that

$$|\phi(x)|\leqslant C$$

for all $x\in G$ and all $\phi\in K$ and for any $\varepsilon>0$ there exists $\delta>0$ such that

$$|\phi(x)-\phi(y)|<\varepsilon$$

for all $x, y\in G$ with $|x-y|<\delta$ and all $\phi\in K$.

Proof. Let K be bounded and equicontinuous and let (ϕ_n) be a sequence in K. We choose a sequence (x_m) in G that is dense in G. Since the sequence

$(\phi_n(x_m))$ is bounded in $\mathbb{C}$ for each x_m, by the standard diagonalization procedure (see the proof of Theorem 1.6) we can choose a subsequence $(\phi_{n(k)})$ such that $(\phi_{n(k)}(x_m))$ converges in $\mathbb{C}$ as $k \to \infty$ for each x_m. Because the set (x_m) is dense in G, given $\varepsilon > 0$, we can choose $m \in \mathbb{N}$ such that any point $x \in G$ has a distance less than δ from at least one element x_j of the set $x_1, \ldots, x_m$. Next choose $N(\varepsilon) \in \mathbb{N}$ such that

$$|\phi_{n(k)}(x_j) - \phi_{n(l)}(x_j)| < \varepsilon, \qquad k, l \geqslant N(\varepsilon)$$

and all $j = 1, \ldots, m$. From the equicontinuity we obtain

$$|\phi_{n(k)}(x) - \phi_{n(l)}(x)| \leqslant |\phi_{n(k)}(x) - \phi_{n(k)}(x_j)|$$

$$+ |\phi_{n(k)}(x_j) - \phi_{n(l)}(x_j)| + |\phi_{n(l)}(x_j) - \phi_{n(l)}(x)| < 3\varepsilon$$

for all $k, l \geqslant N(\varepsilon)$ and all $x \in G$ which establishes the uniform convergence, that is, convergence in the maximum norm of the subsequence $(\phi_{n(k)})$. Hence K is relatively compact.

Conversely, let K be relatively compact. Given $\varepsilon > 0$ there exist functions $\phi_1, \ldots, \phi_m \in K$ such that

$$\min_{1 \leqslant j \leqslant m} \|\phi - \phi_j\|_\infty < \frac{\varepsilon}{3}$$

for all $\phi \in K$. Otherwise we inductively could construct a sequence (ϕ_n) in K with the property

$$\|\phi_n - \phi_l\|_\infty \geqslant \frac{\varepsilon}{3}, \qquad n \neq l.$$

This implies that the sequence (ϕ_n) does not contain a convergent subsequence which contradicts the relative compactness of K. Since each of the $\phi_1, \ldots, \phi_m$ is uniformly continuous, there exists $\delta > 0$ such that

$$|\phi_j(x) - \phi_j(y)| < \frac{\varepsilon}{3}$$

for all $x, y \in G$ with $|x - y| < \delta$ and all $j = 1, \ldots, m$. Then for all $\phi \in K$, choosing j such that

$$\|\phi - \phi_j\|_\infty = \min_{1 \leqslant i \leqslant m} \|\phi - \phi_i\|_\infty,$$

we obtain

$$|\phi(x) - \phi(y)| \leqslant |\phi(x) - \phi_j(x)| + |\phi_j(x) - \phi_j(y)| + |\phi_j(y) - \phi(y)| < \varepsilon$$

for all $x, y \in G$ with $|x - y| < \delta$. Therefore K is equicontinuous. Finally, the boundedness of the relatively compact set K is trivial.

The proof of Theorem 1.10 with the aid of the Arzelá–Ascoli theorem is left as an exercise.

1.2 THE RIESZ THEORY

We now present the Riesz [1] theory for an operator equation

$$\phi - \mathbf{A}\phi = f$$

of the second kind. We assume that X is a normed space, $\mathbf{A}: X \to X$ is a compact linear operator and we define

$$\mathbf{L} := \mathbf{I} - \mathbf{A}$$

where $\mathbf{I}$ denotes the identity operator.

Theorem 1.13 (First Riesz Theorem). The nullspace of the operator $\mathbf{L}$

$$N(\mathbf{L}) := \{\phi \in X \mid \mathbf{L}u = 0\}$$

is a finite dimensional subspace.

Proof. The nullspace of the bounded linear operator $\mathbf{L}$ is trivially a closed linear subspace of X. Since for all $\phi \in N(\mathbf{L})$ we have $\mathbf{A}\phi = \phi$, the restriction of $\mathbf{A}$ to the nullspace $N(\mathbf{L})$ coincides with the identity operator $\mathbf{A}|_{N(\mathbf{L})} = \mathbf{I}: N(\mathbf{L}) \to N(\mathbf{L})$. $\mathbf{A}$ is compact on X and therefore also compact on closed linear subspaces of X. Hence $N(\mathbf{L})$ is finite dimensional by Theorem 1.9.

Theorem 1.14 (Second Riesz Theorem). The range of the operator $\mathbf{L}$

$$\mathbf{L}(X) := \{\mathbf{L}\phi \mid \phi \in X\}$$

is a closed linear subspace.

Proof. The range of the linear operator $\mathbf{L}$ is clearly a linear subspace. Let f be an element of the closure $\overline{\mathbf{L}(X)}$. Then there exists a sequence (ϕ_n) in X such that

$$\mathbf{L}\phi_n \to f, \qquad n \to \infty.$$

To each ϕ_n we choose an element $\chi_n \in N(\mathbf{L})$ such that

$$\|\phi_n - \chi_n\| \leqslant \inf_{\chi \in N(\mathbf{L})} \|\phi_n - \chi\| + \frac{1}{n}.$$

The sequence (ϕ_n') defined by

$$\phi_n' := \phi_n - \chi_n$$

is bounded. We prove this indirectly, that is, we assume that it is not bounded. Then there exists a subsequence $(\phi'_{n(k)})$ such that

$$\|\phi'_{n(k)}\| \geqslant k, \qquad k \in \mathbb{N}.$$

Now we define

$$\psi_k := \frac{\phi'_{n(k)}}{\|\phi'_{n(k)}\|}, \qquad k \in \mathbb{N}.$$

Since $\|\psi_k\| = 1$, $k \in \mathbb{N}$, there exists a subsequence $(\psi_{k(j)})$ such that

$$\mathbf{A}\psi_{k(j)} \to \psi \in X, \qquad j \to \infty.$$

Furthermore,

$$\|\mathbf{L}\psi_k\| = \frac{\|\mathbf{L}\phi'_{n(k)}\|}{\|\phi'_{n(k)}\|} \leqslant \frac{\|\mathbf{L}\phi_{n(k)}\|}{k} \to 0, \qquad k \to \infty,$$

since the sequence $(\mathbf{L}\phi_n)$ is convergent and therefore bounded. Hence

$$\mathbf{L}\psi_{k(j)} \to 0, \qquad j \to \infty.$$

We now obtain

$$\psi_{k(j)} = \mathbf{L}\psi_{k(j)} + \mathbf{A}\psi_{k(j)} \to \psi, \qquad j \to \infty,$$

and since $\mathbf{L}$ is bounded from the two previous equations we conclude

$$\mathbf{L}\psi = 0.$$

But then since $\chi_{n(k)} + \|\phi'_{n(k)}\|\psi \in N(\mathbf{L})$ for all k we find

$$\begin{aligned}
\|\psi_k - \psi\| &= \frac{1}{\|\phi'_{n(k)}\|}\|\phi_{n(k)} - \left(\chi_{n(k)} + \|\phi'_{n(k)}\|\psi\right)\| \\
&\geqslant \frac{1}{\|\phi'_{n(k)}\|} \inf_{\chi \in N(\mathbf{L})} \|\phi_{n(k)} - \chi\| \\
&\geqslant \frac{1}{\|\phi'_{n(k)}\|}\left(\|\phi_{n(k)} - \chi_{n(k)}\| - \frac{1}{n(k)}\right) \\
&= 1 - \frac{1}{n(k)\|\phi'_{n(k)}\|} \to 1, \qquad k \to \infty,
\end{aligned}$$

which contradicts the fact that $\psi_{k(j)} \to \psi$, $j \to \infty$.

Therefore the sequence (ϕ'_n) is bounded and we can select a subsequence $(\phi'_{n(k)})$ such that $(\mathbf{A}\phi'_{n(k)})$ converges as $k \to \infty$. From $\phi'_{n(k)} = \mathbf{L}\phi'_{n(k)} + \mathbf{A}\phi'_{n(k)}$ we now observe that $(\phi'_{n(k)})$ converges

$$\phi'_{n(k)} \to \phi \in X, \qquad k \to \infty.$$

But then $\mathbf{L}\phi'_{n(k)} \to \mathbf{L}\phi \in X$ and therefore $f = \mathbf{L}\phi \in \mathbf{L}(X)$. Hence $\overline{\mathbf{L}(X)} = \mathbf{L}(X)$.

The iterated operators $\mathbf{L}^n$, $n \geq 1$, defined by $\mathbf{L}^0 := \mathbf{I}$, $\mathbf{L}^n := \mathbf{L}\mathbf{L}^{n-1}$, can be written in the form

$$\mathbf{L}^n = (\mathbf{I} - \mathbf{A})^n = \mathbf{I} - \mathbf{A}_n$$

where

$$\mathbf{A}_n = \sum_{k=1}^{n} (-1)^{k-1} \binom{n}{k} \mathbf{A}^k$$

is compact by Theorems 1.4 and 1.5. Therefore by Theorem 1.13 the nullspaces $N(\mathbf{L}^n)$ are finite dimensional and by Theorem 1.14 the ranges $\mathbf{L}^n(X)$ are closed subspaces.

Theorem 1.15 (Third Riesz Theorem). There exists a uniquely determined nonnegative integer r, called the Riesz number of the operator $\mathbf{A}$, such that

$$\{0\} = N(\mathbf{L}^0) \subsetneqq N(\mathbf{L}^1) \subsetneqq \cdots \subsetneqq N(\mathbf{L}^r) = N(\mathbf{L}^{r+1}) = \cdots,$$

$$X = \mathbf{L}^0(X) \supsetneqq \mathbf{L}^1(X) \supsetneqq \cdots \supsetneqq \mathbf{L}^r(X) = \mathbf{L}^{r+1}(X) = \cdots.$$

Furthermore,

$$X = N(\mathbf{L}^r) \oplus \mathbf{L}^r(X).$$

Proof.

1. Since for any ϕ with $\mathbf{L}^n\phi = 0$ it follows that $\mathbf{L}^{n+1}\phi = 0$, we trivially have

$$\{0\} = N(\mathbf{L}^0) \subset N(\mathbf{L}^1) \subset N(\mathbf{L}^2) \subset \cdots.$$

Assume now that

$$\{0\} = N(\mathbf{L}^0) \subsetneqq N(\mathbf{L}^1) \subsetneqq N(\mathbf{L}^2) \subsetneqq \cdots.$$

Since the nullspaces $N(\mathbf{L}^n)$ are finite dimensional by Theorem 1.13, from the Riesz Lemma 1.8 we conclude that there exists $\phi_n \in N(\mathbf{L}^{n+1})$ such that $\|\phi_n\| = 1$ and

$$\|\phi_n - \phi\| \geq \tfrac{1}{2}$$

for all $\phi \in N(\mathbf{L}^n)$. For $n > m$ we consider

$$\mathbf{A}\phi_n - \mathbf{A}\phi_m = \phi_n - (\phi_m + \mathbf{L}\phi_n - \mathbf{L}\phi_m).$$

Here $\phi_m + \mathbf{L}\phi_n - \mathbf{L}\phi_m \in N(\mathbf{L}^n)$ since

$$\mathbf{L}^n(\phi_m + \mathbf{L}\phi_n - \mathbf{L}\phi_m) = \mathbf{L}^{n-m-1}\mathbf{L}^{m+1}\phi_m + \mathbf{L}^{n+1}\phi_n - \mathbf{L}^{n-m}\mathbf{L}^{m+1}\phi_m = 0.$$

Hence

$$\|\mathbf{A}\phi_n - \mathbf{A}\phi_m\| \geqslant \tfrac{1}{2}, \qquad n > m.$$

Therefore the sequence $(\mathbf{A}\phi_n)$ does not contain a convergent subsequence which is a contradiction to the compactness of $\mathbf{A}$.

Now we know that in the sequence $(N(\mathbf{L}^n))$ there are two successive nullspaces that are equal. Define

$$r := \min\{k \mid N(\mathbf{L}^k) = N(\mathbf{L}^{k+1})\}.$$

We now prove by induction that

$$N(\mathbf{L}^r) = N(\mathbf{L}^{r+1}) = N(\mathbf{L}^{r+2}) = \cdots .$$

Assume that we already have that $N(\mathbf{L}^k) = N(\mathbf{L}^{k+1})$ for some $k \geqslant r$. Then for any $\phi \in N(\mathbf{L}^{k+2})$ we have $\mathbf{L}^{k+1}\mathbf{L}\phi = \mathbf{L}^{k+2}\phi = 0$, that is, $\mathbf{L}\phi \in N(\mathbf{L}^{k+1}) = N(\mathbf{L}^k)$. Hence $\mathbf{L}^{k+1}\phi = \mathbf{L}^k\mathbf{L}\phi = 0$ and thus $\phi \in N(\mathbf{L}^{k+1})$. Therefore $N(\mathbf{L}^{k+2}) \subset N(\mathbf{L}^{k+1})$.

We can summarize our results up to now by the formula

$$\{0\} = N(\mathbf{L}^0) \subsetneqq N(\mathbf{L}^1) \subsetneqq \cdots \subsetneqq N(\mathbf{L}^r) = N(\mathbf{L}^{r+1}) = \cdots .$$

2. Since for any $\psi = \mathbf{L}^{n+1}\phi \in \mathbf{L}^{n+1}(X)$ we can write $\psi = \mathbf{L}^n\mathbf{L}\phi$, it trivially follows that

$$X = \mathbf{L}^0(X) \supset \mathbf{L}^1(X) \supset \mathbf{L}^2(X) \supset \cdots .$$

Assume now that

$$X = \mathbf{L}^0(X) \supsetneqq \mathbf{L}^1(X) \supsetneqq \mathbf{L}^2(X) \supsetneqq \cdots .$$

Since the ranges $\mathbf{L}^n(X)$ are closed subspaces by Theorem 1.14, it follows from the Riesz Lemma 1.8 that there exists $\psi_n \in \mathbf{L}^n(X)$ such that $\|\psi_n\| = 1$ and

$$\|\psi_n - \psi\| \geqslant \tfrac{1}{2}$$

for all $\psi \in \mathbf{L}^{n+1}(X)$. We write $\psi_n = \mathbf{L}^n\phi_n$ and for $m > n$ we consider

$$\mathbf{A}\psi_n - \mathbf{A}\psi_m = \psi_n - (\psi_m + \mathbf{L}\psi_n - \mathbf{L}\psi_m).$$

Here $\psi_m + \mathbf{L}\psi_n - \mathbf{L}\psi_m \in \mathbf{L}^{n+1}(X)$ since $\psi_m + \mathbf{L}\psi_n - \mathbf{L}\psi_m = \mathbf{L}^{n+1}(\mathbf{L}^{m-n-1}\phi_m + \phi_n - \mathbf{L}^{m-n}\phi_m)$. Hence

$$\|\mathbf{A}\psi_n - \mathbf{A}\psi_m\| \geqslant \tfrac{1}{2}, \qquad m > n$$

and we can derive the same contradiction as above.

Therefore in the sequence $(\mathbf{L}^n(X))$ there are two subsequent ranges that are equal. Define

$$q := \min\{k \mid \mathbf{L}^k(X) = \mathbf{L}^{k+1}(X)\}.$$

We now prove by induction that

$$\mathbf{L}^q(X) = \mathbf{L}^{q+1}(X) = \mathbf{L}^{q+2}(X) = \cdots .$$

Assume we already have that $\mathbf{L}^k(X) = \mathbf{L}^{k+1}(X)$ for some $k \geqslant q$. Then for any $\psi = \mathbf{L}^{k+1}\phi \in \mathbf{L}^{k+1}(X)$ we can write $\mathbf{L}^k\phi = \mathbf{L}^{k+1}\phi'$ with some $\phi' \in X$ since $\mathbf{L}^k(X) = \mathbf{L}^{k+1}(X)$. Hence $\psi = \mathbf{L}^{k+2}\phi' \in \mathbf{L}^{k+2}(X)$ and therefore $\mathbf{L}^{k+1}(X) \subset \mathbf{L}^{k+2}(X)$.

Again we can summarize our results by the formula

$$X = \mathbf{L}^0(X) \supsetneqq \mathbf{L}^1(X) \supsetneqq \cdots \supsetneqq \mathbf{L}^q(X) = \mathbf{L}^{q+1}(X) = \cdots$$

3. We now show that $r = q$. Assume that $r > q$ and let $\phi \in N(\mathbf{L}^r)$. Then since $\mathbf{L}^{r-1}\phi \in \mathbf{L}^{r-1}(X) = \mathbf{L}^r(X)$, we observe that we can write $\mathbf{L}^{r-1}\phi = \mathbf{L}^r\phi'$ with some $\phi' \in X$. Since $\mathbf{L}^{r+1}\phi' = \mathbf{L}^r\phi = 0$, we have $\phi' \in N(\mathbf{L}^{r+1}) = N(\mathbf{L}^r)$, that is, $\mathbf{L}^{r-1}\phi = \mathbf{L}^r\phi' = 0$. Thus $\phi \in N(\mathbf{L}^{r-1})$ and hence $N(\mathbf{L}^{r-1}) = N(\mathbf{L}^r)$, which contradicts the definition of r.

Assume now that $r < q$ and let $\psi = \mathbf{L}^{q-1}\phi \in \mathbf{L}^{q-1}(X)$. Since $\mathbf{L}\psi = \mathbf{L}^q\phi \in \mathbf{L}^q(X) = \mathbf{L}^{q+1}(X)$ we can write $\mathbf{L}\psi = \mathbf{L}^{q+1}\phi'$ for some $\phi' \in X$. Then $\mathbf{L}^q(\phi - \mathbf{L}\phi') = \mathbf{L}\psi - \mathbf{L}^{q+1}\phi' = 0$ and since $N(\mathbf{L}^{q-1}) = N(\mathbf{L}^q)$, we conclude that $\mathbf{L}^{q-1}(\phi - \mathbf{L}\phi') = 0$, which implies $\psi = \mathbf{L}^q\phi' \in \mathbf{L}^q(X)$. Therefore $\mathbf{L}^{q-1}(X) = \mathbf{L}^q(X)$, which contradicts the definition of q.

4. Let $\psi \in N(\mathbf{L}^r) \cap \mathbf{L}^r(X)$. Then $\psi = \mathbf{L}^r\phi$ for some $\phi \in X$ and $\mathbf{L}^r\psi = 0$. Hence $\mathbf{L}^{2r}\phi = 0$, which means $\phi \in N(\mathbf{L}^{2r}) = N(\mathbf{L}^r)$. Therefore $\psi = \mathbf{L}^r\phi = 0$.

Let $\phi \in X$ be arbitrary. Then $\mathbf{L}^r\phi \in \mathbf{L}^r(X) = \mathbf{L}^{2r}(X)$ and we can write $\mathbf{L}^r\phi = \mathbf{L}^{2r}\phi'$ for some $\phi' \in X$. Define $\psi := \mathbf{L}^r\phi' \in \mathbf{L}^r(X)$ and $\chi := \phi - \psi$. Then $\mathbf{L}^r\chi = \mathbf{L}^r\phi - \mathbf{L}^{2r}\phi' = 0$, which means $\chi \in N(\mathbf{L}^r)$. Therefore the decomposition $\phi = \chi + \psi$ proves the direct sum $X = N(\mathbf{L}^r) \oplus \mathbf{L}^r(X)$.

Now we derive the fundamental results of the Riesz theory by distinguishing the two cases $r = 0$ and $r > 0$.

Theorem 1.16. Let X be a normed space, $\mathbf{A}: X \to X$ a compact linear operator, and let $\mathbf{I} - \mathbf{A}$ be injective. Then the inverse operator $(\mathbf{I} - \mathbf{A})^{-1}$ exists and is bounded.

Proof. By assumption, the operator $\mathbf{L}$ is injective, that is, $N(\mathbf{L}) = \{0\}$. Therefore $r = 0$ and from Theorem 1.15 we conclude $\mathbf{L}(X) = X$, that is, the operator $\mathbf{L}$ is surjective. Hence the inverse operator $\mathbf{L}^{-1}: X \to X$ exists.

Assume $\mathbf{L}^{-1}$ is not bounded. Then there exists a sequence (ϕ_n) with $\|\phi_n\| = 1$ such that the sequence $f_n := \mathbf{L}^{-1}\phi_n$ is not bounded. Define

$$\psi_n := \frac{\phi_n}{\|f_n\|}, \qquad g_n := \frac{f_n}{\|f_n\|}.$$

Then $\psi_n \to 0$, $n \to \infty$, and $\|g_n\| = 1$. Since $\mathbf{A}$ is compact, we can choose a subsequence $(g_{n(k)})$ such that $\mathbf{A}g_{n(k)} \to g \in X$, $k \to \infty$. Then since

$$\psi_n = g_n - \mathbf{A}g_n$$

we observe that $g_{n(k)} \to g$, $k \to \infty$, and therefore $g \in N(\mathbf{L})$. Hence $g = 0$, which contradicts $\|g_n\| = 1$, $n \in N$.

We can rewrite Theorem 1.16 in terms of the solvability of the operator equation of the second kind as follows.

Corollary 1.17. Let X be a normed space and $\mathbf{A}: X \to X$ a compact linear operator. If the homogeneous equation

$$\phi - \mathbf{A}\phi = 0$$

only has the trivial solution $\phi = 0$, then for all $f \in X$ the inhomogeneous equation

$$\phi - \mathbf{A}\phi = f$$

has a unique solution $\phi \in X$ and this solution depends continuously on f.

Theorem 1.18. Let X be a normed space and $\mathbf{A}: X \to X$ a compact linear operator and assume $\mathbf{I} - \mathbf{A}$ is not injective. Then the nullspace $N(\mathbf{I} - \mathbf{A})$ is finite dimensional and the range $(\mathbf{I} - \mathbf{A})X \subsetneq X$ is a proper closed subspace.

Proof. By assumption, we have $N(\mathbf{L}) \supsetneq \{0\}$. This means $r > 0$ and from Theorem 1.15 we conclude that $\mathbf{L}(X) \subsetneq X$.

Corollary 1.19. If the homogeneous equation

$$\phi - \mathbf{A}\phi = 0$$

has nontrivial solutions, then the inhomogeneous equation

$$\phi - \mathbf{A}\phi = f$$

is either unsolvable or its general solution is of the form

$$\phi = \phi^* + \sum_{k=1}^{m} \alpha_k \phi_k$$

where ϕ^* denotes a particular solution of the inhomogeneous equation, $\phi_1, \ldots, \phi_m$ are linearly independent solutions of the homogeneous equation, and $\alpha_1, \ldots, \alpha_m$ are arbitrary complex numbers.

Corollary 1.20. Theorems 1.16 and 1.18 and their Corollaries 1.17 and 1.19 remain valid when we replace $\mathbf{I} - \mathbf{A}$ by $\mathbf{S} - \mathbf{A}$, where $\mathbf{S}$ is a bounded linear operator that has a bounded inverse $\mathbf{S}^{-1}$.

Proof. This follows immediately from the fact that we can transform the equation

$$\mathbf{S}\phi - \mathbf{A}\phi = f$$

into the equivalent form

$$\phi - \mathbf{S}^{-1}\mathbf{A}\phi = \mathbf{S}^{-1} f$$

where $\mathbf{S}^{-1}\mathbf{A}$ is compact by Theorem 1.5.

The main importance of the results of the Riesz theory for compact operators lies in the fact that we can conclude existence from uniqueness as in the case of finite dimensional linear equations.

We conclude this section with the following theorem.

Theorem 1.21. The projection operator $\mathbf{P}: X \to N(\mathbf{L}')$ defined by the decomposition

$$X = N(\mathbf{L}') \oplus \mathbf{L}'(X)$$

is compact. The operator $\mathbf{L} - \mathbf{P} = \mathbf{I} - \mathbf{A} - \mathbf{P}$ is bijective.

Proof. The nullspace $N(\mathbf{L}')$ is finite dimensional by Theorem 1.13. On $N(\mathbf{L}')$ it is easily verified that

$$|||\psi||| := \inf_{\chi \in \mathbf{L}'(X)} \|\psi + \chi\|$$

defines a norm. In particular, we conclude from $|||\psi||| = 0$ that $\psi = 0$ since the range $\mathbf{L}'(X)$ is closed by Theorem 1.14. Since on a finite dimensional linear space all norms are equivalent, there exists a positive number C such that $\|\psi\| \leqslant C|||\psi|||$, that is,

$$\|\psi\| \leqslant C \inf_{\chi \in \mathbf{L}'(X)} \|\psi + \chi\|$$

for all $\psi \in N(\mathbf{L}^r)$. Then for all $\phi \in X$ we have $\mathbf{P}\phi \in N(\mathbf{L}^r)$ and therefore

$$\|\mathbf{P}\phi\| \leqslant C \inf_{\chi \in \mathbf{L}^r(X)} \|\mathbf{P}\phi + \chi\| \leqslant \|\phi\|$$

since $\phi - \mathbf{P}\phi \in \mathbf{L}^r(X)$. Hence $\mathbf{P}$ is bounded and, since it has finite dimensional range $\mathbf{P}(X) = N(\mathbf{L}^r)$, by Theorem 1.7 it is compact.

It follows from Theorem 1.4 that the operator $\mathbf{A}+\mathbf{P}$ is compact. Let $\phi \in N(\mathbf{L}-\mathbf{P})$, that is,

$$\mathbf{L}\phi - \mathbf{P}\phi = 0.$$

Then $\mathbf{L}^{r+1}\phi = 0$ since $\mathbf{P}\phi \in N(\mathbf{L}^r)$. Therefore $\phi \in N(\mathbf{L}^{r+1}) = N(\mathbf{L}^r)$ and $\mathbf{P}\phi = \phi$, which implies

$$\mathbf{L}\phi = \phi.$$

From this we get by iteration that

$$\phi = \mathbf{L}^r\phi = 0.$$

Therefore $N(\mathbf{L}-\mathbf{P}) = \{0\}$ and from Theorem 1.16, applied to the compact operator $\mathbf{A}+\mathbf{P}$, we conclude that $\mathbf{L}-\mathbf{P}$ is surjective.

1.3 THE FREDHOLM THEORY

In the case of Theorem 1.18 in which the homogeneous equation has nontrivial solutions, the Riesz theory gives no answer to the question whether the nonhomogeneous equation for a given inhomogeneity is solvable or not. This question is settled by the Fredholm theory that we shall develop for dual systems following the analysis of Jörgens [1] and Wendland [1], [2].

Definition 1.22. Let X and Y be normed spaces and $\langle .,. \rangle : X \times Y \to \mathbf{C}$ be a nondegenerate bilinear form, that is,

1. For any $\phi \in X$, $\phi \neq 0$, there exists $\psi \in Y$ such that $\langle \phi, \psi \rangle \neq 0$; and for any $\psi \in Y$, $\psi \neq 0$, there exists $\phi \in X$ such that $\langle \phi, \psi \rangle \neq 0$.
2. For all $\phi_1, \phi_2, \phi \in X$, $\psi_1, \psi_2, \psi \in Y$, $\alpha_1, \alpha_2, \beta_1, \beta_2 \in \mathbf{C}$ we have

$$\langle \alpha_1\phi_1 + \alpha_2\phi_2, \psi \rangle = \alpha_1\langle \phi_1, \psi \rangle + \alpha_2\langle \phi_2, \psi \rangle,$$

$$\langle \phi, \beta_1\psi_1 + \beta_2\psi_2 \rangle = \beta_1\langle \phi, \psi_1 \rangle + \beta_2\langle \phi, \psi_2 \rangle.$$

We call two normed spaces X and Y equipped with a nondegenerate bilinear form a dual system and denote it by $\langle X, Y \rangle$.

Note that as opposed to Jörgens and Wendland, we do not assume the bilinear form to be bounded. A similar theory can be developed with a sesquilinear form instead of a bilinear form.

Theorem 1.23. Let G be as in Theorem 1.10. Then $\langle C(G), C(G)\rangle$ is a dual system with the bilinear form

$$\langle \phi, \psi \rangle := \int_G \phi(x)\psi(x)\,dx, \qquad \phi, \psi \in C(G).$$

Proof. Obvious from Definition 1.22.

Definition 1.24. Let $\langle X, Y\rangle$ be a dual system. Then two operators $\mathbf{A}: X \to X$, $\mathbf{B}: Y \to Y$ are called adjoint if for every $\phi \in X$, $\psi \in Y$,

$$\langle \mathbf{A}\phi, \psi \rangle = \langle \phi, \mathbf{B}\psi \rangle.$$

Theorem 1.25. Let $\langle X, Y\rangle$ be a dual system. If an operator $\mathbf{A}: X \to X$ has an adjoint $\mathbf{B}: Y \to Y$, then $\mathbf{B}$ is uniquely determined and $\mathbf{A}$ and $\mathbf{B}$ are linear.

Proof. Suppose there existed two adjoints to $\mathbf{A}$ and we denote these by $\mathbf{B}_1$ and $\mathbf{B}_2$. Let $\mathbf{B} := \mathbf{B}_1 - \mathbf{B}_2$. Then for every $\psi \in Y$ we have $\langle \phi, \mathbf{B}\psi \rangle = \langle \phi, \mathbf{B}_1\psi \rangle - \langle \phi, \mathbf{B}_2\psi \rangle = \langle \mathbf{A}\phi, \psi \rangle - \langle \mathbf{A}\phi, \psi \rangle = 0$ for all $\phi \in X$. Hence, since $\langle .,. \rangle$ is nondegenerate we have $\mathbf{B}\psi = 0$ for every $\psi \in Y$, that is, $\mathbf{B}_1 = \mathbf{B}_2$. To show that $\mathbf{B}$ is linear we simply observe that for every $\phi \in X$

$$\begin{aligned}\langle \phi, \beta_1 \mathbf{B}\psi_1 + \beta_2 \mathbf{B}\psi_2 \rangle &= \beta_1 \langle \phi, \mathbf{B}\psi_1 \rangle + \beta_2 \langle \phi, \mathbf{B}_2\psi \rangle \\ &= \beta_1 \langle \mathbf{A}\phi, \psi_1 \rangle + \beta_2 \langle \mathbf{A}\phi, \psi_2 \rangle \\ &= \langle \mathbf{A}\phi, \beta_1\psi_1 + \beta_2\psi_2 \rangle \\ &= \langle \phi, \mathbf{B}(\beta_1\psi_1 + \beta_2\psi_2) \rangle,\end{aligned}$$

that is, $\beta_1 \mathbf{B}\psi_1 + \beta_2 \mathbf{B}\psi_2 = \mathbf{B}(\beta_1\psi_1 + \beta_2\psi_2)$. In a similar manner, it is seen that $\mathbf{A}$ is linear.

Theorem 1.26. Let K be a continuous or a weakly singular kernel. Then in the dual system $\langle C(G), C(G)\rangle$ the (compact) integral operators defined by

$$(\mathbf{A}\phi)(x) := \int_G K(x, y)\phi(y)\,dy$$

$$(\mathbf{B}\psi)(x) := \int_G K(y, x)\psi(y)\,dy$$

are adjoint.

Proof. The theorem follows from

$$\begin{aligned}\langle \mathbf{A}\phi, \psi\rangle &= \int_G (\mathbf{A}\phi)(x)\psi(x)\,dx \\ &= \int_G \left\{ \int_G K(x,y)\phi(y)\,dy \right\} \psi(x)\,dx \\ &= \int_G \phi(y) \left\{ \int_G K(x,y)\psi(x)\,dx \right\} dy \\ &= \int_G \phi(y)(\mathbf{B}\psi)(y)\,dy = \langle \phi, \mathbf{B}\psi\rangle.\end{aligned}$$

In the case of a weakly singular kernel the interchanging of the order of integration is justified by the fact that $\mathbf{A}_n\phi \to \mathbf{A}\phi$, $n \to \infty$, uniformly on G, where $\mathbf{A}_n$ is the integral operator with continuous kernel K_n introduced in the proof of Theorem 1.11.

Lemma 1.27. Let $\langle X, Y\rangle$ be a dual system. Then to every set of linearly independent elements $\phi_1, \ldots, \phi_n \in X$ there exists a set $\psi_1, \ldots, \psi_n \in Y$ such that

$$\langle \phi_i, \psi_k \rangle = \delta_{ik}, \qquad i, k = 1, \ldots, n.$$

A similar statement holds with the roles of X and Y interchanged.

Proof. For one linearly independent element $\phi_1 \in X$ the lemma is true since $\langle .,.\rangle$ is nondegenerate. Assume the lemma is proven for $n \geq 1$ linearly independent elements. Let $\phi_1, \ldots, \phi_{n+1}$ be $n+1$ linearly independent elements. By our induction assumption, for every $l = 1, \ldots, n+1$, to the set $\phi_1, \ldots, \phi_{l-1}, \phi_{l+1}, \ldots, \phi_{n+1}$ of n elements in X there exists a set of n elements $\psi_1^{(l)}, \ldots, \psi_{l-1}^{(l)}\psi_{l+1}^{(l)}, \ldots, \psi_{n+1}^{(l)}$ in Y such that

$$\langle \phi_i, \psi_k^{(l)} \rangle = \delta_{ik}, \qquad i, k = 1, \ldots, n+1, \qquad i, k \neq l. \tag{1.4}$$

Then there exists $\chi_l \in Y$ such that

$$\begin{aligned}\alpha_l := \langle \phi_l, \chi_l - \sum_{\substack{k=1\\k\neq l}}^{n+1} \psi_k^{(l)} \langle \phi_k, \chi_l \rangle\rangle \\ = \langle \phi_l - \sum_{\substack{k=1\\k\neq l}}^{n+1} \langle \phi_l, \psi_k^{(l)}\rangle \phi_k, \chi_l\rangle \neq 0\end{aligned}$$

since otherwise

$$\phi_l - \sum_{\substack{k=1\\k\neq l}}^{n+1} \langle \phi_l, \psi_k^{(l)}\rangle \phi_k = 0,$$

a contradiction to the linear independence of the $\phi_1, \ldots, \phi_{n+1}$. Define

$$\psi_l := \frac{1}{\alpha_l}\left\{\chi_l - \sum_{\substack{k=1\\k\neq l}}^{n+1} \psi_k^{(l)}\langle\phi_k, \chi_l\rangle\right\}.$$

Then, obviously, $\langle\phi_l, \psi_l\rangle = 1$ and for $i \neq l$,

$$\langle\phi_i, \psi_l\rangle = \frac{1}{\alpha_l}\left\{\langle\phi_i, \chi_l\rangle - \sum_{\substack{k=1\\k\neq l}}^{n+1} \langle\phi_i, \psi_k^{(l)}\rangle\langle\phi_k, \chi_l\rangle\right\} = 0$$

because of (1.4). Hence we obtain $\psi_1, \ldots, \psi_{n+1}$ such that

$$\langle\phi_i, \psi_k\rangle = \delta_{ik}, \qquad i, k = 1, \ldots, n+1.$$

Theorem 1.28 (First Fredholm Theorem). Let $\langle X, Y\rangle$ be a dual system, and $\mathbf{A}: X \to X$, $\mathbf{B}: Y \to Y$ be compact adjoint operators. Then the nullspaces of the operators $\mathbf{I}-\mathbf{A}$ and $\mathbf{I}-\mathbf{B}$ have the same finite dimension.

Proof. By the first Riesz Theorem 1.13 we have

$$m := \dim N(\mathbf{I}-\mathbf{A}) < \infty, \qquad n := \dim N(\mathbf{I}-\mathbf{B}) < \infty.$$

We assume $m < n$. Choose a basis $\phi_1, \ldots, \phi_m$ for the nullspace $N(\mathbf{I}-\mathbf{A})$ (if $m > 0$) and a basis $\psi_1, \ldots, \psi_n$ for the nullspace $N(\mathbf{I}-\mathbf{B})$. By Lemma 1.27 there exist elements $a_1, \ldots, a_m \in Y$ (if $m > 0$) and $b_1, \ldots, b_n \in X$ such that

$$\langle\phi_i, a_k\rangle = \delta_{ik}, \qquad i, k = 1, \ldots, m,$$
$$\langle b_i, \psi_k\rangle = \delta_{ik}, \qquad i, k = 1, \ldots, n.$$

Define a linear operator $\mathbf{T}: X \to X$ with finite dimensional range by

$$\mathbf{T}\phi := \begin{cases} 0 & \text{if } m = 0, \\ \sum_{i=1}^{m} \langle\phi, a_i\rangle b_i & \text{if } m > 0. \end{cases}$$

Let $\phi \in N(\mathbf{I}-\mathbf{A}+\mathbf{T})$, that is, (if $m > 0$)

$$\phi - \mathbf{A}\phi + \sum_{i=1}^{m} \langle\phi, a_i\rangle b_i = 0.$$

It follows that

$$\begin{aligned}\langle\phi, a_k\rangle &= \langle\phi, \psi_k - \mathbf{B}\psi_k\rangle + \langle\phi, a_k\rangle \\ &= \left\langle\phi - \mathbf{A}\phi + \sum_{i=1}^{m} \langle\phi, a_i\rangle b_i, \psi_k\right\rangle = 0\end{aligned}$$

for $k=1,\dots,m$. Therefore $\phi-\mathbf{A}\phi=0$ and hence we can write

$$\phi=\sum_{i=1}^{m}\alpha_i\phi_i.$$

Now from

$$\langle\phi,a_k\rangle=\sum_{i=1}^{m}\alpha_i\langle\phi_i,a_k\rangle=\alpha_k$$

we obtain $\alpha_k=0$ for $k=1,\dots,m$, and therefore $\phi=0$. Thus we have proved that $N(\mathbf{I}-\mathbf{A}+\mathbf{T})=\{0\}$, that is, $\mathbf{I}-\mathbf{A}+\mathbf{T}$ is injective, which of course also is true for the case $m=0$.

Now we show that $\mathbf{I}-\mathbf{A}+\mathbf{T}$ is surjective. Let $\mathbf{P}: X\to N(\mathbf{L}^r)$ be the projection defined by the decomposition $X=N(\mathbf{L}^r)\oplus\mathbf{L}^r(X)$ and recall that the inverse operator $(\mathbf{I}-\mathbf{A}-\mathbf{P})^{-1}: X\to X$ exists by Theorem 1.21. The operator $\mathbf{T}+\mathbf{P}$ has a finite dimensional range $U:=(\mathbf{T}+\mathbf{P})(X)$. Define on the finite dimensional space U an operator $\mathbf{K}: U\to U$ by

$$\mathbf{K}:=\mathbf{I}+(\mathbf{T}+\mathbf{P})(\mathbf{I}-\mathbf{A}-\mathbf{P})^{-1}=(\mathbf{I}-\mathbf{A}+\mathbf{T})(\mathbf{I}-\mathbf{A}-\mathbf{P})^{-1}$$

Let $g\in N(\mathbf{K})$. Then, since $N(\mathbf{I}-\mathbf{A}+\mathbf{T})=\{0\}$, we have $(\mathbf{I}-\mathbf{A}-\mathbf{P})^{-1}g=0$ and from this $g=0$. Therefore the linear operator $\mathbf{K}$ on the finite dimensional space U is injective and therefore surjective. Thus, given any $f\in X$, the inhomogeneous equation

$$\mathbf{K}g=(\mathbf{T}+\mathbf{P})(\mathbf{I}-\mathbf{A}-\mathbf{P})^{-1}f$$

has a unique solution $g\in U$. Now we set

$$\phi:=(\mathbf{I}-\mathbf{A}-\mathbf{P})^{-1}(f-g)$$

and obtain

$$\begin{aligned}(\mathbf{I}-\mathbf{A}+\mathbf{T})\phi&=f-g+(\mathbf{T}+\mathbf{P})(\mathbf{I}-\mathbf{A}-\mathbf{P})^{-1}(f-g)\\&=f-\mathbf{K}g+(\mathbf{T}+\mathbf{P})(\mathbf{I}-\mathbf{A}-\mathbf{P})^{-1}f=f.\end{aligned}$$

Hence, $\mathbf{I}-\mathbf{A}+\mathbf{T}$ is surjective.

Now, since $\mathbf{I}-\mathbf{A}+\mathbf{T}$ is bijective, the inhomogeneous equation

$$\phi-\mathbf{A}\phi+\mathbf{T}\phi=b_{m+1}$$

has a unique solution ϕ. We now arrive at the contradiction

$$\begin{aligned}1=\langle b_{m+1},\psi_{m+1}\rangle&=\langle\phi-\mathbf{A}\phi+\mathbf{T}\phi,\psi_{m+1}\rangle\\&=\langle\phi-\mathbf{A}\phi,\psi_{m+1}\rangle=\langle\phi,\psi_{m+1}-\mathbf{B}\psi_{m+1}\rangle=0\end{aligned}$$

since $\langle\mathbf{T}\phi,\psi_{m+1}\rangle=0$.

Therefore $m\geqslant n$ and a similar argument shows $n\geqslant m$. Hence $m=n$.

Theorem 1.29 (Second Fredholm Theorem). The nonhomogeneous equation

$$\phi - \mathbf{A}\phi = f \quad | \quad \psi - \mathbf{B}\psi = g$$

is solvable if and only if the condition

$$\langle f, \psi \rangle = 0 \quad | \quad \langle \phi, g \rangle = 0$$

is satisfied for all solutions of the homogeneous adjoint equation

$$\psi - \mathbf{B}\psi = 0 \quad | \quad \phi - \mathbf{A}\phi = 0.$$

Proof. Obviously, it suffices to carry out the proof for the equation $\phi - \mathbf{A}\phi = f$.

Necessity. Let ϕ be a solution of $\phi - \mathbf{A}\phi = f$. Then for all solutions ψ of $\psi - \mathbf{B}\psi = 0$, we obtain

$$\langle f, \psi \rangle = \langle \phi - \mathbf{A}\phi, \psi \rangle = \langle \phi, \psi - \mathbf{B}\psi \rangle = 0.$$

Sufficiency. By the first Fredholm theorem we have

$$m = \dim N(\mathbf{I} - \mathbf{A}) = \dim N(\mathbf{I} - \mathbf{B}) < \infty.$$

In the case $m = 0$, the condition $\langle f, \psi \rangle = 0$ is satisfied for all $f \in X$ and by Corollary 1.17 the equation $\phi - \mathbf{A}\phi = f$ indeed is solvable for all $f \in X$.

In the case $m > 0$, from the proof of the previous theorem, we know that $\mathbf{I} - \mathbf{A} + \mathbf{T}$ is bijective. Hence there exists a unique solution ϕ of the equation

$$\phi - \mathbf{A}\phi + \mathbf{T}\phi = f.$$

Then it follows that

$$\begin{aligned} \langle \phi, a_k \rangle &= \langle \phi, \psi_k - \mathbf{B}\psi_k \rangle + \langle \phi, a_k \rangle \\ &= \langle \phi - \mathbf{A}\phi, \psi_k \rangle + \sum_{i=1}^{m} \langle \phi, a_i \rangle \langle b_i, \psi_k \rangle \\ &= \langle \phi - \mathbf{A}\phi + \mathbf{T}\phi, \psi_k \rangle = \langle f, \psi_k \rangle = 0 \end{aligned}$$

for $k = 1, \ldots, m$ since we are assuming that the solvability condition of the theorem is satisfied. Hence $\mathbf{T}\phi = 0$ and thus ϕ also satisfies the original equation

$$\phi - \mathbf{A}\phi = f.$$

We now summarize our results in the following theorem.

Theorem 1.30 (Fredholm Alternative). Let $\langle X, Y \rangle$ be a dual system and $\mathbf{A}: X \to X$, $\mathbf{B}: Y \to Y$, be compact adjoint operators. Then *either*

$$N(\mathbf{I}-\mathbf{A}) = \{0\} \quad \text{and} \quad N(\mathbf{I}-\mathbf{B}) = \{0\}$$

$$\text{and} \quad (\mathbf{I}-\mathbf{A})(X) = X \quad \text{and} \quad (\mathbf{I}-\mathbf{B})(Y) = Y$$

or

$$\dim N(\mathbf{I}-\mathbf{A}) = \dim N(\mathbf{I}-\mathbf{B}) \in \mathbb{N}$$

$$\text{and} \quad (\mathbf{I}-\mathbf{A})(X) = \{f \in X \mid \langle f, \psi \rangle = 0, \quad \psi \in N(\mathbf{I}-\mathbf{B})\}$$

$$\text{and} \quad (\mathbf{I}-\mathbf{B})(Y) = \{g \in Y \mid \langle \phi, g \rangle = 0, \quad \phi \in N(\mathbf{I}-\mathbf{A})\}.$$

Choosing the dual system introduced by Theorem 1.23 and the integral operator considered in Theorem 1.26, our results include the Fredholm alternative for integral equations of the second kind first obtained by Fredholm [1]. The Schauder theory (Schauder [1], Jörgens [1]) is included by taking $Y = X^*$, the dual space of X, and defining a bilinear form by $\langle \phi, \psi \rangle = \psi(\phi)$ for all elements $\phi \in X$ and bounded linear functionals $\psi \in X^*$. As a consequence of the Hahn–Banach theorem, this bilinear form is nondegenerate.

Finally, we note the following theorem (Kress [2], Jörgens [1]).

Theorem 1.31. The operators $\mathbf{A}$ and $\mathbf{B}$ have Riesz number one if and only if for any pair of bases $\phi_1, \ldots, \phi_m$ and $\psi_1, \ldots, \psi_m$ of the nullspaces $N(\mathbf{I}-\mathbf{A})$ and $N(\mathbf{I}-\mathbf{B})$ the matrix $\langle \phi_i, \psi_k \rangle$, $i, k = 1, \ldots, m$, is nonsingular.

Proof. Obviously,

$$\det \langle \phi_i, \psi_k \rangle = 0$$

is equivalent to the existence of a nontrivial solution λ_i, $i = 1, \ldots, m$, of the homogeneous linear system

$$\sum_{i=1}^{m} \langle \phi_i, \psi_k \rangle \lambda_i = 0, \qquad k = 1, \ldots, m.$$

By the Fredholm alternative, this is equivalent to the fact that for

$$f := \sum_{i=1}^{m} \lambda_i \phi_i \in N(\mathbf{I}-\mathbf{A})$$

with $f \neq 0$ the equation $\phi - \mathbf{A}\phi = f$ has a solution ϕ, that is, $\phi \in N((\mathbf{I}-\mathbf{A})^2)$ but $\phi \notin N(\mathbf{I}-\mathbf{A})$, that is, the Riesz number is greater than one.

1.4 A SINGULAR PERTURBATION PROBLEM

In our investigations of the low frequency behavior of solutions to electromagnetic scattering problems, we shall need to consider the following perturbation problem.

Let X be a Banach space and let $K \subset \mathbf{C}$ be a subset such that $0 \in K$ is an accumulation point of K. Consider a family $\{\mathbf{A}_\kappa: X \to X, \kappa \in K\}$ of compact linear operators and define $\mathbf{L}_\kappa := \mathbf{I} - \mathbf{A}_\kappa$. We assume that for $\kappa \neq 0$ the operator $\mathbf{L}_\kappa$ has a trivial nullspace and therefore is bijective. Then for all $\kappa \neq 0$ and all $f_\kappa \in X$ the equation

$$\mathbf{L}_\kappa \phi_\kappa = f_\kappa$$

has a unique solution $\phi_\kappa = \mathbf{L}_\kappa^{-1} f_\kappa$. We are interested in finding sufficient conditions to guarantee that $\phi_\kappa \to \phi_0$, $\kappa \to 0$, where ϕ_0 is a solution of the limiting equation

$$\mathbf{L}_0 \phi_0 = f_0$$

and it is assumed that

$$\|\mathbf{A}_\kappa - \mathbf{A}_0\| \to 0, \qquad \|f_\kappa - f_0\| \to 0, \qquad \kappa \to 0.$$

Let $\mathbf{P}: X \to N(\mathbf{L}_0^r)$ be the projection defined by the decomposition $X = N(\mathbf{L}_0^r) \oplus \mathbf{L}_0^r(X)$ of Theorem 1.15 and recall that by Theorems 1.16 and 1.21 the operator $\mathbf{P}$ is compact and the inverse operator $(\mathbf{L}_0 - \mathbf{P})^{-1}$ exists and is bounded. Now define the bounded operator

$$\mathbf{L}_0^+ := (\mathbf{L}_0 - \mathbf{P})^{-1}(\mathbf{I} - \mathbf{P}).$$

For any $\phi \in X$ we have $\mathbf{L}_0 \mathbf{P} \phi \in N(\mathbf{L}_0^r)$ and therefore $\mathbf{P L}_0 \mathbf{P} \phi = \mathbf{L}_0 \mathbf{P} \phi$. On the other hand, from the decomposition $\phi = \mathbf{P}\phi + \psi$, $\psi \in \mathbf{L}_0^r(X)$, we can conclude that $\mathbf{L}_0 \phi = \mathbf{L}_0 \mathbf{P} \phi + \mathbf{L}_0 \psi$ and $\mathbf{P L}_0 \phi = \mathbf{P L}_0 \mathbf{P} \phi$ since $\mathbf{L}_0 \psi \in \mathbf{L}_0^{r+1}(X) = \mathbf{L}_0^r(X)$. Thus we have the commutative property $\mathbf{P L}_0 = \mathbf{L}_0 \mathbf{P}$. From this we deduce that $\mathbf{P}(\mathbf{L}_0 - \mathbf{P}) = (\mathbf{L}_0 - \mathbf{P})\mathbf{P}$ and $\mathbf{P}(\mathbf{L}_0 - \mathbf{P})^{-1} = (\mathbf{L}_0 - \mathbf{P})^{-1}\mathbf{P}$. Then $\mathbf{P L}_0^+ = \mathbf{L}_0^+ \mathbf{P} = 0$ and therefore

$$\mathbf{L}_0 \mathbf{L}_0^+ = \mathbf{L}_0^+ \mathbf{L}_0 = \mathbf{I} - \mathbf{P}. \tag{1.5}$$

Note that the latter equations imply that $\mathbf{L}_0^+$ is a generalized inverse of $\mathbf{L}_0$.

Define

$$\mathbf{M}_\kappa := \mathbf{L}_0^+ (\mathbf{L}_0 - \mathbf{L}_\kappa).$$

Since $\|\mathbf{A}_\kappa - \mathbf{A}_0\| \to 0$, $\kappa \to 0$, for sufficiently small κ the operator $\mathbf{I} - \mathbf{M}_\kappa$ has an inverse and the Neumann series

$$(\mathbf{I} - \mathbf{M}_\kappa)^{-1} = \mathbf{I} + \mathbf{M}_\kappa + \mathbf{M}_\kappa^2 + \cdots$$

converges. Now we define

$$\mathbf{F}_\kappa := \mathbf{L}_\kappa + \mathbf{P}(\mathbf{L}_0 - \mathbf{L}_\kappa). \tag{1.6}$$

With the help of (1.5) we find that

$$\mathbf{F}_\kappa = \mathbf{L}_0(\mathbf{I} - \mathbf{M}_\kappa). \tag{1.7}$$

Then for the operator

$$\mathbf{F}_\kappa^+ := (\mathbf{I} - \mathbf{M}_\kappa)^{-1}\mathbf{L}_0^+$$

a simple calculation using (1.5) yields

$$\mathbf{F}_\kappa \mathbf{F}_\kappa^+ = \mathbf{I} - \mathbf{P}$$

$$\mathbf{F}_\kappa^+ \mathbf{F}_\kappa = \mathbf{I} - (\mathbf{I} - \mathbf{M}_\kappa)^{-1}\mathbf{P}. \tag{1.8}$$

Note that these equations imply that $\mathbf{F}_\kappa^+$ is a generalized inverse of $\mathbf{F}_\kappa$.

Now we define

$$\psi_\kappa := \mathbf{F}_\kappa^+ f_\kappa, \qquad \psi_0 := \mathbf{L}_0^+ f_0.$$

Then from

$$\|\psi_\kappa - \psi_0\| \leqslant \|\mathbf{F}_\kappa^+ - \mathbf{L}_0^+\| \|f_\kappa\| + \|\mathbf{L}_0^+\| \|f_\kappa - f_0\|$$

we see that

$$\psi_\kappa \to \psi_0, \qquad \kappa \to 0.$$

From (1.8), we obtain $\mathbf{F}_\kappa \psi_\kappa = (\mathbf{I} - \mathbf{P}) f_\kappa$ and from this, using the definition of $\mathbf{F}_\kappa$, we obtain

$$\mathbf{L}_\kappa \psi_\kappa = (\mathbf{I} - \mathbf{P}) f_\kappa + \mathbf{P}(\mathbf{L}_\kappa - \mathbf{L}_0)\psi_\kappa. \tag{1.9}$$

We now try to represent the solution ϕ_κ of $\mathbf{L}_\kappa \phi_\kappa = f_\kappa$ in the form

$$\phi_\kappa = \psi_\kappa + (\mathbf{I} - \mathbf{M}_\kappa)^{-1}\chi_\kappa$$

where $\chi_\kappa \in N(\mathbf{L}_0')$. By straightforward calculations, after using (1.6) and (1.7) to obtain

$$\mathbf{L}_\kappa(\mathbf{I} - \mathbf{M}_\kappa)^{-1} = \mathbf{L}_0 + \mathbf{P}(\mathbf{L}_\kappa - \mathbf{L}_0)(\mathbf{I} - \mathbf{M}_\kappa)^{-1}$$

from (1.9), we find that ϕ_κ satisfies $\mathbf{L}_\kappa \phi_\kappa = f_\kappa$ if and only if $\chi_\kappa \in N(\mathbf{L}_0')$ is a solution of

$$\mathbf{P}(\mathbf{L}_\kappa - \mathbf{L}_0)(\mathbf{I} - \mathbf{M}_\kappa)^{-1}\chi_\kappa + \mathbf{L}_0\chi_\kappa = \mathbf{P}f_\kappa - \mathbf{P}(\mathbf{L}_\kappa - \mathbf{L}_0)\psi_\kappa. \tag{1.10}$$

Because of its equivalence with the uniquely solvable equation $\mathbf{L}_\kappa\phi_\kappa = f_\kappa$ the linear equation (1.10) has a unique solution χ_κ for all $\kappa \neq 0$ in the finite dimensional space $N(\mathbf{L}_0^r)$.

To establish the convergence of the solutions to $\mathbf{L}_\kappa\phi_\kappa = f_\kappa$ it now suffices to study the behavior of the solutions χ_κ of the finite dimensional equation (1.10). The following theorem states sufficient conditions for convergence.

Theorem 1.32. Assume that the Riesz number of $\mathbf{A}_0$ is one and that there exists a number $s \in \mathbb{N}$ such that

$$\mathbf{A}_0 - \mathbf{A}_\kappa = \mathbf{C}\kappa^s + o(\kappa^s)$$

and

$$\mathbf{P}f_\kappa = g\kappa^s + o(\kappa^s)$$

where $g \in N(\mathbf{L}_0)$ and $\mathbf{C}: X \to X$ is a linear operator such that $\mathbf{PC}: N(\mathbf{L}_0) \to N(\mathbf{L}_0)$ is bijective. Then the unique solution ϕ_κ of $\mathbf{L}_\kappa\phi_\kappa = f_\kappa$ converges to a solution ϕ_0 of $\mathbf{L}_0\phi_0 = f_0$.

Proof. Since $\mathbf{A}_0$ has Riesz number $r = 1$, equation (1.10) reduces to

$$\mathbf{PL}_\kappa(\mathbf{I} - \mathbf{M}_\kappa)^{-1}\chi_\kappa = \mathbf{P}(f_\kappa - \mathbf{L}_\kappa\psi_\kappa).$$

From our assumptions on the limiting behavior of $\mathbf{A}_\kappa$ and $\mathbf{P}f_\kappa$ we conclude that

$$\mathbf{PL}_\kappa(\mathbf{I} - \mathbf{M}_\kappa)^{-1}\big|_{N(\mathbf{L}_0)} = \mathbf{PC}\big|_{N(\mathbf{L}_0)}\kappa^s + o(\kappa^s)$$

and

$$\mathbf{P}(f_\kappa - \mathbf{L}_\kappa\psi_\kappa) = \left(g - \mathbf{PC}(\mathbf{L}_0 - \mathbf{P})^{-1}f_0\right)\kappa^s + o(\kappa^s).$$

From this we conclude that the unique solution χ_κ of (1.10) converges as $\kappa \to 0$ to the unique solution $\chi_0 \in N(\mathbf{L}_0)$ of

$$\mathbf{PC}\chi_0 = \left(g - \mathbf{PC}(\mathbf{L}_0 - \mathbf{P})^{-1}f_0\right).$$

Note: Suppose there exists a second Banach space Y, a nondegenerate bilinear form $\langle .,. \rangle : X \times Y \to \mathbf{C}$, and a compact operator $\mathbf{B}_0: Y \to Y$ that is the adjoint of $\mathbf{A}_0$ with respect to the dual system $\langle X, Y \rangle$. Then the projector $\mathbf{P}$, in the case of Riesz number one, can be expressed in terms of the nullspaces of

$\mathbf{I}-\mathbf{A}_0$ and $\mathbf{I}-\mathbf{B}_0$ as follows. By Theorem 1.31 we can choose a basis $\phi_1,\ldots,\phi_m$ of $N(\mathbf{I}-\mathbf{A}_0)$ and a basis $\psi_1,\ldots,\psi_m$ of $N(\mathbf{I}-\mathbf{B}_0)$ such that

$$\langle \phi_i, \psi_k \rangle = \delta_{ik}, \qquad i,k=1,\ldots,m.$$

Then

$$\mathbf{P}\phi = \sum_{k=1}^{m} \langle \phi, \psi_k \rangle \phi_k, \qquad \phi \in X,$$

since obviously $\mathbf{P}\phi \in N(\mathbf{L}_0)$ and $\phi - \mathbf{P}\phi \in \mathbf{L}_0(X)$ because $\langle \phi - \mathbf{P}\phi, \psi_i \rangle = 0$, $i=1,\ldots,m$. Therefore the assumptions of Theorem 1.32 in this case can be rewritten in the form

$$\langle f_\kappa, \psi_i \rangle = g_i \kappa^s + o(\kappa^s) \tag{1.11}$$

and

$$\langle \mathbf{L}_\kappa \phi_i, \psi_k \rangle = c_{ik} \kappa^s + o(\kappa^s) \tag{1.12}$$

where $g_i = \langle g, \psi_i \rangle \in \mathbf{C}$ and where $c_{ik} = \langle \mathbf{PC}\phi_i, \psi_k \rangle$ is a nonsingular complex $m \times m$-matrix.

The results of this section were first obtained by Kress [4] and cast into a basis-free notation by Kirsch [3]. A detailed discussion of the case where the Riesz number is greater than one including an application to a mixed boundary-value problem was given by Klein [1]. The more general case where the dimension of the nullspaces changes between two nonzero values at the critical point $\kappa = 0$ was considered by Engl and Kress [1] and includes an application to combined transmission and boundary-value problems in electro- and magnetostatics.

1.5 SUCCESSIVE APPROXIMATIONS

In order to solve the integral equations of acoustic and electromagnetic scattering by iterative methods, we shall now investigate the convergence of successive approximations

$$\phi_{n+1} := \mathbf{A}\phi_n + f$$

to solve the equation

$$\phi - \mathbf{A}\phi = f$$

where $\mathbf{A}: X \to X$ is a bounded linear operator in a Banach space X.

Definition 1.33. Let $\mathbf{A}: X \to X$ be a bounded linear operator mapping a Banach space X into itself. Then a complex number λ is called an eigenvalue of

$\mathbf{A}$ if there exists an element $\phi \in X$, $\phi \neq 0$, such that $\mathbf{A}\phi = \lambda\phi$. ϕ is called an eigenelement of $\mathbf{A}$. A complex number λ is called a regular value of $\mathbf{A}$ if $(\lambda\mathbf{I} - \mathbf{A})^{-1}$ exists and is bounded. The set of all regular values of $\mathbf{A}$ is called the resolvent set $\rho(\mathbf{A})$ and $\mathbf{R}(\lambda;\mathbf{A}) := (\lambda\mathbf{I} - \mathbf{A})^{-1}$ is called the resolvent. The complement of $\rho(\mathbf{A})$ is called the spectrum $\sigma(\mathbf{A})$ and

$$r(\mathbf{A}) := \sup_{\lambda \in \sigma(\mathbf{A})} |\lambda|$$

is called the spectral radius of $\mathbf{A}$.

For the spectrum of a compact operator we have the following properties.

Theorem 1.34. Let X be an infinite-dimensional Banach space and let $\mathbf{A}: X \to X$ be a compact linear operator. Then $\lambda = 0$ belongs to the spectrum $\sigma(\mathbf{A})$ and $\sigma(\mathbf{A}) \setminus \{0\}$ consists of at most a countable set of eigenvalues with no point of accumulation except, possibly, $\lambda = 0$.

Proof. Suppose $\lambda = 0$ is a regular value of $\mathbf{A}$, that is, $\mathbf{A}^{-1}$ exists and is bounded. Then $\mathbf{I} = \mathbf{A}^{-1}\mathbf{A}$ is compact by Theorem 1.5 and by Theorem 1.9 we obtain the contradiction that X is finite dimensional. Therefore $\lambda = 0$ belongs to the spectrum $\sigma(\mathbf{A})$.

For $\lambda \neq 0$ we can apply the Riesz theory to the operator $\lambda\mathbf{I} - \mathbf{A}$. Either $N(\lambda\mathbf{I} - \mathbf{A}) = \{0\}$ and $(\lambda\mathbf{I} - \mathbf{A})^{-1}$ exists and is bounded by Theorem 1.16 or $N(\lambda\mathbf{I} - \mathbf{A}) \supsetneqq \{0\}$, which means λ is an eigenvalue. Thus any $\lambda \neq 0$ is either a regular value or an eigenvalue of $\mathbf{A}$.

It remains to show that for each $R > 0$ there exists only a finite number of eigenvalues λ with $|\lambda| \geq R$. Assume, on the contrary, that we have a sequence (λ_n) of distinct eigenvalues satisfying $|\lambda_n| \geq R$. Choose eigenelements ϕ_n such that $\mathbf{A}\phi_n = \lambda_n\phi_n$ and define finite dimensional subspaces

$$U_n := \operatorname{span}(\phi_1, \ldots, \phi_n).$$

It is readily verified that eigenelements corresponding to distinct eigenvalues are linearly independent.

We have $U_{n-1} \subsetneqq U_n$ and by the Riesz Lemma 1.8 we can choose a sequence (ψ_n) such that $\psi_n \in U_n$ with $\|\psi_n\| = 1$ and

$$\|\psi_n - \psi\| \geq \tfrac{1}{2}$$

for all $\psi \in U_{n-1}$. Writing

$$\psi_n = \sum_{k=1}^{n} \alpha_{nk}\phi_k$$

we obtain

$$\lambda_n\psi_n - \mathbf{A}\psi_n = \sum_{k=1}^{n-1} (\lambda_n - \lambda_k)\alpha_{nk}\phi_k \in U_{n-1}.$$

Therefore, for $m < n$ we obtain

$$\begin{aligned} \mathbf{A}\psi_n - \mathbf{A}\psi_m &= [\lambda_n\psi_n - (\lambda_n\psi_n - \mathbf{A}\psi_n - \mathbf{A}\psi_m)] \\ &= \lambda_n(\psi_n - \psi) \end{aligned}$$

where $\psi := (1/\lambda_n)[\lambda_n\psi_n - \mathbf{A}\psi_n - \mathbf{A}\psi_m] \in U_{n-1}$. Hence

$$\|\mathbf{A}\psi_n - \mathbf{A}\psi_m\| \geqslant \frac{|\lambda_n|}{2} \geqslant \frac{R}{2}$$

for $m < n$. Therefore the sequence $(\mathbf{A}\psi_n)$ does not contain a convergent subsequence which contradicts the compactness of $\mathbf{A}$.

Theorem 1.35. Let $\mathbf{A}: X \to X$ be a bounded linear operator mapping the Banach space X into itself. Then the Neumann series

$$(\lambda\mathbf{I} - \mathbf{A})^{-1} = \sum_{n=0}^{\infty} \lambda^{-n-1}\mathbf{A}^n$$

converges in the uniform operator norm for all $\lambda > r(\mathbf{A})$.

Proof. Consider the power series

$$\sum_{n=0}^{\infty} \mu^n\mathbf{A}^n$$

which obviously converges in the uniform operator norm for all $\mu \in \mathbf{C}$ such that $|\mu|\,\|\mathbf{A}\| < 1$, since then we have a convergent geometric series as a majorant. As in advanced calculus, it can be shown that

$$a := \left[\limsup_{n\to\infty} \|\mathbf{A}^n\|^{1/n}\right]^{-1}$$

is the radius of convergence of this power series, that is, the series converges in the uniform operator norm for all $\mu \in \mathbf{C}$ with $|\mu| < a$ and diverges for all $\mu \in \mathbf{C}$ with $|\mu| > a$. Therefore

$$\mathbf{S}(\lambda) := \sum_{n=0}^{\infty} \lambda^{-n-1}\mathbf{A}^n$$

is uniformly convergent for all $|\lambda| > a^{-1}$ and defines a bounded linear operator. We obviously have $(\lambda\mathbf{I} - \mathbf{A})\mathbf{S}(\lambda) = \mathbf{I} = \mathbf{S}(\lambda)(\lambda\mathbf{I} - \mathbf{A})$ and therefore $\lambda \in \rho(\mathbf{A})$ and $\mathbf{S}(\lambda) = \mathbf{R}(\lambda; \mathbf{A})$. Hence $r(\mathbf{A}) \leqslant a^{-1}$.

Now let $\lambda_0 \in \rho(\mathbf{A})$. Then for all λ with

$$|\lambda - \lambda_0|\,\|\mathbf{R}(\lambda_0; \mathbf{A})\| < 1$$

the series

$$\mathbf{T}(\lambda) := \sum_{n=0}^{\infty} (\lambda_0 - \lambda)^n[\mathbf{R}(\lambda_0; \mathbf{A})]^{n+1}$$

is uniformly convergent and defines a bounded linear operator. We see that

$$(\lambda \mathbf{I}-\mathbf{A})\mathbf{T}(\lambda) = [(\lambda-\lambda_0)\mathbf{I}+(\lambda_0\mathbf{I}-\mathbf{A})]\mathbf{T}(\lambda)$$

$$= -\sum_{n=0}^{\infty}(\lambda_0-\lambda)^{n+1}[\mathbf{R}(\lambda_0;\mathbf{A})]^{n+1}$$

$$+\sum_{n=0}^{\infty}(\lambda_0-\lambda)^{n}[\mathbf{R}(\lambda_0;\mathbf{A})]^{n} = \mathbf{I}$$

and similarly $\mathbf{T}(\lambda)(\lambda\mathbf{I}-\mathbf{A})=\mathbf{I}$. Therefore $\lambda \in \rho(\mathbf{A})$ and $\mathbf{T}(\lambda)=\mathbf{R}(\lambda;\mathbf{A})$. In particular, this means that the resolvent set is open and that the resolvent $\mathbf{R}(\lambda;\mathbf{A})$ is an analytic mapping from the resolvent set into the Banach space of bounded linear operators on X equipped with the uniform operator norm.

In particular $\mathbf{R}(\lambda;\mathbf{A})$ is analytic for all λ with $|\lambda| > r(\mathbf{A})$ and as in classical analytic function theory, we can expand $\mathbf{R}(\lambda;\mathbf{A})$ into a uniquely determined Laurent expansion

$$\mathbf{R}(\lambda;\mathbf{A}) = \sum_{n=0}^{\infty}\lambda^{-n-1}\mathbf{A}_n$$

with bounded linear operators $\mathbf{A}_n$ such that the series converges uniformly for all $|\lambda| > r(\mathbf{A})$ with respect to the uniform operator norm. For $|\lambda| > a^{-1}$ we already know that $\mathbf{R}(\lambda;\mathbf{A})$ is given by the Neumann series. Therefore from the uniqueness of the Laurent expansion we conclude that the Neumann series is the Laurent expansion and hence converges for all $|\lambda| > r(\mathbf{A})$.

Now we are in the position to obtain the main result of this section.

Theorem 1.36. Let $\mathbf{A}: X \to X$ be a bounded linear operator mapping the Banach space X into itself with spectral radius $r(\mathbf{A}) < 1$. Then for all $f \in X$ the successive approximations

$$\phi_{n+1} := \mathbf{A}\phi_n + f, \qquad n = 0, 1, 2, \ldots$$

with arbitrary $\phi_0 \in X$ converge to the unique solution ϕ of

$$\phi - \mathbf{A}\phi = f.$$

Proof. From Theorem 1.35 we have

$$(\mathbf{I}-\mathbf{A})^{-1} = \sum_{k=0}^{\infty}\mathbf{A}^k.$$

By induction, it is readily seen that

$$\phi_n = \mathbf{A}^n\phi_0 + \sum_{k=0}^{n}\mathbf{A}^k f, \qquad n = 1, 2, \ldots.$$

Therefore

$$\phi_n \to \sum_{k=0}^{\infty} \mathbf{A}^k f = (\mathbf{I}-\mathbf{A})^{-1} f, \qquad n \to \infty.$$

Theorem 1.37. Let $K \subset \mathbf{C}$ be a subset such that $0 \in K$ is an accumulation point of K and let $\{\mathbf{A}_\kappa : X \to X, \kappa \in K\}$ be a family of bounded linear operators such that $\|\mathbf{A}_\kappa - \mathbf{A}_0\| \to 0$, $\kappa \to 0$, and assume that $\mathbf{A}_0$ has spectral radius $r(\mathbf{A}_0) < 1$. Then for sufficiently small $\kappa \in K$ the equation $\phi - \mathbf{A}_\kappa \phi = f$ can be solved by successive approximations.

Proof. Our proof is based on Kleinman and Wendland [1]. Choose $\lambda_0 \in (r(\mathbf{A}_0), 1)$. Then $\mathbf{R}(\lambda; \mathbf{A}_0) = (\lambda \mathbf{I} - \mathbf{A}_0)^{-1}$ exists and is bounded for all $\lambda \in \mathbf{C}$ with $|\lambda| \geqslant \lambda_0$. Since $\mathbf{R}(\lambda; \mathbf{A}_0)$ is analytic and $\|\mathbf{R}(\lambda; \mathbf{A}_0\| \to 0$, $\lambda \to \infty$ (this follows from the Neumann series expansion of $\mathbf{R}(\lambda; \mathbf{A}_0)$), we have

$$C := \max_{|\lambda| \geqslant \lambda_0} \|(\lambda \mathbf{I} - \mathbf{A}_0)^{-1}\| < \infty.$$

Since $\|\mathbf{A}_\kappa - \mathbf{A}_0\| \to 0$, $\kappa \to 0$, there exists $\kappa_0 > 0$ such that

$$\|(\lambda \mathbf{I} - \mathbf{A}_0)^{-1}(\mathbf{A}_\kappa - \mathbf{A}_0)\| \leqslant q < 1$$

for all $\kappa \in K$ with $|\kappa| \leqslant \kappa_0$ and all $\lambda \in \mathbf{C}$ with $|\lambda| \geqslant \lambda_0$. But then

$$(\mathbf{I} - \mathbf{B}_\kappa)^{-1} = \sum_{n=0}^{\infty} \mathbf{B}_\kappa^n$$

converges where

$$\mathbf{B}_\kappa := (\lambda \mathbf{I} - \mathbf{A}_0)^{-1}(\mathbf{A}_\kappa - \mathbf{A}_0).$$

Finally, since

$$(\lambda \mathbf{I} - \mathbf{A}_\kappa) = (\lambda \mathbf{I} - \mathbf{A}_0)\left[\mathbf{I} - (\lambda \mathbf{I} - \mathbf{A}_0)^{-1}(\mathbf{A}_\kappa - \mathbf{A}_0)\right],$$

we see that

$$(\lambda \mathbf{I} - \mathbf{A}_\kappa)^{-1} = (\mathbf{I} - \mathbf{B}_\kappa)^{-1}(\lambda \mathbf{I} - \mathbf{A}_0)^{-1}$$

exists and is bounded. Therefore $r(\mathbf{A}_\kappa) < \lambda_0$ for all $\kappa \in K$ with $|\kappa| \leqslant \kappa_0$ and the theorem now follows from Theorem 1.36.

For a more detailed study of the spectral theory for bounded linear operators the reader is referred to Jörgens [1].

2

REGULARITY PROPERTIES OF SURFACE POTENTIALS

As the title of this book indicates, the first step in our analysis of the scattering of acoustic and electromagnetic waves by an obstacle is to reformulate the boundary-value problems of scattering theory as boundary integral equations. This will be accomplished by representing the solution of the boundary-value problem as a surface potential with respect to a given density, and then using the continuity properties of such potentials to arrive at the sought-after integral equation. Hence, in order to proceed with this objective, it is necessary to examine the regularity properties of surface potentials defined on closed surfaces. For the sake of brevity as well as practical importance, we shall restrict ourselves to surfaces in $\mathbb{R}^3$ that are twice continuously differentiable. The extension of these results to Lyapunov surfaces in $\mathbb{R}^n$ is straightforward (cf. Günter [1], Mikhlin [1]), although the problem of the scattering of waves by domains with corners presents new difficulties due to the loss of compactness of the associated integral operators (cf. Kleinman and Wendland [1], Wendland [3]).

The plan of this chapter is as follows. We first consider the differential geometry of closed surfaces in $\mathbb{R}^3$ and the concept of Hölder continuity and spaces of Hölder continuous functions defined on subsets of $\mathbb{R}^3$. These results are then used to study weakly singular operators in the space of continuous functions and Hölder continuous functions defined on closed surfaces in $\mathbb{R}^3$, with particular emphasis being placed on the compactness properties of these operators. We then turn our attention to single- and double-layer potentials and establish results on the continuity and differentiability properties of these potentials. After stating the corresponding regularity properties for vector potentials, we conclude the chapter by examining the regularity properties of the integral operators that will later appear in our study of scattering problems in acoustic and electromagnetic wave propagation.

2.1 GEOMETRY OF SURFACES

For the rest of this book, we shall let D denote a bounded open region in $\mathbb{R}^3$. The boundary of D, denoted by ∂D, is assumed to consist of a finite number of disjoint, closed bounded surfaces belonging to the class C^2 and we assume that the complement $\mathbb{R}^3 \setminus \bar{D}$ is connected. For the purposes of this chapter, it suffices to consider the case in which the boundary ∂D has just one component. Our results can be extended to the case of boundaries consisting of more than one component in an obvious way.

The property "∂D belongs to class C^2" means that for each point $z \in \partial D$ there exists a three-dimensional neighborhood V_z of z such that the intersection $\partial D \cap V_z$ can be mapped bijectively onto some open domain $U \subset \mathbb{R}^2$ and that this mapping is twice continuously differentiable. We describe this mapping in the form

$$x(u) = \left(x^1(u^1, u^2), x^2(u^1, u^2), x^3(u^1, u^2)\right)$$

where $u = (u^1, u^2) \in U$. The image of the parameter domain U under such a mapping is called a surface element. The whole boundary ∂D is obtained by patching together a finite number of surface elements.

Since the mapping is bijective, the two vectors

$$x_{,j} := \frac{\partial x}{\partial u^j}, \qquad j = 1, 2,$$

are linearly independent. They represent vectors tangent to the surface at the point x, that is, they span the tangent plane at this point. The first fundamental tensor of differential geometry is given by

$$g_{jk} := (x_{,j}, x_{,k}), \qquad j, k = 1, 2.$$

Since the tangent vectors are linearly independent, the symmetric matrix g_{jk} is positive definite and its determinant satisfies

$$g := \det(g_{jk}) = x_{,1}^2 x_{,2}^2 - (x_{,1}, x_{,2})^2 = |[x_{,1}, x_{,2}]|^2 > 0.$$

A curve C on the surface ∂D can be described by a parametric representation $x(u(t))$ where t denotes the arclength on C. Then the tangent vector τ to this curve at the point x is given by

$$\tau = \frac{du^j}{dt} x_{,j}$$

where we use the convention to sum over equal subscripts and superscripts. The line element at the point x is given by

$$dt^2 = \left(x_{,j} du^j, x_{,k} du^k\right) = g_{jk}\, du^j\, du^k,$$

the surface element at the point x by

$$ds = |[x_{,1}\,du^1, x_{,2}\,du^2]| = \sqrt{g}\,du^1\,du^2,$$

and the unit normal vector ν to ∂D at the point x by

$$\nu(x) = \frac{[x_{,1}, x_{,2}]}{|[x_{,1}, x_{,2}]|} = \frac{1}{\sqrt{g}}[x_{,1}, x_{,2}].$$

At each point of the surface we have three linearly independent vectors $x_{,1}$, $x_{,2}$, and ν. From $(\nu,\nu)=1$ we observe that $(\partial\nu/\partial u^j, \nu)=0$, $j=1,2$, and therefore we can write

$$\frac{\partial \nu}{\partial u^j} = -b_j^k x_{,k}, \qquad j=1,2,$$

for some matrix b_j^k, $j,k=1,2$. Then, using $(\nu, x_{,j})=0$, $j=1,2$, we obtain

$$(\nu, x_{,j,k}) = -\left(\frac{\partial \nu}{\partial u^k}, x_{,j}\right) = b_{jk}, \qquad j,k=1,2,$$

where $b_{jk} := g_{jr} b_k^r$ denotes the second fundamental tensor of differential geometry. As is easily verified, the two scalars

$$H := \tfrac{1}{2} b_i^i = \tfrac{1}{2}(b_1^1 + b_2^2)$$

and

$$K := \det(b_k^j) = b_1^1 b_2^2 - b_2^1 b_1^2$$

are independent of the choice of the coordinate system u^1, u^2. The quantity H is called the mean curvature and K the Gaussian curvature of the surface.

Let g^{jk} denote the inverse of the matrix g_{jk}, that is,

$$g^{jr} g_{rk} = \delta_k^j, \qquad j,k=1,2.$$

For a continuously differentiable function ϕ defined on a surface element (i.e., the function $\phi(x(u))$ is continuously differentiable in the parameter domain U), the surface gradient is defined independently of the choice of the coordinate system by

$$\operatorname{Grad}\phi = g^{jk}\frac{\partial \phi}{\partial u^j} x_{,k}.$$

The direction of $\operatorname{Grad}\phi$ is given by the direction of maximal increase of the function ϕ and the modulus of $\operatorname{Grad}\phi$ is the value of this increase. The directional derivative with respect to a curve C on the surface is given by

$$\frac{\partial \phi}{\partial t} = (\tau, \operatorname{Grad}\phi) = \frac{\partial \phi}{\partial u^j}\frac{du^j}{dt}.$$

Theorem 2.1. Let $S \subset \partial D$ be a connected surface contained in ∂D with boundary ∂S of class C^2. Let ν_0 denote the unit normal vector to the boundary curve ∂S that is perpendicular to the surface normal ν and directed to the exterior of S, and let $\tau := [\nu, \nu_0]$ denote the unit tangent vector. Let the function ϕ be continuously differentiable on S and continuous on $\bar{S}$. Then

$$\int_S \operatorname{Grad} \phi \, ds = \int_{\partial S} \phi \nu_0 \, dt - 2 \int_S \phi H \nu \, ds \tag{2.1}$$

and

$$\int_S [\nu, \operatorname{Grad} \phi] \, ds = \int_{\partial S} \phi \tau \, dt. \tag{2.2}$$

Proof. We first assume that S is a surface element. Then using Gauss' theorem in the parameter domain U, we obtain

$$\begin{aligned} 2\int_S \phi H \nu \, ds &= \int_U \phi b_j^j [x_{,1}, x_{,2}] \, du^1 \, du^2 \\ &= \int_U \phi \left\{ \left[\frac{\partial \nu}{\partial u^2}, x_{,1} \right] - \left[\frac{\partial \nu}{\partial u^1}, x_{,2} \right] \right\} du^1 \, du^2 \\ &= \int_{\partial U} \phi \{ [x_{,1}, \nu] \, du^1 + [x_{,2}, \nu] \, du^2 \} \\ &\quad - \int_U \left\{ \frac{\partial \phi}{\partial u^2} [\nu, x_{,1}] - \frac{\partial \phi}{\partial u^1} [\nu, x_{,2}] \right\} du^1 \, du^2. \end{aligned}$$

A straightforward calculation now shows that

$$[x_{,j}, \nu] \, du^j = [\tau, \nu] \, dt = \nu_0 \, dt$$

and

$$\frac{\partial \phi}{\partial u^2} [\nu, x_{,1}] - \frac{\partial \phi}{\partial u^1} [\nu, x_{,2}] = \sqrt{g} \, g^{jk} \frac{\partial \phi}{\partial u^j} x_{,k} = \sqrt{g} \operatorname{Grad} \phi.$$

Thus we conclude that

$$2 \int_S \phi H \nu \, ds = \int_{\partial S} \phi \nu_0 \, dt - \int_S \operatorname{Grad} \phi \, ds.$$

Similarly, we find that

$$\int_S [\nu, \operatorname{Grad} \phi]\, ds = \int_U g^{jk} \frac{\partial \phi}{\partial u^j} [[x_{,1}, x_{,2}], x_{,k}]\, du^1\, du^2$$

$$= \int_U \left(\frac{\partial \phi}{\partial u^1} x_{,2} - \frac{\partial \phi}{\partial u^2} x_{,1} \right) du^1\, du^2$$

$$= \int_{\partial U} \phi x_{,j}\, du^j = \int_{\partial S} \phi \tau\, dt.$$

Hence we have established (2.1) and (2.2) for surface elements. For an arbitrary surface S the theorem now follows by patching surface elements together and observing that boundary integrals over neighboring surface elements cancel.

We shall need the following result in our investigation of the regularity properties of single- and double-layer potentials.

Theorem 2.2. For the twice continuously differentiable surface ∂D there exists a positive constant L such that

$$|(\nu(y), x-y)| \leqslant L|x-y|^2 \tag{2.3}$$

and

$$|\nu(x) - \nu(y)| \leqslant L|x-y| \tag{2.4}$$

for all $x, y \in \partial D$.

Proof. Let z be an arbitrary point on ∂D. Then we can choose a neighborhood V_z and a parametric representation for $\partial D \cap V_z$ with the parameter domain U being the closed interior of a circle in $\mathbb{R}^2$. On $U \times U$ define the matrix

$$G_{jk}(x, y) := g_{jk}(x) + (x - y, x_{,j,k}), \qquad j, k = 1, 2,$$

with image points $u = (u^1, u^2)$, $v = (v^1, v^2) \in U$ for $x, y \in \partial D \cap V_z$. Since the matrix $g_{jk}(x)$ is positive definite on U, we can assume U to be small enough such that $G_{jk}(x, y)$ is positive definite on $U \times U$. Then from the continuity of $G_{jk}(x, y)\xi^j\xi^k$ on the compact set $U \times U \times \Omega$, we conclude that

$$\gamma := \inf_{\substack{x, y \in \partial D \cap V_z \\ \xi = (\xi^1, \xi^2) \in \Omega}} G_{jk}(x, y)\xi^j\xi^k > 0$$

and

$$\Gamma := \sup_{\substack{x, y \in \partial D \cap V_z \\ \xi = (\xi^1, \xi^2) \in \Omega}} G_{jk}(x, y)\xi^j\xi^k < \infty,$$

where Ω denotes the unit circle in $\mathbb{R}^2$.

Now use Taylor's formula to expand

$$(x-y)^2 = G_{jk}(\bar{x}, y)(u^j - v^j)(u^k - v^k)$$

where the image point $\bar{u}$ of $\bar{x}$ lies on the straight line connecting u and v. Hence we have the estimate

$$\gamma|u-v|^2 \leqslant |x-y|^2 \leqslant \Gamma|u-v|^2. \tag{2.5}$$

A further application of Taylor's formula yields

$$(\nu(y), x-y) = B_{jk}(\bar{\bar{x}}, y)(u^j - v^j)(u^k - v^k)$$

with the matrix $B_{jk}(x, y) = (\nu(y), x_{,j,k})$ and where the image point $\bar{\bar{u}}$ of $\bar{\bar{x}}$ again lies on the straight line connecting u and v. As above we have the estimate

$$|(\nu(y), x-y)| \leqslant B|u-v|^2 \tag{2.6}$$

where

$$B := \sup_{\substack{x, y \in \partial D \cap V_z \\ \xi = (\xi^1, \xi^2) \in \Omega}} |B_{jk}(x, y)\xi^j \xi^k| < \infty.$$

Combining (2.5) and (2.6), we have established that for any $z \in \partial D$ there exists a neighborhood V_z such that

$$|(\nu(y), x-y)| \leqslant C_z|x-y|^2$$

for all $x, y \in \partial D \cap V_z$, where C_z is a constant depending on z.

Since ∂D is compact, we can select a finite number of points $z_1, \ldots, z_m$ such that $\cup_{l=1}^m \partial D \cap V_{z_l} = \partial D$. Then we have

$$|(\nu(y), x-y)| \leqslant C|x-y|^2$$

for all $x, y \in \partial D$ with $|x-y| \leqslant \delta$ where

$$\delta := \min_{l=1,\ldots,m} \operatorname{diam} V_{z_l}$$

and

$$C := \max_{l=1,\ldots,m} C_{z_l}.$$

Finally, noting that for $|x-y| \geqslant \delta$ we have $|(\nu(y), x-y)| \leqslant d \leqslant d|x-y|^2/\delta^2$ where $d := \operatorname{diam} \partial D$, we observe that (2.3) is satisfied with $L = \max(C, d/\delta^2)$.

Equation (2.4) follows similarly by using the mean value theorem to estimate $\nu(x)-\nu(y)$.

We conclude this section with the remark that we can introduce parallel surfaces ∂D_h to ∂D by the representation

$$x = z + h\nu(z), \qquad z \in \partial D, \tag{2.7}$$

where the parameter h denotes the distance of ∂D_h from the generating surface ∂D. Since ∂D is assumed to be of class C^2, we observe that ∂D_h is of class C^1 and straightforward calculations show that

$$g(u_1, u_2; h) = g(u_1, u_2)[1 - 2hH + h^2K]^2.$$

This verifies that the surfaces are well defined provided the parameter h is restricted to be sufficiently small to ensure that the invariant $1-2hH+h^2K$ remains positive. It also shows that we can use (2.7) as a three-dimensional coordinate system in a neighborhood of ∂D. For a more detailed analysis see Martensen [1].

2.2 HÖLDER CONTINUITY

The space of uniformly Hölder continuous functions is of basic importance in the investigation of the regularity properties of single- and double-layer potentials.

Definition 2.3. Let G be a bounded closed subset of $\mathbb{R}^3$. By $C^{0,\alpha}(G)$, $0 < \alpha \leqslant 1$, we denote the linear space of all complex-valued functions ϕ defined on G satisfying

$$|\phi(x) - \phi(y)| \leqslant C|x-y|^\alpha$$

where C is a positive constant depending on ϕ but not on x and y. If G is unbounded, then by $\phi \in C^{0,\alpha}(G)$ we mean that ϕ is bounded and the above inequality is satisfied.

The space $C^{0,\alpha}(G)$ is called a Hölder space or a space of uniformly Hölder continuous functions. In our subsequent analysis we shall meet such Hölder spaces in the cases in which G is either the bounded domain $\bar{D}$, the unbounded domain $\mathbb{R}^3 \setminus D$, or the boundary ∂D.

Obviously, if $\phi \in C^{0,\alpha}(G)$, $0 < \alpha \leqslant 1$, then ϕ is uniformly continuous on G.

Theorem 2.4. The Hölder space $C^{0,\alpha}(G)$ is a Banach space with the norm

$$\|\phi\|_\alpha := \sup_{x\in G} |\phi(x)| + \sup_{\substack{x,y\in G\\ x\neq y}} \frac{|\phi(x)-\phi(y)|}{|x-y|^\alpha}.$$

Proof. It is clear that

$$|\phi|_\alpha := \sup_{\substack{x, y \in G \\ x \neq y}} \frac{|\phi(x) - \phi(y)|}{|x - y|^\alpha}$$

defines a seminorm on $C^{0,\alpha}(G)$. Then $\|\cdot\|_\alpha$ is a norm, since $\|\phi\|_\infty := \sup_{x \in G}|\phi(x)|$ defines a norm.

It remains to be shown that $C^{0,\alpha}(G)$ is complete. Let (ϕ_n) denote a Cauchy sequence in $C^{0,\alpha}(G)$. Then obviously (ϕ_n) is also a Cauchy sequence in $C(G)$ and therefore there exists a function $\phi \in C(G)$ such that $\|\phi_n - \phi\|_\infty \to 0$, $n \to \infty$. Since (ϕ_n) is a Cauchy sequence in $C^{0,\alpha}(G)$, given $\varepsilon > 0$, there exists $N(\varepsilon) \in \mathbb{N}$ such that

$$|\phi_n - \phi_m|_\alpha < \varepsilon, \qquad n, m \geqslant N(\varepsilon),$$

that is,

$$|[\phi_n(x) - \phi_m(x)] - [\phi_n(y) - \phi_m(y)]| \leqslant \varepsilon |x - y|^\alpha$$

for all $n, m \geqslant N(\varepsilon)$ and all $x, y \in G$. Since $\phi_n \to \phi$, $n \to \infty$, uniformly on G, by letting $m \to \infty$ we have

$$|[\phi_n(x) - \phi(x)] - [\phi_n(y) - \phi(y)]| \leqslant \varepsilon |x - y|^\alpha$$

for all $n \geqslant N(\varepsilon)$ and all $x, y \in G$. From this we conclude that $\phi \in C^{0,\alpha}(G)$ and $|\phi_n - \phi|_\alpha \leqslant \varepsilon$, $n \geqslant N(\varepsilon)$, which implies $\|\phi_n - \phi\|_\alpha \to 0$, $n \to \infty$.

If $\alpha < \beta$ then clearly any function $\phi \in C^{0,\beta}(G)$ is also contained in $C^{0,\alpha}(G)$. For this imbedding we have the following compactness property.

Theorem 2.5. Let $0 < \alpha < \beta \leqslant 1$ and let G be compact. Then the imbedding operators

$$\mathbf{I}^\beta : C^{0,\beta}(G) \to C(G)$$

$$\mathbf{I}^{\alpha,\beta} : C^{0,\beta}(G) \to C^{0,\alpha}(G)$$

are compact.

Proof. Let K be a bounded set in $C^{0,\beta}(G)$, that is, $\|\phi\|_\beta \leqslant C$ for all $\phi \in K$. Then obviously we have

$$|\phi(x)| \leqslant C, \qquad x \in G$$

and

$$|\phi(x) - \phi(y)| \leqslant C|x - y|^\beta, \qquad x, y \in G, \tag{2.8}$$

for all $\phi \in K$, which implies that K is bounded and equicontinuous. Therefore, by thc Arzelá–Ascoli theorem (Theorem 1.12), the set K is relatively compact in $C(G)$, which, in particular, means that $\mathbf{I}^\beta$: $C^{0,\beta}(G) \to C(G)$ is compact.

It remains to be verified that K is relatively compact in $C^{0,\alpha}(G)$. From (2.8) we have that for all $\phi, \psi \in K$

$$\begin{aligned}
&|[\phi(x)-\psi(x)]-[\phi(y)-\psi(y)]| \\
&\quad = |[\phi(x)-\psi(x)]-[\phi(y)-\psi(y)]|^{\alpha/\beta} \\
&\qquad \times |[\phi(x)-\psi(x)]-[\phi(y)-\psi(y)]|^{1-\alpha/\beta} \\
&\quad \leqslant (2C)^{\alpha/\beta}|x-y|^\alpha(2\|\phi-\psi\|_\infty)^{1-\alpha/\beta}, \qquad x, y \in G,
\end{aligned}$$

which implies that

$$|\phi-\psi|_\alpha \leqslant (2C)^{\alpha/\beta}2^{1-\alpha/\beta}\|\phi-\psi\|_\infty^{1-\alpha/\beta}.$$

But from this we can conclude that any sequence taken from K and converging in $C(G)$ also converges in $C^{0,\alpha}(G)$.

We note that in a similar way we can introduce the Hölder space $C^{1,\alpha}(G)$, $0<\alpha\leqslant 1$, of uniformly Hölder continuously differentiable functions, that is, functions for which $\operatorname{grad}\phi$ (or $\operatorname{Grad}\phi$ in the case of $G=\partial D$) satisfy

$$|\operatorname{grad}\phi(x)-\operatorname{grad}\phi(y)| \leqslant C|x-y|^\alpha$$

and for which ϕ and $\operatorname{grad}\phi$ are bounded in case G is unbounded. The norm in this case is given by

$$\begin{aligned}
\|\phi\|_{1,\alpha} := &\sup_{x\in G}|\phi(x)| + \sup_{x\in G}|\operatorname{grad}\phi(x)| \\
&+ \sup_{\substack{x,y\in G\\ x\neq y}} \frac{|\operatorname{grad}\phi(x)-\operatorname{grad}\phi(y)|}{|x-y|^\alpha}.
\end{aligned}$$

Then the properties given in Theorems 2.4 and 2.5 remain true for $C^{1,\alpha}(G)$.

2.3 WEAKLY SINGULAR INTEGRAL OPERATORS ON SURFACES

In the Banach space $C(\partial D)$ of complex-valued continuous functions defined on the surface ∂D equipped with the maximum norm $\|\phi\|_\infty := \max_{x\in\partial D}|\phi(x)|$ we consider the integral operator $\mathbf{A}$: $C(\partial D) \to C(\partial D)$ defined by

$$(\mathbf{A}\phi)(x) := \int_{\partial D} K(x,y)\phi(y)\,ds(y), \qquad x \in \partial D \tag{2.9}$$

where K is a continuous or weakly singular kernel. A kernel K is said to be weakly singular if K is defined and continuous for all $x, y \in \partial D$, $x \neq y$, and there exist positive constants M and $\alpha \in (0,2]$ such that for all $x, y \in \partial D$, $x \neq y$, we have

$$|K(x, y)| \leqslant M|x-y|^{\alpha-2}. \tag{2.10}$$

Analogous to Theorems 1.10 and 1.11 we can prove the following.

Theorem 2.6. The integral operator $\mathbf{A}$ with continuous or weakly singular kernel is a compact operator on $C(\partial D)$.

Proof. The only major difference in the proof as compared with Theorems 1.10 and 1.11 arises in the verification of the existence of the integral in (2.9) defining the operator $\mathbf{A}$ as an improper integral in the case of a weakly singular kernel. By writing $(\nu(x), \nu(y)) = 1 - (\nu(x), \nu(x) - \nu(y))$, we observe from Theorem 2.2 that there exists a number $R \in (0, 1]$ such that

$$(\nu(x), \nu(y)) \geqslant \tfrac{1}{2} \tag{2.11}$$

for all $x, y \in \partial D$ with $|x-y| \leqslant R$. Furthermore, we can assume that R is sufficiently small such that

$$S_{x,R} := \{y \in \partial D \mid |y-x| < R\}$$

is connected for each $x \in \partial D$. Then condition (2.11) implies that $S_{x,R}$ can be bijectively projected into the tangent plane to ∂D at the point x. By using polar coordinates (ρ, θ) in the tangent plane, we now have the estimate

$$\begin{aligned}\left|\int_{S_{x,R}} K(x,y)\phi(y)\,ds(y)\right| &\leqslant M\|\phi\|_\infty \int_{S_{x,R}} |x-y|^{\alpha-2}\,ds(y)\\ &\leqslant 2M\|\phi\|_\infty \int_0^{2\pi}\int_0^R \rho^{\alpha-2}\rho\,d\rho\,d\theta\\ &= 4\pi M\|\phi\|_\infty \frac{R^\alpha}{\alpha}.\end{aligned}$$

Here we have used the facts that $|x-y| \geqslant \rho$, that the surface element

$$ds(y) = \frac{\rho\,d\rho\,d\theta}{(\nu(x), \nu(y))}$$

can be estimated with the aid of (2.11) by $ds(y) \leqslant 2\rho\,d\rho\,d\theta$, and that the projection of $S_{x,R}$ into the tangent plane is contained in the interior of the circle of radius R and center x. Furthermore,

$$\left|\int_{\partial D \setminus S_{x,R}} K(x,y)\phi(y)\,ds(y)\right| \leqslant M\|\phi\|_\infty \int_{\partial D \setminus S_{x,R}} R^{\alpha-2}\,ds(y)$$

$$< M\|\phi\|_\infty R^{\alpha-2}|\partial D|$$

where $|\partial D|$ denotes the surface area of ∂D. Hence, for all $x \in \partial D$ the integral (2.9) exists as an improper integral and we have

$$\|\mathbf{A}\phi\|_\infty \leqslant C\|\phi\|_\infty \tag{2.12}$$

where the constant C is defined by $C := (4\pi(R^\alpha/\alpha) + R^{\alpha-2}|\partial D|)M$.

We shall now impose further conditions on the kernel in order to ensure the compactness of the integral operator $\mathbf{A}$ on the Hölder space $C^{0,\beta}(\partial D)$.

Theorem 2.7. Let G be a closed domain containing ∂D in its interior. Assume the function K is defined and continuous for all $x \in G$, $y \in \partial D$, $x \neq y$, and assume there exist positive constants M and $\alpha \in (0,2]$ such that for all $x \in G$, $y \in \partial D$, $x \neq y$, we have

$$|K(x,y)| \leqslant M|x-y|^{\alpha-2}. \tag{2.13}$$

Assume further that there exists $m \in \mathbb{N}$ such that

$$|K(x_1,y) - K(x_2,y)| \leqslant M \sum_{j=1}^{m} |x_1-y|^{\alpha-2-j}|x_1-x_2|^j \tag{2.14}$$

for all $x_1, x_2 \in G$, $y \in \partial D$ with $2|x_1 - x_2| \leqslant |x_1 - y|$. Then the generalized potential u defined by

$$u(x) := \int_{\partial D} K(x,y)\phi(y)\,ds(y), \qquad x \in G, \tag{2.15}$$

with density $\phi \in C(\partial D)$ belongs to the Hölder space $C^{0,\beta}(G)$ for all $\beta \in (0,\alpha]$ if $0 < \alpha < 1$, for all $\beta \in (0,1)$ if $\alpha = 1$, and for all $\beta \in (0,1]$ if $1 < \alpha < 2$ and

$$\|u\|_{\beta,G} \leqslant C_\beta \|\phi\|_{\infty,\partial D} \tag{2.16}$$

for some constant C_β depending on β.

Proof. By the arguments used in the proof of Theorem 2.6 the function u is well defined as an improper integral for $x \in \partial D$.

Choose a positive number h_0 such that the parallel surfaces (2.7) are well defined for all $|h| \leqslant h_0$ and define the set D_{h_0} by

$$D_{h_0} := \{x = z + h\nu(z), \qquad z \in \partial D, \qquad |h| \leqslant h_0\}.$$

Then, analogous to (2.12), we can easily show that

$$|u(x)| \leqslant C\|\phi\|_\infty \tag{2.17}$$

for all $x \in D_{h_0}$.

To establish the uniform Hölder continuity, let $x_1, x_2 \in D_{h_0}$ with $0 < |x_1 - x_2| < R/4$. Both x_1, x_2 may lie on ∂D. Now choose uniquely determined points $z_1, z_2 \in \partial D$ such that $x_j = z_j + h_j \nu(z_j), j = 1, 2$. As is easily seen by using Theorem 2.2, we have the estimate

$$|z_1 - z_2| \leqslant 2|x_1 - x_2| \tag{2.18}$$

provided R and h_0 are chosen small enough depending on the constant L. We now set

$$r := 4|x_1 - x_2| \tag{2.19}$$

and using (2.13) we find as in Theorem 2.6 that

$$\left|\int_{S_{z_1,r}} [K(x_1, y) - K(x_2, y)]\phi(y)\, ds(y)\right|$$

$$\leqslant M\|\phi\|_\infty \left\{\int_{S_{z_1,r}} |x_1 - y|^{\alpha-2} ds(y) + \int_{S_{z_2,2r}} |x_2 - y|^{\alpha-2} ds(y)\right\}$$

$$\leqslant C_1\|\phi\|_\infty |x_1 - x_2|^\alpha \tag{2.20}$$

for some constant C_1 depending on M and α. Here we have used the fact that by (2.18) $S_{z_1,r} \subset S_{z_2,2r}$. Using condition (2.14), we have the estimate

$$\left|\int_{S_{z_1,R} \setminus S_{z_1,r}} [K(x_1, y) - K(x_2, y)]\phi(y)\, ds(y)\right|$$

$$\leqslant M\|\phi\|_\infty \sum_{j=1}^{m} |x_1 - x_2|^j \int_{S_{z_1,R} \setminus S_{z_1,r}} |x_1 - y|^{\alpha-2-j} ds(y)$$

$$\leqslant 4\pi M\|\phi\|_\infty \sum_{j=1}^{m} |x_1 - x_2|^j \int_{r/4}^{R} \rho^{\alpha-1-j} d\rho$$

where we have used the fact that the projection of $S_{z_1,R} \setminus S_{z_1,r}$ into the tangent plane at z_1 is contained in the annulus with radii $r/4$ and R. We now note that

$$\int_{r/4}^{R} \rho^{\alpha-1-j}\,d\rho \leqslant \begin{cases} \dfrac{1}{j-\alpha}|x_1-x_2|^{\alpha-j} & \text{if} \quad j>\alpha \\ \log\dfrac{R}{|x_1-x_2|} & \text{if} \quad j=\alpha \\ \dfrac{1}{\alpha-j}R^{\alpha-j} & \text{if} \quad j<\alpha \end{cases}$$

and if $\beta \in (0,1)$, $|x_1-x_2|<1$, we have

$$|x_1-x_2|\log\frac{1}{|x_1-x_2|} \leqslant \frac{1}{1-\beta}|x_1-x_2|^{\beta}.$$

Hence

$$\left|\int_{S_{z_1,R}\setminus S_{z_1,r}} [K(x_1,y)-K(x_2,y)]\phi(y)\,ds(y)\right|$$

$$\leqslant \begin{cases} C_2\|\phi\|_\infty |x_1-x_2|^{\alpha}, & \alpha<1, \\ C_2\|\phi\|_\infty |x_1-x_2|^{\beta}, & \alpha=1, \qquad 0<\beta<1, \\ C_2\|\phi\|_\infty |x_1-x_2|, & \alpha>1, \end{cases} \tag{2.21}$$

for some constant C_2 depending on m, M, R, α, and β. Finally, again using (2.14), we have the estimate

$$\left|\int_{\partial D\setminus S_{z_1,R}} [K(x_1,y)-K(x_2,y)]\phi(y)\,ds(y)\right|$$

$$\leqslant M\|\phi\|_\infty \sum_{j=1}^{m} |x_1-x_2|^j \int_{\partial D\setminus S_{z_1,R}} |x_1-y|^{\alpha-2-j}\,ds(y)$$

$$\leqslant C_3\|\phi\|_\infty |x_1-x_2| \tag{2.22}$$

for some constant C_3 depending on m, M, R, α, and $|\partial D|$. Combining (2.20), (2.21), and (2.22), we obtain

$$|u(x_1)-u(x_2)| \leqslant (C_1+C_2+C_3)|x_1-x_2|^{\beta}\|\phi\|_\infty$$

for all $x_1, x_2 \in D_{h_0}$ with $|x_1-x_2| \leqslant R/4$. Finally, if $|x_1-x_2| \geqslant R/4$, with the aid of (2.17) we can trivially obtain the estimate

$$|u(x_1)-u(x_2)| \leqslant 2C\left(\frac{4}{R}\right)^{\beta}|x_1-x_2|^{\beta}\|\phi\|_\infty .$$

Thus we find that

$$|u(x_1)-u(x_2)| \leqslant C_0|x_1-x_2|^{\beta}\|\phi\|_{\infty}$$

for all $x_1, x_2 \in D_{h_0}$, where C_0 is some constant depending on ∂D, α, and β. The inequality (2.16) now follows from this in a trivial manner.

Remark 2.8. If the kernel K is defined and continuous for all $x, y \in \partial D$, $x \neq y$, and satisfies conditions (2.13) and (2.14) on ∂D, then the potential u defined by (2.15) with density $\phi \in C(\partial D)$ belongs to the Hölder space $C^{0,\beta}(\partial D)$ and

$$\|u\|_{\beta, \partial D} \leqslant C_{\beta}\|\phi\|_{\infty, \partial D}. \tag{2.23}$$

Corollary 2.9. Let K be a weakly singular kernel satisfying condition (2.14) for all $x_1, x_2 \in \partial D$, $y \in \partial D$, with $2|x_1-x_2| \leqslant |x_1-y|$. Then the integral operator $\mathbf{A}$: $C^{0,\beta}(\partial D) \to C^{0,\beta}(\partial D)$ defined by (2.9) is compact for all $\beta \in (0, \alpha]$ if $0<\alpha<1$, for all $\beta \in (0,1)$ if $\alpha=1$ and for all $\beta \in (0,1]$ if $1<\alpha<2$.

Proof. Inequality (2.23) shows that the integral operator $\mathbf{A}$: $C(\partial D) \to C^{0,\beta}(\partial D)$ is bounded. Then we combine Theorem 1.5 and the imbedding Theorem 2.5 to obtain the compactness of $\mathbf{A}$: $C^{0,\beta}(\partial D) \to C^{0,\beta}(\partial D)$.

We conclude this section with a lemma that we shall prove by the same techniques we used in Theorem 2.7.

Lemma 2.10. Assume the function K to be defined and continuous for all $x \in D_{h_0}$, $y \in \partial D$, $x \neq y$, and assume that there exists a positive constant M such that for all $x \in D_{h_0}$, $y \in \partial D$, $x \neq y$, we have

$$|K(x,y)| \leqslant M|x-y|^{-2}. \tag{2.24}$$

Furthermore, assume there exists $m \in \mathbb{N}$ such that

$$|K(x_1,y)-K(x_2,y)| \leqslant M\sum_{j=1}^{m}|x_1-y|^{-2-j}|x_1-x_2|^j \tag{2.25}$$

for all $x_1, x_2 \in D_{h_0}$, $y \in \partial D$, with $2|x_1-x_2| \leqslant |x_1-y|$, and that

$$\left|\int_{\partial D \setminus S_{z,r}} K(x,y)\, ds(y)\right| \leqslant M \tag{2.26}$$

for all $z \in \partial D$ and $x = z + h\nu(z) \in D_{h_0}$ and all $0<r<R$. Now define

$$u(x) := \int_{\partial D} K(x,y)[\phi(y)-\phi(z)]\, ds(y), \qquad x \in D_{h_0}, \tag{2.27}$$

with density $\phi \in C^{0,\alpha}(\partial D)$, $0<\alpha<1$. Then u belongs to $C^{0,\alpha}(D_{h_0})$ and

$$\|u\|_{\alpha, D_{h_0}} \leqslant C\|\phi\|_{\alpha, \partial D} \tag{2.28}$$

for some constant C.

Proof. Using (2.18) we observe that $|K(x,y)[\phi(y)-\phi(z)]| \leqslant 2^{\alpha}M|\phi|_{\alpha}$ $|x-y|^{\alpha-2}$ which establishes the existence of the improper integral in (2.27). Then analogous to (2.20), (2.21), and (2.22), we have

$$\left|\int_{S_{z_1,r}} \{K(x_1,y)[\phi(y)-\phi(z_1)]-K(x_2,y)[\phi(y)-\phi(z_2)]\}\,ds(y)\right|$$

$$\leqslant C_1|\phi|_{\alpha}|x_1-x_2|^{\alpha},$$

$$\left|\int_{S_{z_1,R}\setminus S_{z_1,r}} [K(x_1,y)-K(x_2,y)][\phi(y)-\phi(z_2)]\,ds(y)\right|$$

$$\leqslant C_2|\phi|_{\alpha}|x_1-x_2|^{\alpha}$$

and

$$\left|\int_{\partial D\setminus S_{z_1,R}} [K(x_1,y)-K(x_2,y)][\phi(y)-\phi(z_2)]\,ds(y)\right|$$

$$\leqslant C_3|\phi|_{\alpha}|x_1-x_2|$$

for some constants C_1, C_2, and C_3 depending on M, m, α, and ∂D. Note that the logarithmic term does not appear in the second inequality because $0<\alpha<1$. Finally, because of our assumptions (2.26) and using (2.18), we have

$$\left|[\phi(z_1)-\phi(z_2)]\int_{\partial D\setminus S_{z_1,r}} K(x_1,y)\,ds(y)\right|$$

$$\leqslant C_4|\phi|_{\alpha}|x_1-x_2|^{\alpha}$$

since $\phi\in C^{0,\alpha}(\partial D)$. The last four inequalities can be combined to yield

$$|u(x_1)-u(x_2)| \leqslant (C_1+C_2+C_3+C_4)|\phi|_{\alpha}|x_1-x_2|^{\alpha}$$

and the proof is now completed as in Theorem 2.7.

Remark 2.11. Analogous to Remark 2.8, we can state a variant of Lemma 2.10 for a kernel K defined only on ∂D, that is,

$$\|u\|_{\alpha,\partial D} \leqslant C\|\phi\|_{\alpha,\partial D}. \tag{2.29}$$

2.4 SINGLE- AND DOUBLE-LAYER POTENTIALS

Let k be a complex number such that

$$\operatorname{Im} k \geqslant 0.$$

As is readily verified, the function

$$\Phi(x, y) := \frac{1}{4\pi} \frac{e^{ik|x-y|}}{|x-y|}, \qquad x, y \in \mathbb{R}^3, \qquad x \neq y, \tag{2.30}$$

is a solution to the Helmholtz equation

$$\Delta \Phi + k^2 \Phi = 0$$

with respect to x for any fixed y. Because of its polelike singularity at $x = y$, the function Φ is called a fundamental solution to the Helmholtz equation.

Given a function $\phi \in C(\partial D)$, the function

$$u(x) := \int_{\partial D} \Phi(x, y) \phi(y)\, ds(y), \qquad x \in \mathbb{R}^3 \setminus \partial D, \tag{2.31}$$

is called the acoustic single-layer potential with density ϕ. Since for $x \in \mathbb{R}^3 \setminus \partial D$ we can differentiate under the integral sign, we see that u is a solution of the Helmholtz equation and therefore, as we shall see later (Theorem 3.5), analytic in $\mathbb{R}^3 \setminus \partial D$. In the following, we shall investigate properties of surface potentials for points on the boundary.

Since

$$|\Phi(x, y)| \leqslant \frac{1}{4\pi|x-y|}, \qquad x \neq y,$$

the kernel Φ is weakly singular with $\alpha = 1$ and hence the single-layer potential is well defined for all points $x \in \partial D$. Using

$$\left| \frac{1}{|x_1 - y|} - \frac{1}{|x_2 - y|} \right| \leqslant \frac{|x_1 - x_2|}{|x_1 - y|\,|x_2 - y|} \leqslant \frac{2|x_1 - x_2|}{|x_1 - y|^2},$$

for $2|x_1 - x_2| \leqslant |x_2 - y|$ and

$$|e^{ik|x_1 - y|} - e^{ik|x_2 - y|}| \leqslant k|x_1 - x_2|$$

we observe that Φ satisfies (2.14) with $m = 1$. Therefore, applying Theorem 2.7, we immediately obtain the following theorem.

Theorem 2.12. The single-layer potential u with continuous density ϕ is uniformly Hölder continuous throughout $\mathbb{R}^3$ and

$$\|u\|_{\alpha,\mathbb{R}^3} \leq C_\alpha \|\phi\|_{\infty,\partial D} \tag{2.32}$$

for all $0 < \alpha < 1$ for some constant C_α depending on ∂D and α.

Given a function $\psi \in C(\partial D)$, the function

$$v(x) := \int_{\partial D} \frac{\partial \Phi(x,y)}{\partial \nu(y)} \psi(y)\, ds(y), \qquad x \in \mathbb{R}^3 \setminus \partial D \tag{2.33}$$

is called the acoustic double-layer potential with density ψ. We assume the unit normal ν to be directed into the exterior domain $\mathbb{R}^3 \setminus \bar{D}$. We note that the double-layer potential v is a solution of the Helmholtz equation and, therefore, analytic in $\mathbb{R}^3 \setminus \partial D$. In the following, we shall distinguish by indices $+$ and $-$ the limits obtained by approaching the boundary ∂D from inside $\mathbb{R}^3 \setminus \bar{D}$ and D, respectively, that is,

$$v_+(x) = \lim_{\substack{y \to x \\ y \in \mathbb{R}^3 \setminus \bar{D}}} v(y), \qquad v_-(x) = \lim_{\substack{y \to x \\ y \in D}} v(y), \qquad x \in \partial D.$$

For the next three theorems, it suffices to carry out the proof in the potential theoretic case $k = 0$ with fundamental solution

$$\Phi_0(x,y) = \frac{1}{4\pi |x-y|}, \qquad x, y \in \mathbb{R}^3, \qquad x \neq y. \tag{2.34}$$

The extension to arbitrary $k \neq 0$ follows from the fact that the difference of the double-layer potential with kernel Φ and Φ_0 and continuous density ψ is uniformly Hölder continuous throughout $\mathbb{R}^3$ by Theorem 2.7. The function

$$\begin{aligned} K(x,y) &:= \frac{\partial \Phi(x,y)}{\partial \nu(y)} - \frac{\partial \Phi_0(x,y)}{\partial \nu(y)} \\ &= \frac{(\nu(y), x-y)}{4\pi |x-y|^3} \left[e^{ik|x-y|} - ik|x-y| e^{ik|x-y|} - 1 \right] \end{aligned}$$

satisfies conditions (2.13) and (2.14) with $\alpha = 2$ and $m = 1$ on all of $\mathbb{R}^3$.

Theorem 2.13. The double-layer potential v with continuous density ψ can be continuously extended from $\mathbb{R}^3 \setminus \bar{D}$ to $\mathbb{R}^3 \setminus D$ and from D to $\bar{D}$ with limiting values

$$v_\pm(x) = \int_{\partial D} \frac{\partial \Phi(x,y)}{\partial \nu(y)} \psi(y)\, ds(y) \pm \frac{1}{2}\psi(x), \qquad x \in \partial D, \tag{2.35}$$

where the integral exists as an improper integral.

Proof. As already pointed out, we need only consider the case $k=0$. From

$$\frac{\partial\Phi_0(x,y)}{\partial\nu(y)}=\frac{(\nu(y),x-y)}{4\pi|x-y|^3} \tag{2.36}$$

and Theorem 2.2, we observe that

$$\left|\frac{\partial\Phi_0(x,y)}{\partial\nu(y)}\right|\leqslant\frac{L}{4\pi|x-y|},\qquad x,y\in\partial D,\qquad x\neq y. \tag{2.37}$$

Hence the integral in (2.35) exists as an improper integral.

We first prove Theorem 2.13 for the double-layer potential

$$w(x)=\int_{\partial D}\frac{\partial\Phi_0(x,y)}{\partial\nu(y)}\,ds(y),\qquad x\in\mathbb{R}^3\setminus\partial D \tag{2.38}$$

with constant density $\psi=1$. Using Gauss' theorem, we readily see that

$$w(x)=\begin{cases}0, & x\in\mathbb{R}^3\setminus\bar{D},\\ -1, & x\in D.\end{cases} \tag{2.39}$$

A further application of Gauss' theorem shows that

$$\begin{aligned}\int_{\partial D}\frac{\partial\Phi_0(x,y)}{\partial\nu(y)}\,ds(y)&=\lim_{r\to 0}\int_{H_{x,r}}\frac{\partial\Phi_0(x,y)}{\partial\nu(y)}\,ds(y)\\&=\lim_{r\to 0}\frac{-1}{4\pi r^2}\int_{H_{x,r}}ds(y),\qquad x\in\partial D,\end{aligned}$$

where $H_{x,r}$ denotes that part of the surface of the sphere $\Omega_{x,r}$ of radius r and center x that is contained in D and where ν denotes the exterior unit normal to this sphere. With the help of Theorem 2.2, it can be seen that

$$\int_{H_{x,r}}ds(y)=2\pi r^2+O(r^3)$$

uniformly on ∂D. Hence

$$\int_{\partial D}\frac{\partial\Phi_0(x,y)}{\partial\nu(y)}\,ds(y)=-\frac{1}{2},\qquad x\in\partial D, \tag{2.40}$$

which concludes the proof of Theorem 2.13 in the special case of a constant density.

For arbitrary continuous density in $D_{h_0}\setminus\partial D$ we first write v in the form

$$v(x)=\psi(z)w(x)+u(x),\qquad x=z+h\nu(z),\qquad x\in D_{h_0}\setminus\partial D \tag{2.41}$$

where

$$u(x) = \int_{\partial D} \frac{\partial \Phi_0(x,y)}{\partial \nu(y)} [\psi(y) - \psi(z)]\, ds(y). \tag{2.42}$$

In order to prove the theorem we must show that u is continuous on D_{h_0}. By (2.37) and Theorem 2.6, the integral in (2.42) exists as an improper integral for $x \in \partial D$ and represents a continuous function on ∂D. Therefore, it suffices to show that

$$\lim_{x \to z} u(x) = \lim_{h \to 0} u(z + h\nu(z)) = u(z), \qquad z \in \partial D,$$

uniformly on ∂D.

Using Theorem 2.2, we have the estimate

$$\begin{aligned} |x-y|^2 &= |z-y|^2 + 2(z-y, x-z) + |x-z|^2 \\ &\geq \tfrac{1}{2}\{|z-y|^2 + |x-z|^2\} \end{aligned}$$

provided h_0 is sufficiently small. Then, writing

$$4\pi \frac{\partial \Phi_0(x,y)}{\partial \nu(y)} = \frac{(\nu(y), z-y)}{|x-y|^3} + \frac{(\nu(y), x-z)}{|x-y|^3},$$

for $r < R$, we obtain by projecting onto the tangent plane that

$$\begin{aligned} \int_{S_{z,r}} \left| \frac{\partial \Phi_0(x,y)}{\partial \nu(y)} \right| ds(y) &\leq C_1 \left\{ \int_0^r d\rho + |x-z| \int_0^\infty \frac{\rho\, d\rho}{(\rho^2 + |x-z|^2)^{3/2}} \right\} \\ &= C_1(r+1) \leq C_1(R+1) \end{aligned} \tag{2.43}$$

where C_1 denotes some constant depending on ∂D. From the mean value theorem we see that

$$\left| \frac{\partial \Phi_0(x,y)}{\partial \nu(y)} - \frac{\partial \Phi_0(z,y)}{\partial \nu(y)} \right| \leq C_2 \frac{|x-z|}{|z-y|^3}$$

for $2|x - z| \leq |z - y|$ and therefore

$$\int_{\partial D \setminus S_{z,r}} \left| \frac{\partial \Phi_0(x,y)}{\partial \nu(y)} - \frac{\partial \Phi_0(z,y)}{\partial \nu(y)} \right| ds(y) \leq C_3 \frac{|x-z|}{r^3} \tag{2.44}$$

for some constants C_2 and C_3. Now we can combine (2.43) and (2.44) to obtain

$$|u(x) - u(z)| \leq C \left\{ \sup_{|y-z| \leq r} |\psi(y) - \psi(z)| + \frac{|x-z|}{r^3} \right\}$$

for some constant C. Given $\varepsilon > 0$ we can choose $r > 0$ such that

$$|\psi(y) - \psi(z)| < \frac{\varepsilon}{2C}$$

for all $y, z \in \partial D$ with $|y - z| < r$ since ψ is uniformly continuous on ∂D. Then taking $\delta < (\varepsilon/2C)r^3$, we see that

$$|u(x) - u(z)| < \varepsilon$$

for all $|x - z| < \delta$ and the theorem is now proved.

Corollary 2.14. For the double-layer potential v with continuous density ψ, we have the jump relation

$$v_+ - v_- = \psi \quad \text{on} \quad \partial D. \tag{2.45}$$

Theorem 2.15. The direct values of the double-layer potential

$$v(x) := \int_{\partial D} \frac{\partial \Phi(x, y)}{\partial \nu(y)} \psi(y)\, ds(y), \qquad x \in \partial D, \tag{2.46}$$

with continuous density ψ represent a uniformly Hölder continuous function on ∂D with

$$\|v\|_{\alpha, \partial D} \leqslant C_\alpha \|\psi\|_{\infty, \partial D} \tag{2.47}$$

for $0 < \alpha < 1$ and some constant C_α depending on ∂D and α.

Proof. The theorem will follow from Remark 2.8 if we can show that the kernel (2.36) satisfies conditions (2.13) and (2.14) of Theorem 2.7 with $\alpha = 1$ and $m = 2$ for x, y restricted to the boundary ∂D. But condition (2.13) is already verified by (2.37), and (2.14) follows from the decomposition

$$4\pi\left[\frac{\partial \Phi_0(x_1, y)}{\partial \nu(y)} - \frac{\partial \Phi_0(x_2, y)}{\partial \nu(y)}\right] = \left\{\frac{1}{|x_1 - y|^3} - \frac{1}{|x_2 - y|^3}\right\}(\nu(y), x_2 - y) + \frac{(\nu(y), x_1 - x_2)}{|x_1 - y|^3} \tag{2.48}$$

if we use the mean value theorem to estimate the first term on the right-hand side and Theorem 2.2 and the inequality

$$\begin{aligned} |(\nu(y), x_1 - x_2)| &\leqslant |(\nu(y) - \nu(x_1), x_1 - x_2)| + |(\nu(x_1), x_1 - x_2)| \\ &\leqslant L\{|y - x_1|\,|x_1 - x_2| + |x_1 - x_2|^2\} \end{aligned}$$

to estimate the second term.

Theorem 2.16. The double-layer potential v with uniformly Hölder continuous density $\psi \in C^{0,\alpha}(\partial D)$, $0<\alpha<1$, is uniformly Hölder continuous in $\mathbb{R}^3 \setminus D$ and in $\bar{D}$ with

$$\|v\|_{\alpha,\mathbb{R}^3\setminus D} \leqslant C_\alpha \|\psi\|_{\alpha,\partial D}, \qquad \|v\|_{\alpha,\bar{D}} \leqslant C_\alpha \|\psi\|_{\alpha,\partial D} \tag{2.49}$$

where C_α is some constant depending on ∂D and α.

Proof. In the decomposition (2.41), the first term obviously has the properties stated in the theorem in the domain D_{h_0}. For the second part, we apply Lemma 2.10 and observe that the kernel (2.36) satisfies conditions (2.24) and (2.25) with $m=1$ for $x \in \mathbb{R}^3$ and $y \in \partial D$. Furthermore, from (2.43) it is seen that condition (2.26) is also satisfied. Hence the second term in (2.41) is uniformly Hölder continuous in D_{h_0}. The extension to all of $\mathbb{R}^3$ now follows from the analyticity of v in $\mathbb{R}^3 \setminus \partial D$.

2.5 DERIVATIVES OF SINGLE- AND DOUBLE-LAYER POTENTIALS

We begin our investigation of the differentiability of surface potentials at the boundary with the following theorem.

Theorem 2.17. The first derivatives of a single-layer potential u with uniformly Hölder continuous density $\phi \in C^{0,\alpha}(\partial D)$, $0<\alpha<1$, can be uniformly extended in a Hölder continuous fashion from $\mathbb{R}^3 \setminus \bar{D}$ into $\mathbb{R}^3 \setminus D$ and from D into $\bar{D}$ with limiting values

$$\operatorname{grad} u_\pm(x) = \int_{\partial D} \operatorname{grad}_x \Phi(x,y)\phi(y)\,ds(y) \mp \tfrac{1}{2}\nu(x)\phi(x), \qquad x \in \partial D, \tag{2.50}$$

where the integral exists as a Cauchy principal value. Furthermore, we have the estimates

$$\|\operatorname{grad} u\|_{\alpha,\mathbb{R}^3\setminus D} \leqslant C_\alpha \|\phi\|_{\alpha,\partial D},$$

$$\|\operatorname{grad} u\|_{\alpha,\bar{D}} \leqslant C_\alpha \|\phi\|_{\alpha,\partial D} \tag{2.51}$$

for some constant C_α depending on ∂D and α.

Proof. We first treat the case of a single-layer potential

$$w(x) = \int_{\partial D} \Phi(x,y)\,ds(y)$$

with constant density $\phi = 1$. Using the symmetry relation

$$\operatorname{grad}_x \Phi(x, y) = -\operatorname{grad}_y \Phi(x, y) \tag{2.52}$$

(by the subscripts we indicate differentiation with respect to x and y), we can write $\operatorname{grad} w = U + V$ where

$$U(x) := -\int_{\partial D} \operatorname{Grad}_y \Phi(x, y)\, ds(y), \qquad x \in \mathbb{R}^3 \setminus \partial D,$$

and

$$V(x) := -\int_{\partial D} \nu(y) \frac{\partial \Phi(x, y)}{\partial \nu(y)}\, ds(y), \qquad x \in \mathbb{R}^3 \setminus \partial D.$$

Obviously, V represents a double-layer potential with density $\nu \in C^{0,1}(\partial D)$. Therefore, we can apply Theorems 2.13 and 2.16 to deduce the uniform Hölder continuity of V, including the estimates (2.51) and the limiting values

$$V_{\pm}(x) = -\int_{\partial D} \nu(y) \frac{\partial \Phi(x, y)}{\partial \nu(y)}\, ds(y) \mp \frac{1}{2} \nu(x), \qquad x \in \partial D.$$

Using the integral identity (2.1), we can rewrite U in the form

$$U(x) = 2\int_{\partial D} H(y) \nu(y) \Phi(x, y)\, ds(y), \qquad x \in \mathbb{R}^3 \setminus \partial D$$

that is, as a single-layer potential with continuous density $2H\nu$. We can now apply Theorem 2.12 to deduce the uniform Hölder continuity of U, including the estimates (2.51) and the limiting values

$$U_{\pm}(x) = 2\int_{\partial D} H(y) \nu(y) \Phi(x, y)\, ds(y), \qquad x \in \partial D.$$

Thus, by combining these results, the proof can be completed by showing that

$$\lim_{r \to 0} \int_{\partial D \setminus S_{x,r}} \operatorname{Grad}_y \Phi(x, y)\, ds(y) = -2\int_{\partial D} H(y) \nu(y) \Phi(x, y)\, ds(y). \tag{2.53}$$

From the integral identity (2.1) we have

$$\int_{\partial D \setminus S_{x,r}} \operatorname{Grad}_y \Phi(x, y)\, ds(y) = \int_{|y - x| = r} \Phi(x, y) \nu_0(y)\, dt(y)$$

$$- 2\int_{\partial D \setminus S_{x,r}} H(y) \nu(y) \Phi(x, y)\, ds(y).$$

But

$$\int_{|y-x|=r} \Phi(x,y)\nu_0(y)\,dt(y) = \frac{e^{ikr}}{4\pi r}\int_{|y-x|=r} \nu_0(y)\,dt(y)$$

$$= -\frac{e^{ikr}}{2\pi r}\int_{S_{x,r}} H(y)\nu(y)\,ds(y),$$

and using the estimate

$$\left|\int_{S_{x,r}} H(y)\nu(y)\,ds(y)\right| \leqslant Cr^2$$

valid for some constant C depending on ∂D, we now obtain (2.53).

The extension to arbitrary densities now follows analogously to the proof of Theorem 2.16 by writing

$$\operatorname{grad} u(x) = \phi(z)\operatorname{grad} w(x) + W(x)$$

and applying Lemma 2.10 to

$$W(x) := \int_{\partial D} \operatorname{grad}_x \Phi(x,y)[\phi(y)-\phi(z)]\,ds(y).$$

Corollary 2.18. For the single-layer potential u with uniformly Hölder continuous density ϕ we have the jump relation

$$\operatorname{grad} u_+ - \operatorname{grad} u_- = -\nu\phi \quad \text{on} \quad \partial D. \tag{2.54}$$

As can be seen from counterexamples in Günter [1], for a single-layer potential with merely continuous density the gradient on the boundary, in general, does not exist. However, for the normal derivative, we can prove the following theorem.

Theorem 2.19. For the single-layer potential u with continuous density ϕ we have

$$\frac{\partial u_\pm}{\partial \nu}(x) = \int_{\partial D} \frac{\partial \Phi(x,y)}{\partial \nu(x)}\phi(y)\,ds(y) \mp \frac{1}{2}\phi(x), \qquad x \in \partial D, \tag{2.55}$$

where

$$\frac{\partial u_\pm}{\partial \nu}(x) = \lim_{\substack{h\to 0\\ h>0}} \frac{\partial u(x \pm h\nu(x))}{\partial \nu(x)}$$

is to be understood in the sense of uniform convergence and where the integral in (2.55) exists as an improper integral.

Proof. Let v denote the double-layer potential with density ϕ. Then for $x \in D_{h_0} \setminus \partial D$, with $x = z + h\nu(z)$, we can write

$$(\nu(z), \operatorname{grad} u(x)) + v(x) = \int_{\partial D} (\nu(y) - \nu(z), \operatorname{grad}_y \Phi(x, y)) \phi(y)\, ds(y)$$

where we have made use of (2.52). From Theorem 2.7, the right-hand side is seen to be continuous and conditions (2.13) and (2.14) are satisfied with $\alpha = 1$ and $m = 1$. The proof is now completed by applying Theorem 2.13 to the double-layer potential v.

Corollary 2.20. For the single-layer potential u with continuous density ϕ we have the jump relation

$$\frac{\partial u_+}{\partial \nu} - \frac{\partial u_-}{\partial \nu} = -\phi \quad \text{on} \quad \partial D. \tag{2.56}$$

We conclude our analysis of surface potentials with three results on the derivatives of double-layer potentials. It again suffices to carry out the proofs in the case $k = 0$. The extension to arbitrary $k \neq 0$ follows from the observation that the difference of the gradients of double-layer potentials with kernels Φ and Φ_0 and continuous density ψ is uniformly Hölder continuous throughout $\mathbb{R}^3$ by Theorem 2.7. The components of

$$\begin{aligned} K(x, y) :&= \operatorname{grad}_x \frac{\partial \Phi(x, y)}{\partial \nu(y)} - \operatorname{grad}_x \frac{\partial \Phi_0(x, y)}{\partial \nu(y)} \\ &= (x - y) \frac{(\nu(y), y - x)}{4\pi |x - y|^5} \left[(3 - 3ik|x - y| - k^2|x - y|^2) e^{ik|x - y|} - 3 \right] \\ &\quad + \frac{\nu(y)}{4\pi |x - y|^3} \left[(1 - ik|x - y|) e^{ik|x - y|} - 1 \right] \end{aligned} \tag{2.57}$$

satisfy conditions (2.13) and (2.14) with $\alpha = 1$ and $m = 1$ on all of $\mathbb{R}^3$.

As again can be seen from counterexamples given in Günter [1], for a double-layer potential with continuous density the derivatives on the boundary in general do not exist. But we can prove the following continuity property of the normal derivative.

Theorem 2.21. For the double-layer potential v with continuous density ψ we have

$$\frac{\partial v_+}{\partial \nu} = \frac{\partial v_-}{\partial \nu} \quad \text{on} \quad \partial D \tag{2.58}$$

in the sense that

$$\lim_{\substack{h \to 0 \\ h > 0}} \frac{\partial}{\partial \nu(x)} v(x + h\nu(x)) - \frac{\partial}{\partial \nu(x)} v(x - h\nu(x)) = 0, \qquad x \in \partial D,$$

uniformly for x on ∂D.

Proof. Setting $x_{\pm} = x \pm h\nu(x)$, and using (2.39), we can write

$$\frac{\partial}{\partial \nu(x)} v(x_+) - \frac{\partial}{\partial \nu(x)} v(x_-)$$

$$= \frac{1}{4\pi} \int_{\partial D} (\psi(y) - \psi(x)) \left\{ \left(\frac{1}{|x_+ - y|^3} - \frac{1}{|x_- - y|^3} \right) (\nu(x), \nu(y)) \right.$$

$$- 3 \left(\frac{(\nu(y), x_+ - y)(\nu(x), x_+ - y)}{|x_+ - y|^5} \right.$$

$$\left. \left. - \frac{(\nu(y), x_- - y)(\nu(x), x_- - y)}{|x_- - y|^5} \right) \right\} ds(y). \tag{2.59}$$

We now note the identities

$$|x_+ - y|^2 - |x_- - y|^2 = 4h(\nu(x), x - y)$$

and

$$\frac{1}{|x_+ - y|^3} - \frac{1}{|x_- - y|^3}$$

$$= \frac{4h(\nu(x), y - x)\{|x_+ - y|^2 + |x_+ - y|\,|x_- - y| + |x_- - y|^2\}}{|x_+ - y|^3 |x_- - y|^3 \{|x_+ - y| + |x_- - y|\}}$$

and use Theorem 2.2 to obtain the estimate

$$\left| \frac{1}{|x_+ - y|^3} - \frac{1}{|x_- - y|^3} \right| \leqslant C \frac{h}{|x_+ - y|^3}$$

for some constant C. Analogously, using $(\nu(y), x_+ - y)(\nu(x), x_+ - y) - (\nu(y), x_- - y)(\nu(x), x_- - y) = 2h[(\nu(y), \nu(x))(\nu(x), x - y) + (\nu(y), x - y)]$, we find the same estimate is true for the second term in the integrand of (2.59). The uniform existence of the limit can now be established analogously to the proof of Theorem 2.13 (compare with the estimate (2.43)).

Our last two theorems deal with sufficient conditions for the differentiability of double-layer potentials on the boundary.

Theorem 2.22. The direct values of the double-layer potential v given by (2.46) with uniformly Hölder continuous density $\psi \in C^{0,\alpha}(\partial D)$, $0<\alpha<1$, represent a uniformly Hölder continuously differentiable function on ∂D with

$$\|\operatorname{Grad} v\|_{\alpha,\partial D} \leqslant C_\alpha \|\psi\|_{\alpha,\partial D} \tag{2.60}$$

where C_α is some constant depending on ∂D and α.

Proof. Let $x \in \partial D$ be an arbitrary point and let $\tau(x)$ be a unit vector in the tangent-plane at x. Choose a curve C of class C^2 on ∂D passing through x with unit tangent vector $\tau(x)$. Let x_h be a point on C such that the arc between x_h and x has length h. Then

$$x_h = x + h\tau(x) + O(h^2)$$

and using this we see that

$$\begin{aligned}\frac{4\pi}{h}&\left\{\frac{\partial\Phi_0(x_h,y)}{\partial\nu(y)} - \frac{\partial\Phi_0(x,y)}{\partial\nu(y)}\right\}\\ &= \frac{1}{h}\left[\frac{(\nu(y),x_h-y)}{|x_h-y|^3} - \frac{(\nu(y),x-y)}{|x-y|^3}\right]\\ &= \frac{(\tau(x),\nu(y))}{|x-y|^3} - 3\frac{(\nu(y),x-y)(\tau(x),x-y)}{|x-y|^5} + O\left(\frac{|h|}{|x-y|^3}\right).\end{aligned}$$

Thus, in view of (2.40), we expect that

$$\begin{aligned}u(x) := \frac{1}{4\pi}\int_{\partial D}&[\psi(y)-\psi(x)]\\ &\cdot\left[\frac{(\tau(x),\nu(y))}{|x-y|^3} - 3\frac{(\nu(y),x-y)(\tau(x),x-y)}{|x-y|^5}\right]ds(y)\end{aligned} \tag{2.61}$$

represents the derivative $(\tau(x),\operatorname{Grad} v(x))$. To prove this, we first show that the integral in (2.61) exists. By writing $(\tau(x),\nu(y)) = (\tau(x),\nu(y)-\nu(x))$ and using Theorem 2.2 we can verify that the kernel of (2.61) satisfies the conditions of Remark 2.11. Therefore the integral exists as an improper integral and represents a uniformly Hölder continuous function satisfying an estimate of the form (2.29).

It remains to be shown that (2.61) indeed defines the derivative of v. To do this, we have to estimate the difference

$$\delta := \frac{1}{h}\left[v(x_h) - v(x)\right] - u(x).$$

By subdividing the integral into three parts as in the proofs of Theorems 2.7 and 2.10, we find that

$$\int_{S_{x,r}} \{\ldots\} = O\left(\frac{r^{\alpha+1}}{|h|} + r^\alpha\right)$$

$$\int_{S_{x,R}\setminus S_{x,r}} \{\ldots\} = O\left(|h| r^{\alpha-1}\right)$$

$$\int_{\partial D\setminus S_{x,R}} \{\ldots\} = O(|h|)$$

provided $|h| \leqslant r/4$. We now choose $r = 4|h|$ and obtain $\delta = O(|h|^\alpha)$ from which the differentiability of v follows

Theorem 2.23. The first derivatives of a double-layer potential v with uniformly Hölder continuously differentiable density $\psi \in C^{1,\alpha}(\partial D)$, $0 < \alpha < 1$, can be uniformly extended in a Hölder continuous fashion from $\mathbb{R}^3 \setminus \bar{D}$ into $\mathbb{R}^3 \setminus D$ and from D into $\bar{D}$ with limiting values

$$\begin{aligned} \operatorname{grad} v_\pm(x) = k^2 \int_{\partial D} \Phi(x, y)\nu(y)\psi(y)\, ds(y) \\ - \int_{\partial D} \left[\operatorname{grad}_x \Phi(x, y), \left[\operatorname{Grad} \psi(y), \nu(y)\right]\right] ds(y) \\ \pm \tfrac{1}{2} \operatorname{Grad} \psi(x), \qquad x \in \partial D, \end{aligned} \tag{2.62}$$

where the second integral exists as a Cauchy principal value. Furthermore, we have the inequality

$$\begin{aligned} \|\operatorname{grad} v\|_{\alpha, \mathbb{R}^3\setminus \partial D} \leqslant C_\alpha \|\psi\|_{1,\alpha,\partial D}, \\ \|\operatorname{grad} v\|_{\alpha, \bar{D}} \leqslant C_\alpha \|\psi\|_{1,\alpha,\partial D} \end{aligned} \tag{2.63}$$

for some constant C_α depending on ∂D and α.

Proof. Using (2.52), we can write the double-layer potential in the form

$$v(x) = -\operatorname{div} \int_{\partial D} \Phi(x, y)\nu(y)\psi(y)\, ds(y), \qquad x \in \mathbb{R}^3 \setminus \partial D,$$

and use the vector identity $\operatorname{curl}\operatorname{curl} A = -\Delta A + \operatorname{grad}\operatorname{div} A$ to obtain

$$\operatorname{grad} v(x) = k^2 \int_{\partial D} \Phi(x,y)\nu(y)\psi(y)\,ds(y)$$

$$-\operatorname{curl}\operatorname{curl} \int_{\partial D} \Phi(x,y)\nu(y)\psi(y)\,ds(y), \qquad x \in \mathbb{R}^3 \setminus \partial D.$$

Since $\psi \in C^{1,\alpha}(\partial D)$, we can use (2.52) and (2.2) to deduce that

$$\operatorname{curl} \int_{\partial D} \Phi(x,y)\nu(y)\psi(y)\,ds(y)$$

$$= \int_{\partial D} \left[\nu(y), \psi(y)\operatorname{Grad}_y \Phi(x,y)\right] ds(y)$$

$$= -\int_{\partial D} \Phi(x,y)[\nu(y), \operatorname{Grad}\psi(y)]\,ds(y), \qquad x \in \mathbb{R}^3 \setminus \partial D.$$

We now have

$$\operatorname{grad} v(x) = k^2 \int_{\partial D} \Phi(x,y)\nu(y)\psi(y)\,ds(y)$$

$$+ \int_{\partial D} [\operatorname{grad}_x \Phi(x,y), [\nu(y), \operatorname{Grad}\psi(y)]]\,ds(y), \qquad x \in \mathbb{R}^3 \setminus \partial D$$

and the proof is concluded by using Theorems 2.12 and 2.17.

2.6 VECTOR POTENTIALS

Given a vector field $a \in C(\partial D)$, we now consider the vector potential

$$A(x) := \int_{\partial D} \Phi(x,y) a(y)\,ds(y), \qquad x \in \mathbb{R}^3 \setminus \partial D. \tag{2.64}$$

We note that we have already encountered vector potentials in the proof of Theorem 2.23. In particular, we can conclude the following theorem from our previous analysis.

Theorem 2.24. The first derivatives of the cartesian components of a vector potential A with uniformly Hölder continuous density $a \in C^{0,\alpha}(\partial D)$, $0 < \alpha < 1$, can be uniformly extended in a Hölder continuous fashion from $\mathbb{R}^3 \setminus \bar{D}$ into

$\mathbb{R}^3\setminus D$ and from D into $\bar{D}$ with limiting values

$$\operatorname{curl} A_{\pm}(x) = \int_{\partial D} \operatorname{curl}_x\{\Phi(x,y)a(y)\}\,ds(y)$$
$$\mp \tfrac{1}{2}[\nu(x), a(x)], \qquad x \in \partial D, \tag{2.65}$$

$$\operatorname{div} A_{\pm}(x) = \int_{\partial D} \operatorname{div}_x\{\Phi(x,y)a(y)\}\,ds(y)$$
$$\mp \tfrac{1}{2}(\nu(x), a(x)), \quad x \in \partial D, \tag{2.66}$$

where the integrals exist as Cauchy principal values. Furthermore, we have the estimates

$$\|\operatorname{curl} A\|_{\alpha, \mathbb{R}^3\setminus D} \leqslant C_\alpha \|a\|_{\alpha, \partial D},$$
$$\|\operatorname{curl} A\|_{\alpha, \bar{D}} \leqslant C_\alpha \|a\|_{\alpha, \partial D}, \tag{2.67}$$

$$\|\operatorname{div} A\|_{\alpha, \mathbb{R}^3\setminus D} \leqslant C_\alpha \|a\|_{\alpha, \partial D},$$
$$\|\operatorname{div} A\|_{\alpha, \bar{D}} \leqslant C_\alpha \|a\|_{\alpha, \partial D} \tag{2.68}$$

for some constant C_α depending on ∂D and α.

Corollary 2.25. For the vector potential A with uniformly Hölder continuous density a, the jump relations

$$\operatorname{curl} A_+ - \operatorname{curl} A_- = -[\nu, a] \tag{2.69}$$

$$\operatorname{div} A_+ - \operatorname{div} A_- = -(\nu, a) \quad \text{on} \quad \partial D \tag{2.70}$$

are valid.

The following theorem is the analog of Theorem 2.19 for the normal derivative of a single-layer potential.

Theorem 2.26. For the vector potential A with continuous tangential density a, we have

$$[\nu(x), \operatorname{curl} A_{\pm}(x)] = \int_{\partial D} [\nu(x), \operatorname{curl}_x\{\Phi(x,y)a(y)\}]\,ds(y) \pm \tfrac{1}{2}a(x),$$
$$x \in \partial D, \tag{2.71}$$

where

$$[\nu(x), \operatorname{curl} A_{\pm}(x)] = \lim_{\substack{h\to 0\\ h>0}} [\nu(x), \operatorname{curl} A(x \pm h\nu(x))], \qquad x \in \partial D,$$

has to be understood in the sense of uniform convergence on ∂D and where the integral in (2.71) exists as an improper integral.

Proof. The proof follows in the same manner as Theorem 2.19 after observing that by Theorem 2.2 the kernel

$$[\nu(x), \operatorname{curl}_x\{\Phi(x,y)a(y)\}] = \operatorname{grad}_x\Phi(x,y)(\nu(x)-\nu(y), a(y)) - a(y)\frac{\partial\Phi(x,y)}{\partial\nu(x)}$$

has the same regularity properties as the kernel of the double-layer potential. It is essential that a is a tangential vector, that is, $(a,\nu)=0$ on ∂D.

Corollary 2.27. For the vector potential A with continuous tangential density a we have the jump relation

$$[\nu, \operatorname{curl} A_+] - [\nu, \operatorname{curl} A_-] = a \quad \text{on} \quad \partial D. \tag{2.72}$$

We conclude this section by developing a sufficient condition for $\operatorname{div} A$ to be differentiable on the boundary ∂D. For this we have to introduce the concept of the surface divergence of a continuous tangential field.

Definition 2.28. Let S_n be a sequence of surfaces contained in ∂D with boundary ∂S_n of class C^2 and with surface area $|S_n|$. Let the sequence S_n converge to a point x of ∂D in the sense that for every $\varepsilon > 0$ there exists a $N(\varepsilon) \in \mathbb{N}$ such that for all $n \geqslant N(\varepsilon)$ the subset S_n is contained in the sphere with radius ε and center x. Let ν_0 denote the unit normal vector to the boundary curve ∂S_n that is perpendicular to the surface normal ν and directed to the exterior of S_n. Let a be a continuous tangential field defined on ∂D. Then, if

$$\lim_{S_n \to x} \frac{1}{|S_n|} \int_{\partial S_n} (\nu_0, a)\, dt$$

exists and is independent of the sequence S_n, we denote this limit by $\operatorname{Div} a$ and call it the surface divergence of a at the point x.

We note the following form of Gauss' theorem from Müller [5]. Let a and $\operatorname{Div} a$ be continuous on ∂D and let ϕ be a continuously differentiable function on ∂D. Then

$$\int_{\partial D} \phi \operatorname{Div} a\, ds + \int_{\partial D} (\operatorname{Grad} \phi, a)\, ds = 0 \tag{2.73}$$

and, in particular,

$$\int_{\partial D} \operatorname{Div} a\, ds = 0. \tag{2.74}$$

Let E be a continuous vector field defined either in $D_{h_0}^+ := \{x = z + h\nu(z) \mid z \in \partial D,\ 0 \leqslant h \leqslant h_0\}$ or in $D_{h_0}^- := \{x = z - h\nu(z) \mid z \in \partial D,\ 0 \leqslant h \leqslant h_0\}$ and let E be continuously differentiable in the interior of $D_{h_0}^+$ or $D_{h_0}^-$ such that curl E can be extended continuously into $D_{h_0}^+$ or $D_{h_0}^-$. Then from Stokes' theorem

$$\int_{\partial S_n} (\nu_0, \nu, E)\, dt = -\int_{\partial S_n} (\tau, E)\, dt = -\int_{S_n} (\nu, \operatorname{curl} E)\, ds$$

where $\tau := [\nu, \nu_0]$ denotes the unit tangent vector to the boundary ∂S_n, we can conclude that $\operatorname{Div}[\nu, E]$ exists and is given by the formula

$$\operatorname{Div}[\nu, E] = -(\nu, \operatorname{curl} E). \tag{2.75}$$

Theorem 2.29. The divergence of a vector potential A with continuous tangential density a possessing a continuous surface divergence $\operatorname{Div} a$ can be expressed in the form of a single-layer potential

$$\operatorname{div} A(x) = \int_{\partial D} \Phi(x, y) \operatorname{Div} a(y)\, ds(y), \qquad x \in \mathbb{R}^3 \setminus \partial D. \tag{2.76}$$

Proof. Use (2.52) and (2.73).

2.7 INTEGRAL OPERATORS FOR BOUNDARY-VALUE PROBLEMS

We now conclude this chapter by expressing our regularity results in terms of the integral operators that we shall later use in our study of boundary-value problems in scattering theory.

We introduce integral operators $\mathbf{K}, \mathbf{K}'\colon C(\partial D) \to C(\partial D)$ by

$$(\mathbf{K}\psi)(x) := 2\int_{\partial D} \frac{\partial \Phi(x, y)}{\partial \nu(y)} \psi(y)\, ds(y), \qquad x \in \partial D \tag{2.77}$$

$$(\mathbf{K}'\phi)(x) := 2\int_{\partial D} \frac{\partial \Phi(x, y)}{\partial \nu(x)} \phi(y)\, ds(y), \qquad x \in \partial D. \tag{2.78}$$

By interchanging the order of integration, it is easily seen that $\mathbf{K}$ and $\mathbf{K}'$ are adjoint with respect to the dual system $\langle C(\partial D), C(\partial D)\rangle$ defined by

$$\langle \psi, \phi \rangle := \int_{\partial D} \psi \phi\, ds. \tag{2.79}$$

Theorem 2.30. The operators $\mathbf{K}, \mathbf{K}'$ are compact in $C(\partial D)$ and $C^{0,\alpha}(\partial D)$ for $0 < \alpha < 1$. $\mathbf{K}$ and $\mathbf{K}'$ map $C(\partial D)$ into $C^{0,\alpha}(\partial D)$, and $\mathbf{K}$ maps $C^{0,\alpha}(\partial D)$ into $C^{1,\alpha}(\partial D)$.

Proof. By using Theorem 2.2 (cf. the proof of Theorem 2.15), we see that

$$\frac{\partial\Phi(x,y)}{\partial\nu(y)}=\frac{(\nu(y),x-y)}{|x-y|^2}(1-ik|x-y|)\Phi(x,y)$$

and

$$\frac{\partial\Phi(x,y)}{\partial\nu(x)}=\frac{(\nu(x),y-x)}{|x-y|^2}(1-ik|x-y|)\Phi(x,y)$$

are weakly singular kernels satisfying condition (2.14) with $m=2$. The compactness of **K** and **K′** now follows from Theorem 2.6 and Corollary 2.9. The fact that **K** and **K′** map $C(\partial D)$ into $C^{0,\alpha}(\partial D)$ follows from Theorem 2.7 and the result that **K** takes $C^{0,\alpha}(\partial D)$ into $C^{1,\alpha}(\partial D)$ is a consequence of Theorem 2.22.

We now introduce the integral operator **S**: $C(\partial D)\to C(\partial D)$ by

$$(\mathbf{S}\phi)(x):=2\int_{\partial D}\Phi(x,y)\phi(y)\,ds(y),\qquad x\in\partial D. \tag{2.80}$$

The operator **S** is obviously self-adjoint, that is, $\langle\mathbf{S}\phi,\psi\rangle=\langle\phi,\mathbf{S}\psi\rangle$ for all $\phi,\psi\in C(\partial D)$ and from Theorems 2.12 and 2.17 it follows that **S** has the same compactness and mapping properties as **K**.

By $\mathfrak{N}(\partial D)$ we denote the linear space of all continuous functions ψ with the property that the double-layer potential v with density ψ has continuous normal derivatives on both sides of ∂D. By Theorem 2.21, both normal derivatives coincide, and by Theorem 2.23 the set $\mathfrak{N}(\partial D)$ is not empty since $C^{1,\alpha}(\partial D)\subset\mathfrak{N}(\partial D)$. Now define the operator **T**: $\mathfrak{N}(\partial D)\to C(\partial D)$ by

$$(\mathbf{T}\psi)(x):=2\frac{\partial}{\partial\nu(x)}\int_{\partial D}\frac{\partial\Phi(x,y)}{\partial\nu(y)}\psi(y)\,ds(y),\qquad x\in\partial D. \tag{2.81}$$

Using Green's theorem and the jump relation (2.45), it can easily be seen that the operator **T** is self-adjoint, that is, $\langle\mathbf{T}\psi,\phi\rangle=\langle\psi,\mathbf{T}\phi\rangle$ for all $\psi,\phi\in\mathfrak{N}(\partial D)$. The operator **T** is unbounded, but since the kernel defined by (2.57) satisfies conditions (2.13) and (2.14) with $\alpha=1$ and $m=1$ for $x,y\in\partial D$ it follows that the operator $\mathbf{T}-\mathbf{T}_0$ is compact in $C(\partial D)$ and in $C^{0,\alpha}(\partial D)$, $0<\alpha<1$, where $\mathbf{T}_0$ denotes the operator (2.81) with Φ replaced by Φ_0 (see (2.34)). We summarize our results in the following theorem.

Theorem 2.31. The operators **S** and $\mathbf{T}-\mathbf{T}_0$ are compact in $C(\partial D)$ and compact in $C^{0,\alpha}(\partial D)$ for $0<\alpha<1$. **S** and $\mathbf{T}-\mathbf{T}_0$ map $C(\partial D)$ into $C^{0,\alpha}(\partial D)$, and **S** maps $C^{0,\alpha}(\partial D)$ into $C^{1,\alpha}(\partial D)$.

The operators **K**, **K′**, **S**, and **T** occur in the treatment of acoustic scattering problems. We shall now define the operators that are relevant for electromagnetic scattering problems.

We first introduce the normed subspaces

$$\mathfrak{T}(\partial D):=\{a\colon \partial D\to \mathbf{C}^3 \mid (a,\nu)=0,\ a\in C(\partial D)\}$$

of continuous tangential fields and

$$\mathfrak{T}^{0,\alpha}(\partial D):=\{a\in \mathfrak{T}(\partial D) \mid a\in C^{0,\alpha}(\partial D)\}$$

of uniformly Hölder continuous tangential fields, $0<\alpha\leqslant 1$, and define the integral operators $\mathbf{M},\mathbf{M}'\colon \mathfrak{T}(\partial D)\to \mathfrak{T}(\partial D)$ by

$$(\mathbf{M}a)(x):=2\int_{\partial D}[\nu(x),\operatorname{curl}_x\{\Phi(x,y)a(y)\}]\,ds(y),\qquad x\in\partial D \tag{2.82}$$

$$\mathbf{M}'b:=[\nu,\mathbf{M}[\nu,b]]. \tag{2.83}$$

By interchanging the order of integration, it is easily seen that $\mathbf{M}$ and $\mathbf{M}'$ are adjoint with respect to the dual system $\langle \mathfrak{T}(\partial D),\mathfrak{T}(\partial D)\rangle$ defined by

$$\langle a,b\rangle:=\int_{\partial D}(a,b)\,ds. \tag{2.84}$$

By using the decomposition made in the proof of Theorem 2.26, we can prove the following theorem which is analogous to Theorem 2.30.

Theorem 2.32. The operators $\mathbf{M},\mathbf{M}'$ are compact in $\mathfrak{T}(\partial D)$ and $\mathfrak{T}^{0,\alpha}(\partial D)$ for $0<\alpha<1$, and map $\mathfrak{T}(\partial D)$ into $\mathfrak{T}^{0,\alpha}(\partial D)$.

Finally, we introduce the subspaces

$$\mathfrak{S}(\partial D):=\{a\in \mathfrak{T}(\partial D) \mid \operatorname{Div} a\in C(\partial D)\}$$

and

$$\mathfrak{S}_\perp(\partial D):=\{a\in \mathfrak{T}(\partial D) \mid \operatorname{Div}[\nu,a]\in C(\partial D)\}$$

of continuous tangential fields possessing continuous surface divergence and

$$\mathfrak{S}^{0,\alpha}(\partial D):=\{a\in \mathfrak{T}^{0,\alpha}(\partial D) \mid \operatorname{Div} a\in C^{0,\alpha}(\partial D)\}$$

and

$$\mathfrak{S}_\perp^{0,\alpha}(\partial D):=\{a\in \mathfrak{T}^{0,\alpha}(\partial D) \mid \operatorname{Div}[\nu,a]\in C^{0,\alpha}(\partial D)\}$$

of Hölder continuous tangential fields with Hölder continuous surface divergence, $0<\alpha\leqslant 1$, and define the operator $\mathbf{N}\colon \mathfrak{S}_\perp(\partial D)\to \mathfrak{T}(\partial D)$ by

$$(\mathbf{N}a)(x):=2\left[\nu(x),\operatorname{curl}_x\operatorname{curl}_x\int_{\partial D}\Phi(x,y)[\nu(y),a(y)]\,ds(y)\right],\qquad x\in\partial D. \tag{2.85}$$

The fact that the operator **N** is well defined follows from Theorems 2.17 and 2.29, and the identity

$$\operatorname{curl}\operatorname{curl}\int_{\partial D}\Phi(x,y)[\nu(y),a(y)]\,ds(y)=k^2\int_{\partial D}\Phi(x,y)[\nu(y),a(y)]\,ds(y)$$

$$+\operatorname{grad}\int_{\partial D}\Phi(x,y)\operatorname{Div}[\nu(y),a(y)]\,ds(y),\qquad x\in\mathbb{R}^3\setminus\partial D. \tag{2.86}$$

Using Gauss' theorem and the jump relation (2.72), it is easily seen that **N** is self-adjoint, that is, $\langle \mathbf{N}a, b\rangle = \langle a, \mathbf{N}b\rangle$ for all $a, b \in \mathfrak{S}_{\perp}(\partial D)$. The operator **N** is unbounded, but by using the fact that for a continuous vector field b, the kernel given by

$$K(x,y)=\operatorname{grad}_x\operatorname{div}_x\Phi(x,y)b(y)-\operatorname{grad}_x\operatorname{div}_x\Phi_0(x,y)b(y)$$

$$=\operatorname{grad}_x\frac{\partial\Phi_0(x,y)}{\partial b(y)}-\operatorname{grad}_x\frac{\partial\Phi(x,y)}{\partial b(y)},$$

satisfies conditions (2.13) and (2.14) with $\alpha=1$ and $m=1$ for all $x, y\in\partial D$ it follows from Theorem 2.6 and Corollary 2.9 that the following theorem is true, where $\mathbf{N}_0$ denotes the operator (2.85) with Φ replaced by Φ_0.

Theorem 2.33. The operator $\mathbf{N}-\mathbf{N}_0$ is compact in $\mathfrak{T}(\partial D)$ and $\mathfrak{T}^{0,\alpha}(\partial D)$ for $0<\alpha<1$ and maps $\mathfrak{T}(\partial D)$ into $\mathfrak{T}^{0,\alpha}(\partial D)$.

3

BOUNDARY-VALUE PROBLEMS FOR THE SCALAR HELMHOLTZ EQUATION

This chapter is probably the most important chapter of this book because essentially all of what follows is either based on or motivated by the results we are now about to discuss. Our basic aim is to show how the existence of a unique solution to exterior boundary-value problems for the Helmholtz equation can be established by the method of integral equations defined over the boundary of the scattering obstacle. We will discuss in subsequent chapters the generalization of these results to the case of Maxwell's equations as well as their application to the investigation of the inverse scattering problem. The main advantage of the use of boundary integral equation methods to study exterior boundary-value problems for the Helmholtz equation lies in the fact that this approach reduces a problem defined over an unbounded domain to one defined on a bounded domain of lower dimension, that is, the boundary of the scattering obstacle. This fact is, of course, crucial from the point of view of numerical analysis. However, the gains inherent in the method of boundary integral equations are not achieved without cost. This is basically due to the fact that the straightforward use of potential theory to formulate boundary integral equations for the classical boundary-value problems of scattering theory lead to equations that are not uniquely solvable at the eigenvalues of certain interior boundary-value problems. Hence a major task is to derive analytic methods for overcoming this difficulty. The numerical implementation of these methods presents further problems because, in general, the eigenvalues are not known in advance, and this necessitates the use of more sophisticated methods to deal with the problem of interior eigenvalues, in particular, the study of strongly singular integral equations and their regularization.

The broad plan of this chapter is as follows. We first discuss the physical origins of exterior boundary-value problems for the Helmholtz equation, the concept of a radiation condition, and the asymptotic behavior of solutions to the Helmholtz equation satisfying such a radiation condition. This discussion enables us to introduce the concept of a far-field pattern and to derive the basic uniqueness results for the exterior Dirichlet and Neumann problems. We then turn to the problem of the existence of solutions to these two boundary-value problems, and first establish this by looking for a solution in the form of either a double- or single-layer potential. As mentioned above, this leads to the problem of the unique solvability at interior eigenvalues, and we shall see that this problem arises regardless of whether we reformulate the boundary-value problems as an integral equation of the first or second kind. Motivated by this analysis, we consider various modifications of the above approach in order to obtain integral equations that are uniquely solvable for all values of the wave number. The first of these methods, due to Jones [1] and Ursell [1], [2], leads to weakly singular integral equations, whereas the second approach, due to Leis [2], Brakhage and Werner [1], and Panich [1], leads, in the case of the Neumann problem, to a strongly singular integral equation that needs to be regularized. After completing this discussion of the Dirichlet and Neumann problems, we then show how these results can be extended to treat the impedance and transmission problems. We conclude the chapter by showing how Green's representation theorem can be used to derive integral equations for the boundary-value problems of scattering theory that are adjoint to those obtained by the use of potential theory and to briefly discuss the use of this representation theorem to derive the so-called null-field method for solving acoustic scattering problems.

For an interesting survey of topics closely related to those of this chapter we refer the reader to Dolph [1].

Throughout this chapter, we shall assume that our problems are three dimensional; the minor modifications needed to treat the two-dimensional case are discussed in Section 3.10.

3.1 TIME-HARMONIC ACOUSTIC SCATTERING

Consider acoustic wave propagation in a homogeneous isotropic medium in $\mathbb{R}^3$ with density ρ, speed of sound c, and damping coefficient γ. The wave motion can be determined from a velocity potential $U=U(x,t)$ from which the velocity field v is obtained by

$$v=\frac{1}{\rho}\operatorname{grad} U$$

and the pressure p by

$$p-p_0=-\frac{\partial U}{\partial t}-\gamma U$$

where p_0 denotes the pressure of the undisturbed medium. In the linearized theory, the velocity potential U satisfies the dissipative wave equation

$$\frac{\partial^2 U}{\partial t^2} + \gamma \frac{\partial U}{\partial t} - c^2 \Delta U = 0 \tag{3.1}$$

and hence for time-harmonic acoustic waves of the form $U(x,t) = u(x)e^{-i\omega t}$ with frequency $\omega > 0$, we deduce that the space dependent part u satisfies the reduced wave equation or Helmholtz equation

$$\Delta u + k^2 u = 0 \tag{3.2}$$

where the wave number $k \neq 0$ is given by $k^2 = \omega(\omega + i\gamma)/c^2$. We choose the sign of k such that

$$\operatorname{Im} k \geqslant 0. \tag{3.3}$$

Therefore the mathematical description of the scattering of time-harmonic waves by an obstacle D leads to boundary-value problems for the Helmholtz equation. Prescribing the values of u on the boundary of the obstacle (i.e., the Dirichlet problem) physically corresponds to prescribing the pressure of the acoustic wave. In particular, consider the scattering of a given incoming acoustic wave u^i by an obstacle D. Then the total acoustic wave is of the form $u = u^i + u^s$ where u^s denotes the scattered wave and for a sound-soft obstacle the total pressure must vanish on the boundary, that is, $u^s = -u^i$ on the boundary. Similarly, prescribing the normal derivative of u on the boundary (i.e., the Neumann problem) physically corresponds to prescribing the normal component of the velocity of the acoustic wave, that is, to acoustic scattering from a sound-hard obstacle.

A boundary condition that presents a more realistic treatment of the acoustic properties of the obstacle D as compared with the Dirichlet or Neumann boundary condition is given by

$$(\nu, v) + \chi(p - p_0) = 0,$$

that is, the normal velocity on the boundary is proportional to the excess pressure on the boundary. The coefficient χ is called the acoustic impedance of the obstacle D and is, in general, a space dependent function defined on the boundary ∂D. This impedance condition leads to a boundary-value problem for the velocity potential u of the form

$$\frac{\partial u}{\partial \nu} + \lambda u = 0$$

where $\lambda = i\chi\rho(\omega + i\gamma)$.

3.2 GREEN'S REPRESENTATION THEOREM AND SOMMERFELD'S RADIATION CONDITION

We begin our analysis by establishing the basic property that any solution to the Helmholtz equation can be represented as the combination of a single- and a double-layer acoustic surface potential.

For any domain G with boundary ∂G of class C^2, we introduce the linear space $\mathfrak{R}(G)$ of all complex-valued functions $u \in C^2(G) \cap C(\bar{G})$ for which the normal derivative on the boundary exists in the sense that the limit

$$\frac{\partial u}{\partial \nu}(x) = \lim_{\substack{h \to 0 \\ h > 0}} (\nu(x), \operatorname{grad} u(x - h\nu(x)) \qquad x \in \partial G,$$

exists uniformly on ∂G. Here we assume the normal ν to be directed into the exterior of G. We note that the assumption $u, v \in \mathfrak{R}(G)$ suffices to guarantee the validity of the first Green's theorem

$$\int_G u \,\Delta v \, dx = \int_{\partial G} u \frac{\partial v}{\partial \nu} ds - \int_G (\operatorname{grad} u, \operatorname{grad} v) \, dx \tag{3.4}$$

and the second Green's theorem

$$\int_G (u \,\Delta v - v \,\Delta u) \, dx = \int_{\partial G} \left(u \frac{\partial v}{\partial \nu} - v \frac{\partial u}{\partial \nu} \right) ds \tag{3.5}$$

for a bounded domain G with C^2 boundary ∂G. This follows by first integrating over the parallel surfaces introduced by (2.7) and then passing to the limit ∂G.

Recall that by D we denote a bounded region in $\mathbb{R}^3$ with the boundary ∂D consisting of a finite number of disjoint, closed, bounded surfaces belonging to the class C^2. The exterior $\mathbb{R}^3 \setminus D$ is assumed to be connected, whereas D itself may have more than one component. We assume the normal ν to ∂D to be directed into the exterior of D. Finally, recall the fundamental solution

$$\Phi(x, y) = \frac{1}{4\pi} \frac{e^{ik|x-y|}}{|x-y|}$$

of the Helmholtz equation in $\mathbb{R}^3$ that we first introduced in (2.30).

Theorem 3.1. Let $u \in \mathfrak{R}(D)$ be a solution to the Helmholtz equation

$$\Delta u + k^2 u = 0 \quad \text{in} \quad D.$$

Then

$$\int_{\partial D} \left\{ u(y) \frac{\partial \Phi(x, y)}{\partial \nu(y)} - \frac{\partial u}{\partial \nu}(y) \Phi(x, y) \right\} ds(y) = \begin{cases} -u(x), & x \in D, \\ 0, & x \in \mathbb{R}^3 \setminus \bar{D}. \end{cases}$$

Proof. We choose an arbitrary fixed point $x \in D$ and circumscribe it with a sphere $\Omega_{x,r} := \{y \in \mathbb{R}^3 \mid |x-y| = r\}$. We assume the radius r to be small enough such that $\Omega_{x,r} \subset D$ and direct the unit normal ν to $\Omega_{x,r}$ into the interior of $\Omega_{x,r}$. Now we apply the second Green's theorem (3.5) to the functions $u(y)$ and $\Phi(x, y)$ in the region $\{y \in D \mid |x-y| > r\}$ to obtain

$$\int_{\partial D + \Omega_{x,r}} \left\{ u(y) \frac{\partial \Phi(x,y)}{\partial \nu(y)} - \frac{\partial u}{\partial \nu}(y) \Phi(x,y) \right\} ds(y) = 0. \tag{3.6}$$

Since on $\Omega_{x,r}$ we have

$$\Phi(x,y) = \frac{e^{ikr}}{4\pi r}, \qquad \operatorname{grad}_y \Phi(x,y) = \left(\frac{1}{r} - ik \right) \frac{e^{ikr}}{4\pi r} \nu(y)$$

a straightforward calculation, using the mean value theorem, shows that

$$\lim_{r \to 0} \int_{\Omega_{x,r}} \left\{ u(y) \frac{\partial \Phi(x,y)}{\partial \nu(y)} - \frac{\partial u}{\partial \nu}(y) \Phi(x,y) \right\} ds(y) = u(x)$$

whence from (3.6) the representation theorem is established for $x \in D$.

The statement for $x \in \mathbb{R}^3 \setminus \bar{D}$ readily follows from Green's theorem applied to the functions $u(y)$ and $\Phi(x, y)$ in the region D.

Straightforward calculations show that

$$\left(\frac{x}{|x|}, \operatorname{grad}_x \Phi(x,y) \right) - ik\Phi(x,y) = O\left(\frac{1}{|x|^2} \right), \qquad |x| \to \infty$$

and

$$\left(\frac{x}{|x|}, \operatorname{grad}_x \frac{\partial \Phi(x,y)}{\partial \nu(y)} \right) - ik \frac{\partial \Phi(x,y)}{\partial \nu(y)} = O\left(\frac{1}{|x|^2} \right), \qquad |x| \to \infty$$

uniformly for all directions $x/|x|$ and uniformly for all y contained in the bounded set ∂D. From this we conclude the following.

Theorem 3.2. Both the single-layer acoustic potential defined by (2.31) and the double-layer acoustic potential defined by (2.33) satisfy the Sommerfeld radiation condition

$$\left(\frac{x}{|x|}, \operatorname{grad} u(x) \right) - iku(x) = o\left(\frac{1}{|x|} \right), \qquad |x| \to \infty$$

uniformly for all directions $x/|x|$.

As we shall soon see, the Sommerfeld radiation condition completely characterizes the behavior of solutions to the Helmholtz equation at infinity.

Theorem 3.3. Let $u \in \mathfrak{R}(\mathbb{R}^3 \setminus \bar{D})$ be a solution to the Helmholtz equation

$$\Delta u + k^2 u = 0 \quad \text{in} \quad \mathbb{R}^3 \setminus \bar{D}$$

satisfying the Sommerfeld radiation condition

$$\left(\frac{x}{|x|}, \operatorname{grad} u(x) \right) - iku(x) = o\left(\frac{1}{|x|} \right), \qquad |x| \to \infty, \tag{3.7}$$

uniformly for all directions $x/|x|$. Then

$$\int_{\partial D} \left\{ u(y) \frac{\partial \Phi(x,y)}{\partial \nu(y)} - \frac{\partial u}{\partial \nu}(y) \Phi(x,y) \right\} ds(y) = \begin{cases} 0, & x \in D, \\ u(x), & x \in \mathbb{R}^3 \setminus \bar{D}. \end{cases}$$

Proof. We first show that

$$\int_{|y|=R} |u|^2 \, ds = O(1), \qquad R \to \infty. \tag{3.8}$$

To accomplish this, we first observe that from the radiation condition (3.7) it follows that

$$\begin{aligned} 0 &= \lim_{R \to \infty} \int_{|y|=R} \left| \frac{\partial u}{\partial \nu} - iku \right|^2 ds \\ &= \lim_{R \to \infty} \int_{|y|=R} \left\{ \left| \frac{\partial u}{\partial \nu} \right|^2 + |k|^2 |u|^2 + 2 \operatorname{Im}\left(ku \frac{\partial \bar{u}}{\partial \nu} \right) \right\} ds \end{aligned} \tag{3.9}$$

where ν denotes the outward unit normal to the sphere $\Omega_R := \{ y \in \mathbb{R}^3 \mid |y| = R \}$. We take R large enough so that $\Omega_R \subset \mathbb{R}^3 \setminus \bar{D}$ and apply the first Green's theorem (3.4) in the domain $D_R := \{ y \in \mathbb{R}^3 \setminus \bar{D} \mid |y| < R \}$ to obtain

$$k \int_{|y|=R} u \frac{\partial \bar{u}}{\partial \nu} ds = k \int_{\partial D} u \frac{\partial \bar{u}}{\partial \nu} ds - \bar{k} |k|^2 \int_{D_R} |u|^2 \, dy + k \int_{D_R} |\operatorname{grad} u|^2 \, dy.$$

Now we substitute the imaginary part of the last equation into (3.9) and find that

$$\begin{aligned} &\lim_{R \to \infty} \left\{ \int_{|y|=R} \left\{ \left| \frac{\partial u}{\partial \nu} \right|^2 + |k|^2 |u|^2 \right\} ds + 2 \operatorname{Im}(k) \int_{D_R} \{ |k|^2 |u|^2 + |\operatorname{grad} u|^2 \} \, dy \right\} \\ &= -2 \operatorname{Im}\left(k \int_{\partial D} u \frac{\partial \bar{u}}{\partial \nu} ds \right). \end{aligned} \tag{3.10}$$

All four terms on the left-hand side of (3.10) are nonnegative since $\operatorname{Im}(k) \geqslant 0$. Hence these terms must be individually bounded as $R \to \infty$ since their sum tends to a finite limit. Equation (3.8) follows immediately.

We now note the identity

$$\int_{|y|=R} \left\{ u(y) \frac{\partial \Phi(x,y)}{\partial \nu(y)} - \frac{\partial u}{\partial \nu}(y) \Phi(x,y) \right\} ds(y)$$

$$= \int_{|y|=R} u(y) \left\{ \frac{\partial \Phi(x,y)}{\partial \nu(y)} - ik\Phi(x,y) \right\} ds(y)$$

$$- \int_{|y|=R} \Phi(x,y) \left\{ \frac{\partial u}{\partial \nu}(y) - iku(y) \right\} ds(y) =: I_1 + I_2$$

and apply Schwarz's inequality to each of the integrals I_1 and I_2. From the radiation condition

$$\frac{\partial \Phi(x,y)}{\partial \nu(y)} - ik\Phi(x,y) = O\left(\frac{1}{R^2}\right), \qquad y \in \Omega_R,$$

for the fundamental solution and (3.8) we see that $I_1 = O(1/R)$ as $R \to \infty$. The radiation condition (3.7) and $\Phi(x,y) = O(1/R)$, $y \in \Omega_R$, yield $I_2 = o(1)$ for $R \to \infty$. Hence

$$\lim_{R \to \infty} \int_{|y|=R} \left\{ u(y) \frac{\partial \Phi(x,y)}{\partial \nu(y)} - \frac{\partial u}{\partial \nu}(y) \Phi(x,y) \right\} ds(y) = 0.$$

The proof is now completed as in Theorem 3.1 by applying the second Green's theorem in the domain $\{y \in D_R \mid |x-y| > r\}$ if $x \in \mathbb{R}^3 \setminus \bar{D}$ or D_R if $x \in D$.

Remark 3.4. It is obvious that any solution of the Helmholtz equation satisfying the Sommerfeld radiation condition (3.7) automatically satisfies

$$u(x) = O\left(\frac{1}{|x|}\right), \qquad |x| \to \infty, \tag{3.11}$$

uniformly for all directions $x/|x|$.

Note that it is not necessary to impose this additional condition for the representation theorem to be valid.

Physically, the fundamental solution $\Phi(x,y)$ describes an outgoing spherical wave of the form

$$\frac{e^{i(k|x-y| - \omega t)}}{4\pi |x-y|}.$$

Thus the radiation condition (3.7), first introduced by Sommerfeld [1], mathematically characterizes outgoing waves. Similarly, using the fundamental solution

$$\frac{e^{-ik|x-y|}}{4\pi|x-y|},$$

it is possible to develop an analogous theory of incoming waves characterized by the condition

$$\left(\frac{x}{|x|}, \operatorname{grad} u(x)\right) + iku(x) = o\left(\frac{1}{|x|}\right), \qquad |x| \to \infty.$$

From the representation Theorem 3.1, we immediately conclude that solutions to the Helmholtz equation are analytic functions of their independent variables.

Theorem 3.5. Any two times continuously differentiable solution to the Helmholtz equation is analytic.

Proof. We make use of the fact that any holomorphic function of several complex variables, that is, a function satisfying the Cauchy–Riemann equations with respect to each of the complex variables, is also analytic, that is, it has a local power series expansion and vice versa (Gilbert [1]). Our theorem now follows from the observation that the fundamental solution $\Phi(x, y)$ is an analytic function of the cartesian coordinates x_1, x_2, x_3 of x and the fact that the integrands in the representation Theorem 3.1 are continuous with respect to y if x is contained in a compact subset of D. Therefore the Cauchy–Riemann equations for u can be verified by differentiating with respect to x under the integral sign.

It follows from Theorem 3.5 that any solution to the Helmholtz equation that vanishes in an open subset of its domain of definition must vanish identically.

Another direct consequence of the representation Theorem 3.3 is the following expansion theorem due to Atkinson [1] and Wilcox [1], [2].

Theorem 3.6. Let $u \in C^2(\mathbf{R}^3 \setminus \bar{D})$ be a solution of the Helmholtz equation satisfying the Sommerfeld radiation condition. Let R_0 be such that $\Omega_{R_0} := \{x \in \mathbf{R}^3 \mid |x| = R_0\} \subset \mathbf{R}^3 \setminus \bar{D}$ and let r, θ, ϕ denote the spherical coordinates of x. Then u has an expansion of the form

$$u(x) = \frac{e^{ikr}}{r} \sum_{n=0}^{\infty} \frac{F_n(\theta, \phi)}{r^n} \tag{3.12}$$

that is valid for all $r \geq R_0$ and that converges absolutely and uniformly with respect to the variables r, θ, ϕ. The series can be differentiated term by term with respect to r, θ, ϕ any number of times and the resulting series all converge absolutely and uniformly.

Proof. From the representation Theorem 3.3 applied to the exterior of a sphere Ω_R with appropriately chosen radius $R < R_0$ we have

$$u(x)=\int_{|y|=R}\left\{u(y)\frac{\partial\Phi(x,y)}{\partial\nu(y)}-\frac{\partial u}{\partial\nu}(y)\Phi(x,y)\right\}ds(y)$$

for $|x| \geqslant R_0$. Introducing spherical coordinates (r, θ, ϕ) and (R, θ', ϕ') for x and y, respectively, we can rewrite this equation in the form

$$u(x)=\frac{R^2}{4\pi}\int_{\Omega}\left\{u(y)\left(\frac{y}{|y|},\frac{y-x}{\rho}\right)\frac{d}{d\rho}\frac{e^{ik\rho}}{\rho}-\frac{\partial u}{\partial\nu}(y)\frac{e^{ik\rho}}{\rho}\right\}d\omega \quad (3.13)'$$

where Ω denotes the unit sphere, $d\omega = \sin\theta' d\theta' d\phi'$ is the surface element on the unit sphere, and

$$\rho = |x-y| = \left(R^2 - 2rR\cos\gamma + r^2\right)^{1/2},$$

$$\cos\gamma = \cos\theta\cos\theta' + \sin\theta\sin\theta'\cos(\phi-\phi').$$

Consider first the function

$$u_1(x) := \int_{\Omega}\frac{e^{ik\rho}}{\rho}f(\theta',\phi')\,d\omega$$

where $f\colon \Omega \to \mathbf{C}$ is continuous. Put $w := R/r$. Then

$$\frac{e^{ik(\rho-r)}}{\rho}=\frac{we^{ikR[(1-2w\cos\gamma+w^2)^{1/2}-1]/w}}{R\left(1-2w\cos\gamma+w^2\right)^{1/2}}$$

where the square root is the branch having the value $+1$ at $w=0$. Since $(1-2w\cos\gamma+w^2)^{1/2}$ is an analytic function of w for $w<1$, we see that for $w<1$ the expansion

$$\frac{e^{ik(\rho-r)}}{\rho}=\sum_{n=1}^{\infty}a_n(\gamma)w^n$$

is valid. This series is absolutely and uniformly convergent for $r \geqslant R_0$ and $\gamma \in [0, 2\pi]$, it can be differentiated term by term with respect to r and γ any number of times, and the resulting series all converge absolutely and uniformly. Multiplying by $f(\theta',\phi')$ and integrating over Ω, we obtain an expansion

$$u_1(x)=\frac{e^{ikr}}{r}\sum_{n=0}^{\infty}\frac{f_n(\theta,\phi)}{r^n}$$

with the convergence properties indicated in the theorem.

The function u is represented in equation (3.13) as a sum of integrals u_1 and u_2 where

$$u_2(x):=\int_{\Omega}\frac{1}{\rho}(R-r\cos\gamma)\frac{d}{d\rho}\frac{e^{ik\rho}}{\rho}g(\theta',\phi')\,d\omega$$

with a continuous function $g\colon \Omega\to\mathbf{C}$. The integral defining u_2 is similar in structure to u_1 and may be treated in the same way. Hence the expansion theorem for u follows.

Corollary 3.7. Every solution u of the Helmholtz equation satisfying the radiation condition has the asymptotic behavior

$$u(x)=\frac{e^{ikr}}{r}F_0(\theta,\phi)+O\left(\frac{1}{r^2}\right). \tag{3.14}$$

The function $F_0\colon \Omega\to\mathbf{C}$ is called the far-field pattern or radiation pattern of u.

Corollary 3.8. The coefficients F_n in the expansion (3.12) are recursively determined in terms of the far-field pattern F_0 by the formula

$$2iknF_n=n(n-1)F_{n-1}+\mathbf{B}F_{n-1},\qquad n=1,2,3,\ldots,$$

where

$$\mathbf{B}:=\frac{1}{\sin\theta}\frac{\partial}{\partial\theta}\left(\sin\theta\frac{\partial}{\partial\theta}\right)+\frac{1}{\sin^2\theta}\frac{\partial^2}{\partial\phi^2}$$

is Beltrami's operator for the sphere.

Proof. The expansion (3.12) has to satisfy the Helmholtz equation. Differentiating the series term by term in spherical coordinates we find

$$\Delta u+k^2u=\frac{e^{ikr}}{r}\sum_{n=1}^{\infty}\frac{1}{r^{n+1}}\{-2iknF_n+n(n-1)F_{n-1}+\mathbf{B}F_{n-1}\}.$$

Since $\Delta u + k^2 u = 0$ the coefficients of the powers of $1/r$ in this series must vanish. This gives the recursion formula.

The following corollary establishes a one-to-one correspondence between solutions to the Helmholtz equation satisfying the radiation condition and their far-field pattern.

Corollary 3.9. Let $u \in C^2(\mathbb{R}^3 \setminus \bar{D})$ be a solution to the Helmholtz equation satisfying the Sommerfeld radiation condition for which the far-field pattern vanishes identically. Then $u = 0$ in $\mathbb{R}^3 \setminus \bar{D}$.

Proof. Since $F_0 = 0$ we see from Corollary 3.8 and the expansion (3.12) that $u = 0$ in the exterior of a suitably chosen sphere. By the analyticity of u (Theorem 3.5) we can now conclude that $u = 0$ in $\mathbb{R}^3 \setminus \bar{D}$.

3.3 THE DIRICHLET AND NEUMANN BOUNDARY-VALUE PROBLEMS: UNIQUENESS THEOREMS

We shall consider the following boundary-value problems for the Helmholtz equation. Recall again that D is a bounded region in $\mathbb{R}^3$ with C^2 boundary ∂D.

Interior Dirichlet Problem

Find a function $u \in C^2(D) \cap C(\bar{D})$ satisfying the Helmholtz equation in D and the boundary condition

$$u = f \quad \text{on} \quad \partial D \tag{3.15}$$

where f is a given continuous function.

Interior Neumann Problem

Find a function $u \in \mathfrak{R}(D)$ (i.e., a function $u \in C^2(D) \cap C(\bar{D})$ possessing normal derivatives in the sense of uniform convergence) satisfying the Helmholtz equation in D and the boundary condition

$$\frac{\partial u}{\partial \nu} = g \quad \text{on} \quad \partial D \tag{3.16}$$

where g is a given continuous function.

Exterior Dirichlet Problem

Find a function $u \in C^2(\mathbb{R}^3 \setminus \bar{D}) \cap C(\mathbb{R}^3 \setminus D)$ satisfying the Helmholtz equation in $\mathbb{R}^3 \setminus \bar{D}$, the Sommerfeld radiation condition at infinity, and the boundary condition

$$u = f \quad \text{on} \quad \partial D \tag{3.17}$$

where f is a given continuous function.

Exterior Neumann Problem

Find a function $u \in \mathfrak{R}(\mathbb{R}^3 \setminus \bar{D})$ satisfying the Helmholtz equation in $\mathbb{R}^3 \setminus \bar{D}$, the Sommerfeld radiation condition at infinity, and the boundary condition

$$\frac{\partial u}{\partial \nu} = g \quad \text{on} \quad \partial D \tag{3.18}$$

where g is a given continuous function.

In the following analysis we shall make use of the fact that any solution u to the homogeneous Dirichlet problem, that is, any solution with identically vanishing boundary values $u = 0$ on ∂D, is automatically continuously differentiable up to the boundary. A proof of this property will be given later in Theorem 3.27.

Theorem 3.10. Let $\operatorname{Im} k > 0$. Then the interior Dirichlet and Neumann problems have at most one solution.

Proof. We have to show that solutions to the homogeneous boundary value problems vanish identically. Let $u \in \mathfrak{R}(D)$ be a solution to the Helmholtz equation. Then from the first Green's theorem (3.4) we obtain

$$\int_{\partial D} \bar{u} \frac{\partial u}{\partial \nu} \, ds = \int_D \{|\operatorname{grad} u|^2 - k^2 |u|^2\} \, dx.$$

Since the left-hand side of this equation vanishes if either $u = 0$ or $\partial u / \partial \nu = 0$ on ∂D, and $\operatorname{Im} k > 0$, we see that

$$\int_D |u|^2 \, dx = 0.$$

Hence $u = 0$ in D.

For k real, we in general do not have uniqueness for the interior problems. To see this, we note that by separation of variables in spherical coordinates r, θ, ϕ we have that the functions

$$u_n^m(x) = j_n(kr) Y_n^m(\theta, \phi), \tag{3.19}$$

$n=0,1,2,\ldots,\ m=-n,\ldots,n$, are solutions to the Helmholtz equation that are regular at the origin. Here j_n denotes the spherical Bessel function of order n and

$$Y_n^m(\theta,\phi)=\left[\frac{(2n+1)(n-|m|)!}{4\pi(n+|m|)!}\right]^{1/2}P_n^{|m|}(\cos\theta)e^{im\phi}$$

denotes the spherical harmonic of order n where P_n^m denotes the associated Legendre polynomial (Erdélyi et al. [1]). In particular, for $n=0$, we have the special case

$$u_0^0(x)=\frac{\sin kr}{\sqrt{4\pi}\,kr}.$$

If D is the unit ball then from (3.19) we observe that for all values of k that are zeros of the spherical Bessel functions, the homogeneous interior Dirichlet problem has nontrivial solutions. Similarly, the homogeneous interior Neumann problem for the unit ball has nontrivial solutions for all values of k that are zeros of the first derivatives of the spherical Bessel functions. In general, by means of variational methods or by the Hilbert–Schmidt theory of symmetric integral operators, it can be shown that for any bounded domain D there exists a countable set $\mathfrak{D}(D)$ of positive wave numbers k accumulating only at infinity for which the interior Dirichlet problem admits nontrivial solutions. We call these values interior Dirichlet eigenvalues. For the interior Neumann problem, there also exists a countable set $\mathfrak{N}(D)$ of positive wave numbers k accumulating only at infinity for which nontrivial solutions occur. These values we call interior Neumann eigenvalues. For a detailed discussion of these eigenvalue problems, see Courant and Hilbert [1], Leis [3], or Stakgold [2].

The uniqueness results for the exterior problems are based on the following lemma due to Rellich [1].

Lemma 3.11. Let k be positive and $u\in C^2(\mathbb{R}^3\setminus\bar{D})$ a solution to the Helmholtz equation satisfying the Sommerfeld radiation condition and

$$\int_{|x|=R}|u|^2\,ds=o(1),\qquad R\to\infty. \tag{3.20}$$

Then $u=0$ in $\mathbb{R}^3\setminus\bar{D}$.

Proof. From the expansion (3.12) we get

$$\int_{|x|=R}|u|^2\,ds=\int_{\Omega}|F_0|^2\,d\omega+O\left(\frac{1}{R}\right),\qquad R\to\infty,$$

for any radiating solution of the Helmholtz equation. Hence (3.20) implies $F_0=0$, and therefore from Corollary 3.9 we can conclude that $u=0$ in $\mathbb{R}^3\setminus\bar{D}$.

Theorem 3.12. Let $u \in \mathfrak{R}(\mathbf{R}^3 \setminus \bar{D})$ be a solution to the Helmholtz equation satisfying the radiation condition and

$$\operatorname{Im}\left(k \int_{\partial D} u \frac{\partial \bar{u}}{\partial \nu} ds \right) \geqslant 0. \tag{3.21}$$

Then $u = 0$ in $\mathbf{R}^3 \setminus D$.

Proof. If $\operatorname{Im} k > 0$, then from (3.10) we see that

$$\int_{D_R} |u|^2 \, dx \to 0, \qquad R \to \infty.$$

Hence $u = 0$ in $\mathbf{R}^3 \setminus D$. If $\operatorname{Im} k = 0$, then from (3.10) we have

$$\int_{|x| = R} |u|^2 \, ds \to 0, \qquad R \to \infty$$

and $u = 0$ in $\mathbf{R}^3 \setminus D$ by Lemma 3.11.

An immediate consequence of Theorem 3.12 is the following uniqueness result.

Theorem 3.13. The exterior Dirichlet and Neumann problems have at most one solution.

In concluding this section we wish to mention that Lemma 3.10 is actually a special case of a stronger result by Rellich [1] of which we shall now outline a short proof making extensive use of properties of spherical Bessel functions and spherical harmonics.

Lemma 3.14. Let k be positive and let $u \in C^2(\mathbf{R}^3 \setminus \bar{D})$ be a solution of the Helmholtz equation satisfying

$$\int_{|x| = R} |u|^2 \, ds \to 0, \qquad R \to \infty. \tag{3.22}$$

Then $u = 0$ in $\mathbf{R}^3 \setminus \bar{D}$.

Proof. Let (r, θ, ϕ) denote spherical coordinates. For sufficiently large fixed r we can expand u in a uniformly convergent series of spherical harmonics

$$u(x) = \sum_{n=0}^{\infty} \sum_{m=-n}^{n} a_{nm}(r) Y_n^m(\theta, \phi) \tag{3.23}$$

where

$$a_{nm}(r) = \int_0^{2\pi} \int_0^{\pi} u(r, \theta, \phi) \bar{Y}_n^m(\theta, \phi) \sin\theta \, d\theta \, d\phi.$$

Since $u \in C^2(\mathbb{R}^3 \setminus \bar{D})$, we can differentiate under the integral and integrate by parts using $\Delta u + k^2 u = 0$ to conclude that a_{nm} is a solution of Bessel's equation

$$\frac{d^2 a_{nm}}{dr^2} + \frac{2}{r}\frac{da_{nm}}{dr} + \left(k^2 - \frac{n(n+1)}{r^2}\right)a_{nm} = 0,$$

that is,

$$a_{nm}(r) = \alpha_{nm} h_n^{(1)}(kr) + \beta_{nm} h_n^{(2)}(kr)$$

where α_{nm} and β_{nm} are constants and $h_n^{(1,2)}$ denotes a spherical Hankel function of the first and second kind, respectively.

We integrate (3.23) term by term using the orthonormality of the spherical harmonics to obtain

$$\int_{|x|=R} |u|^2\, ds = R^2 \sum_{n=0}^{\infty} \sum_{m=-n}^{n} |a_{nm}(R)|^2.$$

From the assumption (3.22), we see that

$$R^2 |a_{nm}(R)|^2 \to 0, \qquad R \to \infty,$$

$n = 0, 1, 2, \ldots,\ m = -n, \ldots, n$. Hence, using the asymptotic formulae

$$h_n^{(1,2)}(kR) = \frac{1}{kR} e^{\pm i(kR - n\pi/2 - \pi/2)} + O\left(\frac{1}{R^2}\right)$$

we obtain $a_{nm} = 0$ for all $n = 0, 1, 2, \ldots,\ m = -n, \ldots, n$. Therefore $u = 0$ outside a sufficiently large sphere and hence $u = 0$ in $\mathbb{R}^3 \setminus \bar{D}$ by analyticity (Theorem 3.5).

3.4 THE EXISTENCE OF SOLUTIONS TO THE DIRICHLET AND NEUMANN PROBLEMS

We shall show in this section that analogous to the potential theoretic case $k = 0$, we can reduce the boundary-value problems for the Helmholtz equation to integral equations of the second kind by seeking the solution in the form of an appropriate surface potential. However, in contrast to the case of Laplace's equation, it is now necessary to take into consideration the fact that the interior boundary-value problems are in general not uniquely solvable.

Theorem 3.15. The double-layer potential

$$u(x) = \int_{\partial D} \frac{\partial \Phi(x, y)}{\partial \nu(y)} \psi(y)\, ds(y), \qquad x \in \mathbb{R}^3 \setminus \partial D, \qquad (3.24)$$

with continuous density ψ is a solution of the interior Dirichlet problem (3.15) in D provided ψ is a solution of the integral equation

$$\psi(x)-2\int_{\partial D}\frac{\partial\Phi(x,y)}{\partial\nu(y)}\psi(y)\,ds(y)=-2f(x),\qquad x\in\partial D.\tag{3.25}$$

It solves the exterior Dirichlet problem (3.17) in $\mathbb{R}^3\setminus\bar{D}$ provided ψ is a solution of the integral equation

$$\psi(x)+2\int_{\partial D}\frac{\partial\Phi(x,y)}{\partial\nu(y)}\psi(y)\,ds(y)=2f(x),\qquad x\in\partial D.\tag{3.26}$$

Proof. The double-layer potential u obviously satisfies the Helmholtz equation in $\mathbb{R}^3\setminus\partial D$ and the Sommerfeld radiation condition (Theorem 3.2). By Theorem 2.13 we see that u continuously assumes the prescribed boundary values on ∂D if the density ψ solves the integral equation (3.25) or (3.26) for the interior or exterior problem, respectively.

In a similar fashion, Theorem 2.19 implies the following result.

Theorem 3.16. The single-layer potential

$$u(x)=\int_{\partial D}\Phi(x,y)\phi(y)\,ds(y),\qquad x\in\mathbb{R}^3\setminus\partial D,\tag{3.27}$$

with continuous density ϕ is a solution of the interior Neumann problem (3.16) in D provided ϕ is a solution of the integral equation

$$\phi(x)+2\int_{\partial D}\frac{\partial\Phi(x,y)}{\partial\nu(x)}\phi(y)\,ds(y)=2g(x),\qquad x\in\partial D.\tag{3.28}$$

It solves the exterior Neumann problem (3.18) in $\mathbb{R}^3\setminus\bar{D}$ provided ϕ is a solution of the integral equation

$$\phi(x)-2\int_{\partial D}\frac{\partial\Phi(x,y)}{\partial\nu(x)}\phi(y)\,ds(y)=-2g(x),\qquad x\in\partial D.\tag{3.29}$$

The above integral equations for boundary-value problems for the Helmholtz equation—in two dimensions—were first considered by Kupradse [1], [2], [3]. Rigorous proofs for the existence of solutions based on the first and second part of Fredholm's alternative were given by Vekua [1], [2], Weyl [1], Müller [2], and Leis [1]. In the following, we shall describe a slightly modified version of their approach.

Recalling the definitions (2.77) and (2.78) of the compact operators $\mathbf{K},\mathbf{K}'$: $C(\partial D)\to C(\partial D)$, we can rewrite the integral equations in the form

$$\psi-\mathbf{K}\psi=-2f\tag{3.25$'$}$$

and

$$\psi + \mathbf{K}\psi = 2f \tag{3.26'}$$

for the interior and exterior Dirichlet problems and

$$\phi + \mathbf{K}'\phi = 2g \tag{3.28'}$$

and

$$\phi - \mathbf{K}'\phi = -2g \tag{3.29'}$$

for the interior and exterior Neumann problems. We shall now apply the Riesz–Fredholm theory to these equations. Note that the equations belonging to the interior (exterior) Dirichlet and exterior (interior) Neumann problems are adjoint to one another.

We shall soon see that the nullspace of the operator $\mathbf{I}+\mathbf{K}$ corresponds to solutions of the homogeneous interior Neumann problem. Hence, we introduce the linear space

$$U := \left\{ u|_{\partial D} \middle| u \in \mathfrak{R}(D), \quad \Delta u + k^2 u = 0 \text{ in } D, \quad \frac{\partial u}{\partial \nu} = 0 \text{ on } \partial D \right\}.$$

Note that if k is not an interior Neumann eigenvalue then $U = \{0\}$.

Theorem 3.17. $N(\mathbf{I}+\mathbf{K}) = U$.

Proof. Let $\psi \in N(\mathbf{I}+\mathbf{K})$ and define a double-layer potential u by (3.24). Then $u_+ = 0$ on ∂D and from the uniqueness of the solution to the exterior Dirichlet problem (Theorem 3.13) it follows that $u = 0$ in $\mathbb{R}^3 \setminus D$. From Theorem 2.21 we see that $\partial u_- / \partial \nu = 0$ on ∂D, that is, u is a solution to the homogeneous interior Neumann problem. Finally, from Corollary 2.14, we find that $\psi = u_+ - u_- = -u_-$ and hence $\psi \in U$.

Conversely, let $\psi \in U$, that is, $\psi = u|_{\partial D}$ where u is a solution to the homogeneous interior Neumann problem. Then from Theorem 3.1 we see that

$$\int_{\partial D} \frac{\partial \Phi(x, y)}{\partial \nu(y)} \psi(y)\, ds(y) = 0, \qquad x \in \mathbb{R}^3 \setminus \bar{D}.$$

Letting $x \to \partial D$ and using Theorem 2.13 we have $\psi + \mathbf{K}\psi = 0$, that is, $\psi \in N(\mathbf{I}+\mathbf{K})$.

By Fredholm's alternative (Theorem 1.30) we have $\dim N(\mathbf{I}+\mathbf{K}) = \dim N(\mathbf{I}+\mathbf{K}') = m_N$ where $m_N = 0$ if k is not an interior Neumann eigenvalue and where $m_N \in \mathbb{N}$ if k is an eigenvalue. In the second case, we have the following theorem.

Theorem 3.18. Let $\phi_1,\ldots,\phi_{m_N}$ be a basis of $N(\mathbf{I}+\mathbf{K}')$ and define

$$u_j(x):=\int_{\partial D}\Phi(x,y)\phi_j(y)\,ds(y),\qquad x\in\mathbf{R}^3\setminus\partial D,\tag{3.30}$$

$j=1,\ldots,m_N$. Then,

$$\phi_j=-\frac{\partial u_{j+}}{\partial\nu}\quad\text{on}\quad\partial D,\tag{3.31}$$

$j=1,\ldots,m_N$, and the functions

$$\psi_j:=-\bar{u}_{j+}\quad\text{on}\quad\partial D,\tag{3.32}$$

$j=1,\ldots,m_N$, form a basis for $N(\mathbf{I}+\mathbf{K})$. The matrix

$$\langle\psi_j,\phi_l\rangle=\int_{\partial D}\bar{u}_{j+}\frac{\partial u_{l+}}{\partial\nu}\,ds,\qquad j,l=1,\ldots,m_N,$$

is regular and hence by Theorem 1.31 the Riesz number is one.

Proof. Since $\phi_j+\mathbf{K}'\phi_j=0$ we clearly have $\partial u_{j-}/\partial\nu=0$ on ∂D. Using the jump relations for single-layer potentials, we therefore have $\phi_j=-\partial u_{j+}/\partial\nu$ on ∂D and $\bar{u}_{j+}=\bar{u}_{j-}\in U=N(\mathbf{I}+\mathbf{K})$ by Theorem 3.17.

Assume α_j, $j=1,\ldots,m_N$, solve

$$\sum_{j=1}^{m_N}\alpha_j\langle\psi_j,\phi_l\rangle=0,\qquad l=1,\ldots,m_N$$

and define

$$u:=\sum_{j=1}^{m_N}\bar{\alpha}_ju_j.$$

Then

$$\int_{\partial D}\bar{u}_+\frac{\partial u_+}{\partial\nu}\,ds=0$$

and from Theorem 3.12 we can conclude that $u=0$ in $\mathbf{R}^3\setminus D$. In particular, $\partial u_+/\partial\nu=0$ on ∂D and therefore $\sum_{j=1}^{m_N}\bar{\alpha}_j\phi_j=0$. Hence $\alpha_j=0$, $j=1,\ldots,m_N$, since the ϕ_j, $j=1,\ldots,m_N$, are linearly independent.

The above analysis implies that the matrix $\langle\psi_j,\phi_l\rangle$, $j,l=1,\ldots,m_N$, is regular and the ψ_j, $j=1,\ldots,m_N$ are linearly independent, that is, they form a basis for $N(\mathbf{I}+\mathbf{K})$.

Remark 3.19. Since the interior Neumann eigenvalues are real, we can choose the basis of solutions to the homogeneous interior Neumann problem to be real. Therefore, we can select the basis $\phi_1,\dots,\phi_{m_N}$ for $N(\mathbf{I}+\mathbf{K}')$ in Theorem 3.18 such that the $\psi_1,\dots,\psi_{m_N}$ are real valued.

We now are in the position to establish the existence of solutions to the interior Neumann and the exterior Dirichlet problems.

Theorem 3.20. The interior Neumann problem (3.16) is solvable if and only if

$$\int_{\partial D} gu\,ds = 0 \tag{3.33}$$

for all solutions u of the homogeneous interior Neumann problem.

Proof. If k is not an interior Neumann eigenvalue, condition (3.33) is trivially satisfied for any inhomogeneity g. Furthermore, by Theorem 3.17 and Fredholm's alternative we have $N(\mathbf{I}+\mathbf{K}')=\{0\}$ and the integral equation $\phi+\mathbf{K}'\phi=2g$ for the inhomogeneous interior Neumann problem is uniquely solvable for all inhomogeneities g.

If k is an eigenvalue, then Theorem 3.17 shows that condition (3.33) coincides with the solvability condition of Fredholm's alternative for the inhomogeneous equation $\phi+\mathbf{K}'\phi=2g$ and hence a solution to the integral equation exists.

The necessity of condition (3.33) for the solvability of the inhomogeneous interior Neumann problem follows from applying the second Green's theorem (3.5) to solutions of the inhomogeneous and homogeneous problems.

Theorem 3.21. The exterior Dirichlet problem (3.17) is uniquely solvable.

Proof. If k is not an eigenvalue of the interior Neumann problem, the integral equation $\psi+\mathbf{K}\psi=2f$ for the inhomogeneous exterior Dirichlet problem is uniquely solvable by the first part of Fredholm's alternative since by Theorem 3.17 $N(\mathbf{I}+\mathbf{K})=\{0\}$.

If k is an eigenvalue, we seek a solution of (3.17) in the form

$$u(x)=\int_{\partial D}\frac{\partial\Phi(x,y)}{\partial\nu(y)}\psi(y)\,ds(y)-\sum_{j=1}^{m_N}\alpha_j u_j(x),\qquad x\in\mathbb{R}^3\setminus\bar{D}, \tag{3.34}$$

where the u_j, $j=1,\dots,m_N$, are defined by (3.30). According to Remark 3.19, we can assume $u_{j+}|_{\partial D}$ to be real. Using (3.32), we see that u solves the exterior Dirichlet problem provided ψ and the coefficients α_j, $j=1,\dots,m_N$, are chosen such that

$$\psi+\mathbf{K}\psi=2f-2\sum_{j=1}^{m_N}\alpha_j\psi_j. \tag{3.35}$$

By Theorem 3.18, we can determine the coefficients to be the unique solution of the linear system

$$\sum_{j=1}^{m_N} \alpha_j \langle \psi_j, \phi_l \rangle = \langle f, \phi_l \rangle, \qquad l = 1, \ldots, m_N.$$

By the second part of Fredholm's alternative, we can now conclude that the integral equation (3.35) has a solution since its right-hand side satisfies the solvability condition. Note that the solution to the integral equation is not unique.

The nullspace of the operator $\mathbf{I} - \mathbf{K}'$ corresponds to solutions to the homogeneous interior Dirichlet problem. We therefore introduce the linear space

$$V := \left\{ \left. \frac{\partial v}{\partial \nu} \right|_{\partial D} \middle| v \in \mathfrak{R}(D), \quad \Delta v + k^2 v = 0 \text{ in } D, \qquad v = 0 \quad \text{on} \quad \partial D \right\}$$

and note that $V = \{0\}$ if k is not an interior Dirichlet eigenvalue. For the definition of V we have anticipated the fact that solutions to the homogeneous Dirichlet problem automatically belong to $\mathfrak{R}(D)$ by Theorem 3.27.

Theorem 3.22. $N(\mathbf{I} - \mathbf{K}') = V$.

Proof. This is proved in the same manner as Theorem 3.17 with the roles of Dirichlet and Neumann problems and single- and double-layer potentials interchanged.

Note that by Fredholm's alternative $\dim N(\mathbf{I} - \mathbf{K}') = \dim N(\mathbf{I} - \mathbf{K}) = m_D$ where $m_D = 0$ if k is not an interior Dirichlet eigenvalue and where $m_D \in \mathbb{N}$ if k is an eigenvalue. In the second case, we have the following theorem.

Theorem 3.23. Let $\delta_1, \ldots, \delta_{m_D}$ be a basis for $N(\mathbf{I} - \mathbf{K})$ and define

$$v_j(x) := \int_{\partial D} \frac{\partial \Phi(x, y)}{\partial \nu(y)} \delta_j(y)\, ds(y), \qquad x \in \mathbb{R}^3 \setminus \partial D, \tag{3.36}$$

$j = 1, \ldots, m_D$. Then

$$\delta_j = v_{j+} \quad \text{on} \quad \partial D, \tag{3.37}$$

$j = 1, \ldots, m_D$, and the functions

$$\chi_{j+} := \frac{\partial \bar{v}_{j+}}{\partial \nu} \quad \text{on} \quad \partial D, \tag{3.38}$$

$j = 1, \ldots, m_D$, form a basis for $N(\mathbf{I} - \mathbf{K}')$. The matrix

$$\langle \chi_j, \delta_l \rangle = \int_{\partial D} v_{l+} \frac{\partial \bar{v}_{j+}}{\partial \nu} ds, \qquad j, l = 1, \ldots, m_D,$$

is regular and hence by Theorem 1.31 the Riesz number is one.

Proof. This is proved in the same way as Theorem 3.18.

Note that since the interior Dirichlet eigenvalues are real, we can again choose the basis $\delta_1,\dots,\delta_{m_D}$ such that the $\chi_1,\dots,\chi_{m_D}$ are real valued.

Theorem 3.24. The interior Dirichlet problem (3.15) is solvable if and only if

$$\int_{\partial D} f\frac{\partial v}{\partial \nu}\,ds = 0 \tag{3.39}$$

for all solutions v to the homogeneous interior Dirichlet problem.

Proof. This is proved in the same way as Theorem 3.20. In proving the necessity of condition (3.39) by a direct application of the second Green's theorem (3.5) to a solution u of the inhomogeneous and a solution v of the homogeneous problem, we must have $u \in \mathfrak{R}(D)$. This is true by Theorem 3.27 if $f \in C^{1,\alpha}(\partial D)$. For general $f \in C(\partial D)$ define

$$\gamma_j := \int_{\partial D} f\chi_j\,ds,$$

$j=1,\dots,m_D$, where the χ_j are defined by (3.38). Let α_l, $l=1,\dots,m_D$, be the unique solution of

$$\sum_{l=1}^{m_D} \alpha_l\langle\chi_j,\delta_l\rangle = \gamma_j, \qquad j=1,\dots,m_D.$$

Then the function $\tilde{f} := f - \sum_{l=1}^{m_D}\alpha_l\delta_l$ satisfies condition (3.39). Since this condition is sufficient for the solvability of the interior Dirichlet problem, there exists a solution $\tilde{u}$ with boundary-value $\tilde{u} = \tilde{f}$ on ∂D. Then $u^* := u - \tilde{u}$ solves the interior Dirichlet problem with boundary values $u^* = \sum_{l=1}^{m_D}\alpha_l\delta_l$. By Theorem 2.30 the operator $\mathbf{K}$ maps $C(\partial D)$ into $C^{0,\alpha}(\partial D)$ and $C^{0,\alpha}(\partial D)$ into $C^{1,\alpha}(\partial D)$. Therefore, the elements $\delta_1,\dots,\delta_{m_D}$ of the nullspace $N(\mathbf{I}-\mathbf{K})$ automatically belong to $C^{1,\alpha}(\partial D)$ and by our previous argument the solvability condition (3.39) for u^* must be satisfied, that is,

$$\sum_{l=1}^{m_D} \alpha_l\langle\chi_j,\delta_l\rangle = 0, \qquad j=1,\dots,m_D.$$

This implies $\gamma_j = 0$, $j=1,\dots,m_D$, which completes the proof.

Theorem 3.25. The exterior Neumann problem (3.18) is uniquely solvable.

Proof. This is proved in the same way as Theorem 3.21 if a solution is sought in the form

$$u(x) = \int_{\partial D}\Phi(x,y)\phi(y)\,ds(y) + \sum_{j=1}^{m_D}\alpha_j v_j(x), \qquad x \in \mathbf{R}^3\setminus\bar{D}, \tag{3.40}$$

which leads to the integral equation

$$\phi - \mathbf{K}'\phi = -2g + 2\sum_{j=1}^{m_D} \alpha_j \chi_j. \tag{3.41}$$

We now conclude this section by showing that any solution to the Dirichlet problem with boundary values belonging to $C^{1,\alpha}(\partial D)$ is automatically uniformly Hölder continuously differentiable up to the boundary; in particular, the solution belongs to $\mathcal{R}(D)$ or $\mathcal{R}(\mathbf{R}^3\setminus\bar{D})$.

Lemma 3.26. Let G be a bounded domain with diameter d, $u\in C^2(G)\cap C(\bar{G})$ a solution to the Helmholtz equation in G with $u=0$ on ∂G, and assume

$$2|k|^2(e^d-1)<1.$$

Then $u=0$ in G.

Proof. Without loss of generality we assume G is contained in the cube $Q:=\{x=(x_1,x_2,x_3)\in\mathbf{R}^3 \mid 0\leqslant x_1,x_2,x_3\leqslant d\}$. We first show that for any real-valued function $w\in C^2(G)\cap C(\bar{G})$ with $\Delta w\in C(\bar{G})$ and $w=0$ on ∂G we have

$$\|w\|_{\infty,G}\leqslant (e^d-1)\|\Delta w\|_{\infty,G}. \tag{3.42}$$

To show this we consider the function

$$v(x):=(e^d-e^{x_1})\|\Delta w\|_\infty, \qquad x\in G.$$

Then

$$-\Delta v(x)=\|\Delta w\|_\infty e^{x_1}\geqslant \|\Delta w\|_\infty, \qquad x\in G,$$

and hence if we define

$$v_{1,2}:=-v\pm w$$

we see that $\Delta v_{1,2}\geqslant 0$ in G. Since on the boundary ∂G we have $v_{1,2}=-v\leqslant 0$ from the maximum principle (cf. Colton [4]), we can conclude that $v_{1,2}\leqslant 0$ in G. Hence $\|w\|_\infty\leqslant\|v\|_\infty$ from which (3.42) follows.

We now write u and k^2 in terms of their real and imaginary parts, $u=u_1+iu_2$, $k^2=k_1+ik_2$. Since $\Delta u_1=-k_1u_1+k_2u_2$ and $\Delta u_2=-k_1u_2-k_2u_1$, we can apply (3.42) to u_1 and u_2 to find

$$\|u_1\|_\infty+\|u_2\|_\infty\leqslant 2|k|^2(e^d-1)(\|u_1\|_\infty+\|u_2\|_\infty).$$

Since $2|k|^2(e^d-1)<1$ we can now conclude that $u=0$ in G.

Theorem 3.27. Let u be a solution to the interior or the exterior Dirichlet problem with boundary values $f \in C^{1,\alpha}(\partial D)$, $0<\alpha<1$. Then $u \in C^{1,\alpha}(\bar{D})$ or $u \in C^{1,\alpha}(\mathbb{R}^3 \setminus D)$, respectively.

Proof. It suffices to consider the interior problem. We choose an arbitrary point $x_0 \in \partial D$ and show that $u \in C^{1,\alpha}(\bar{D} \cap U)$ where U is some neighborhood of x_0. Let G be a bounded domain contained in D whose boundary is of class C^2 and contains $\partial D \cap U$. We assume G is small enough such that by Lemma 3.26 the only solution to the homogeneous Dirichlet problem in G is the trivial solution. Then we can uniquely solve the interior Dirichlet problem $\Delta v + k^2 v = 0$ in G, $v = u$ on ∂G. We now look for a solution to this problem in the form (3.24) applied to the domain G. Since we have chosen G such that the homogeneous form of equation (3.25) only has the trivial solution, the inhomogeneous equation is uniquely solvable. But for this equation the right-hand side $u|_{\partial G}$ coincides with f on $\partial D \cap U$ and is of class $C^{1,\alpha}$. Hence, by Theorem 2.30 the solution ψ of the integral equation is also of class $C^{1,\alpha}$. The proof now follows from Theorem 2.23.

3.5 BOUNDARY INTEGRAL EQUATIONS OF THE FIRST KIND

It is also possible to reduce the boundary-value problems for the Helmholtz equation to boundary integral equations of the first kind. In particular, for the Dirichlet problem we immediately have the following result:

Theorem 3.28. The single-layer potential

$$u(x) = \int_{\partial D} \Phi(x,y)\phi(y)\,ds(y), \qquad x \in \mathbb{R}^3 \setminus \partial D, \tag{3.43}$$

with continuous density ϕ solves the interior and the exterior Dirichlet problems (3.15) and (3.17) provided ϕ is a solution of the integral equation

$$\int_{\partial D} \Phi(x,y)\phi(y)\,ds(y) = f(x), \qquad x \in \partial D. \tag{3.44}$$

Recalling the definition (2.80) of the compact operator $\mathbf{S}$: $C(\partial D) \to C(\partial D)$, we can rewrite (3.44) in the short form

$$\mathbf{S}\phi = 2f. \tag{3.44'}$$

Since by Theorem 2.19 the single-layer potential with continuous density has continuous normal derivatives on the boundary, the integral equation (3.44′) is only solvable for those functions f for which the solutions to the interior and the exterior Dirichlet problems belong to $\mathfrak{R}(D)$ and $\mathfrak{R}(\mathbb{R}^3 \setminus \bar{D})$, respectively. Recalling the linear subspace $\mathcal{N}(\partial D)$ of all continuous functions $\psi \in C(\partial D)$ for which the double-layer potential v with density ψ has continuous normal derivatives on both sides of ∂D (cf. Section 2.7), we can state the following theorem.

Theorem 3.29. A solution u to the interior (and the exterior) Dirichlet problem with boundary values $u = f$ on ∂D has continuous normal derivatives on ∂D, that is, $u \in \mathcal{R}(D)$ (and $u \in \mathcal{R}(\mathbb{R}^3 \setminus \bar{D})$) if and only if $f \in \mathcal{N}(\partial D)$.

Proof. The necessity that $f \in \mathcal{N}(\partial D)$ follows immediately from Theorem 3.1 and Theorem 2.19.

To show that the condition is also sufficient, we first define the double-layer potential

$$v(x) := \int_{\partial D} \frac{\partial \Phi(x, y)}{\partial \nu(y)} f(y)\, ds(y), \qquad x \in \mathbb{R}^3 \setminus \partial D.$$

In $\mathbb{R}^3 \setminus D$ we can regard v as the unique solution w of the exterior Neumann problem with the continuous normal derivatives $\partial w/\partial \nu = \partial v/\partial \nu$ on ∂D. As shown in Theorems 3.16 and 3.25, this solution w can be represented in the form of a single-layer potential

$$w(x) = \int_{\partial D} \Phi(x, y) \phi(y)\, ds(y), \qquad x \in \mathbb{R}^3 \setminus \partial D,$$

provided k is not an interior Dirichlet eigenvalue. Now consider the function

$$W := \begin{cases} w - v - u & \text{in} \quad D \\ w - v & \text{in} \quad \mathbb{R}^3 \setminus \bar{D}. \end{cases}$$

Then from Corollaries 2.14 and 2.20 and the boundary condition $u = f$ on ∂D we see that $W_+ - W_- = 0$ on ∂D. Since by construction W vanishes in $\mathbb{R}^3 \setminus D$ we conclude that $W = 0$ in D if k is not an interior Dirichlet eigenvalue. Thus $u = w - v$ in D from which we finally get $u \in \mathcal{R}(D)$. This proof can easily be extended to the case where k is an interior eigenvalue by representing the solution to the exterior Neumann problem in the form (3.40). For the exterior problem the proof is carried out in an analogous manner.

Theorem 3.30. For any inhomogeneity $f \in \mathcal{N}(\partial D)$ the integral equation (3.44) of the first kind for the Dirichlet problem has a unique solution provided k is not an interior Dirichlet eigenvalue. If k is an eigenvalue, then the integral equation is solvable if and only if f satisfies the condition (3.39) and, in this case, the solution is not unique.

Proof. We first prove uniqueness. Let $\phi \in C(\partial D)$ be a solution of the homogeneous equation $\mathbf{S}\phi = 0$. Then the single-layer potential u defined by (3.43) solves the homogeneous interior and exterior Dirichlet problem. Hence, from the uniqueness results for the Dirichlet problem and the jump relation of Corollary 2.20, we can conclude that $\phi=0$ provided k is not an interior eigenvalue. If k is an eigenvalue, we see from Theorem 3.1 that $N(\mathbf{S})=V$.

To establish existence, we use Theorems 3.21 and 3.24 and denote by u the solution to the interior Dirichlet problem in D and the solution to the exterior Dirichlet problem in $\mathbf{R}^3\setminus\bar{D}$. Observe that $u\in\mathfrak{R}(D)$ and $u\in\mathfrak{R}(\mathbf{R}^3\setminus\bar{D})$ by Theorem 3.29 and the assumption $f\in\mathfrak{N}(\partial D)$. We now define

$$\phi:=\frac{\partial u_-}{\partial \nu}-\frac{\partial u_+}{\partial \nu} \quad \text{on} \quad \partial D$$

and use Theorems 3.1 and 3.3 to see that

$$u(x)=\int_{\partial D}\Phi(x,y)\phi(y)\,ds(y), \qquad x\in\mathbf{R}^3\setminus\partial D,$$

which implies that ϕ solves the equation $\mathbf{S}\phi=2f$.

Recall now the definition (2.81) of the unbounded operator $\mathbf{T}$: $\mathfrak{N}(\partial D)\to C(\partial D)$. Then if k is not an eigenvalue to the interior Dirichlet or Neumann problem, we see by the existence proof of Theorem 3.30 and our previous analysis of the integral equations (3.25) and (3.26) for the Dirichlet problems that the solution ϕ to the equation (3.44) is given by

$$\phi=-2\mathbf{T}(\mathbf{I}-\mathbf{K})^{-1}(\mathbf{I}+\mathbf{K})^{-1}f.$$

To arrive at this relationship we have used the fact that $(\mathbf{I}-\mathbf{K})^{-1}+(\mathbf{I}+\mathbf{K})^{-1}=2(\mathbf{I}-\mathbf{K})^{-1}(\mathbf{I}+\mathbf{K})^{-1}$. Hence the inverse operator $\mathbf{S}^{-1}$: $\mathfrak{N}(\partial D)\to C(\partial D)$ of $\mathbf{S}$ is given by

$$\mathbf{S}^{-1}=-\mathbf{T}(\mathbf{I}-\mathbf{K})^{-1}(\mathbf{I}+\mathbf{K})^{-1}. \tag{3.45}$$

If we look for a solution to the Neumann problem in the form of a double-layer potential, we are faced with the difficulty that the normal derivative of a double-layer potential with continuous density in general does not exist. Hence we assume the density to belong to $\mathfrak{N}(\partial D)$. We can then state the following theorem.

Theorem 3.31. The double-layer potential

$$u(x)=\int_{\partial D}\frac{\partial\Phi(x,y)}{\partial\nu(y)}\psi(y)\,ds(y), \qquad x\in\mathbf{R}^3\setminus\partial D, \tag{3.46}$$

with density $\psi\in\mathfrak{N}(\partial D)$ solves the interior and the exterior Neumann problems, (3.16) and (3.18), provided ψ is a solution of the singular integral equation

$$\frac{\partial}{\partial\nu(x)}\int_{\partial D}\frac{\partial\Phi(x,y)}{\partial\nu(y)}\psi(y)\,ds(y)=g(x),\qquad x\in\partial D. \tag{3.47}$$

Using our previous operator notation, we can rewrite (3.47) in the form

$$\mathbf{T}\psi=2g \tag{3.47'}$$

and analogous to Theorem 3.30 obtain the following result.

Theorem 3.32. For any inhomogeneity $g\in C(\partial D)$ the integral equation (3.47) has a unique solution provided k is not an interior Neumann eigenvalue. If k is an eigenvalue, then the integral equation is solvable if and only if g satisfies condition (3.33) and, in this case, the solution is not unique.

From the details of the proof, it is easily seen that if k is not an eigenvalue to the interior Dirichlet or Neumann problem, the solution ψ of (3.47) is given by

$$\psi=-2\mathbf{S}(\mathbf{I}-\mathbf{K}')^{-1}(\mathbf{I}+\mathbf{K}')^{-1}g.$$

Hence the inverse operator $\mathbf{T}^{-1}$: $C(\partial D)\to\mathfrak{N}(\partial D)$ of $\mathbf{T}$ is given by

$$\mathbf{T}^{-1}=-\mathbf{S}(\mathbf{I}-\mathbf{K}')^{-1}(\mathbf{I}+\mathbf{K}')^{-1}. \tag{3.48}$$

Traditionally, the use of integral equations of the first kind for studying boundary-value problems in acoustic scattering theory has been neglected due to the lack of a Riesz–Fredholm theory for equations of the first kind and the fact that integral equations of the first kind are improperly posed. The ill-posed nature of such equations can be seen from the facts that the inverse $\mathbf{S}^{-1}$ of the compact operator $\mathbf{S}$ given by (3.45) cannot, in view of Theorems 1.5 and 1.9, be bounded and, in addition, the range $\mathbf{S}(C(\partial D))=\mathfrak{N}(\partial D)$ is not closed in $C(\partial D)$. Hence small perturbations of the right-hand side of the equation $\mathbf{S}\phi=2f$ can cause large perturbations in the solution ϕ or might even render the equation unsolvable if the perturbed right-hand side no longer belongs to $\mathfrak{N}(\partial D)$. Despite these difficulties, within recent years significant advances have been made in the numerical analysis of integral equations of the first kind, particularly through the work of Giroire [1], Nedelec [1], Giroire and Nedelec [1], and Hsiao and Wendland [1] in the limiting potential theoretic case $k=0$, and the interested reader is referred to these references for further details.

3.6 MODIFIED INTEGRAL EQUATIONS

In our analysis of the integral equations (3.26) and (3.29) of the second kind and (3.44) and (3.47) of the first kind for exterior boundary-value problems, we had to distinguish between uniquely and nonuniquely solvable integral equations. Because the solutions of the exterior boundary-value problems are unique for all wave numbers with $\operatorname{Im} k \geqslant 0$, the complication of nonuniqueness for the integral equations at the interior eigenvalues arises from our method of solution rather than from the nature of the problem itself. Therefore it is desirable to develop methods leading to integral equations that are uniquely solvable for all values of the wave numbers. This is particularly important for our later study of the inverse scattering problem because in this case the shape of the domain is unknown and hence the interior eigenvalues are also unknown. However, the formulation of uniquely solvable integral equations is also important for the direct problem, since if attempts are made to obtain numerical approximations to the solution by discretizing the integral equations of the previous sections, then the resulting linear systems will become ill-conditioned in a neighborhood of the interior eigenvalues and for a general domain D we do not know beforehand where these eigenvalues are. Since integral equation methods for exterior boundary-value problems have the advantage of reducing the problems from the unbounded three-dimensional domain $\mathbb{R}^3 \setminus \bar{D}$ to its bounded two-dimensional boundary ∂D and automatically satisfy the radiation condition at infinity, there is an increasing interest in numerical methods based on integral equations, and from this point of view it is important to develop integral equations that are uniquely solvable for all wave numbers.

Leis [2], Brakhage and Werner [1], and Panich [1] independently suggested to try and find the solution of the exterior Dirichlet problem (3.17) in the form of a combined double- and single-layer potential

$$u(x) = \int_{\partial D} \left\{ \frac{\partial \Phi(x, y)}{\partial \nu(y)} - i\eta \Phi(x, y) \right\} \psi(y)\, ds(y), \qquad x \in \mathbb{R}^3 \setminus \partial D, \tag{3.49}$$

where $\eta \neq 0$ is an arbitrary real number such that

$$\eta \operatorname{Re} k \geqslant 0. \tag{3.50}$$

Obviously, (3.49) solves the exterior Dirichlet problem (3.17) provided the density $\psi \in C(\partial D)$ is a solution of the integral equation

$$\psi + \mathbf{K}\psi - i\eta \mathbf{S}\psi = 2f. \tag{3.51}$$

Theorem 3.33. The combined double- and single-layer integral equation (3.51) for the exterior Dirichlet problem is uniquely solvable for all wave numbers satisfying $\operatorname{Im} k \geqslant 0$.

Proof. Since $\mathbf{K} - i\eta\mathbf{S}$ is a compact operator, by Theorem 1.16 it suffices to show that the homogeneous form of equation (3.51) has only the trivial solution $\psi = 0$. Let $\psi \in C(\partial D)$ be a solution to the homogeneous equation $\psi + \mathbf{K}\psi - i\eta\mathbf{S}\psi = 0$. Then u as defined by (3.49) solves the homogeneous exterior Dirichlet problem and, therefore, $u = 0$ in $\mathbb{R}^3 \setminus D$. From the jump relations of Corollaries 2.14 and 2.20, we now have

$$-u_- = \psi, \qquad -\frac{\partial u_-}{\partial \nu} = i\eta\psi \quad \text{on} \quad \partial D,$$

and the first Green's theorem (3.4) implies that

$$i\eta\int_{\partial D} |\psi|^2\, ds = \int_{\partial D} \bar{u}_- \frac{\partial u_-}{\partial \nu}\, ds = \int_D (|\operatorname{grad} u|^2 - k^2 |u|^2)\, dx.$$

The imaginary part of this equation is

$$\eta\int_{\partial D} |\psi|^2\, ds = -2\operatorname{Re} k \operatorname{Im} k \int_D |u|^2\, dx$$

from which we see that $\psi = 0$ because of (3.50) and the fact that $\operatorname{Im} k \geqslant 0$.

A similar approach for the exterior Neumann problem runs into difficulties due to the fact that the normal derivative of a double-layer potential with continuous density in general does not exist on the boundary and even if it does exist, the corresponding integral equation is strongly singular. In this case, results on the existence of a solution requires regularization of the integral equation, that is, transforming the integral equation into a form for which the Riesz theory is applicable. Keeping these comments in mind, we seek the solution of the exterior Neumann problem in the form of a combined single- and double-layer potential

$$u(x) = \int_{\partial D} \left\{ \Phi(x, y) + i\eta \frac{\partial \Phi(x, y)}{\partial \nu(y)} \right\} \phi(y)\, ds(y), \qquad x \in \mathbb{R}^3 \setminus \partial D, \tag{3.52}$$

where $\eta \neq 0$ is chosen as in (3.50) and where we require the density ϕ to belong to $\mathfrak{N}(\partial D)$. Obviously, (3.52) solves the exterior Neumann problem (3.18) provided the density $\phi \in \mathfrak{N}(\partial D)$ is a solution of the singular integral equation

$$\phi - \mathbf{K}'\phi - i\eta\mathbf{T}\phi = -2g. \tag{3.53}$$

Theorem 3.34. The combined single- and double-layer integral equation (3.53) for the exterior Neumann problem is uniquely solvable for all wave numbers satisfying $\operatorname{Im} k \geqslant 0$.

Proof. By using essentially the same argument as in the proof of Theorem 3.33, it can be seen that the homogeneous equation $\phi - \mathbf{K}'\phi - i\eta\mathbf{T}\phi = 0$ has only the trivial solution $\phi = 0$ in $\mathfrak{N}(\partial D)$.

To establish existence we shall regularize the singular integral equation (3.53) by using a slight modification of an approach due to Leis [3]. We start by picking a wave number k_0 that is not an interior eigenvalue to the Dirichlet and Neumann problems, for instance, any k_0 with $\operatorname{Im} k_0 > 0$. In the following, we indicate by a subscript the fact that the parameter k appearing in the operators is set equal to k_0. From (3.48), we recall that the operator $\mathbf{A}_0 := -\mathbf{S}_0(\mathbf{I}-\mathbf{K}'_0)^{-1}(\mathbf{I}+\mathbf{K}'_0)^{-1}$ is the inverse of the operator $\mathbf{T}_0$. Since $\mathbf{S}_0$ is compact, by Theorems 1.5 and 1.16 the operator $\mathbf{A}_0$ is compact.

For an arbitrary k we can use the identity $\mathbf{A}_0\mathbf{T}_0 = \mathbf{T}_0\mathbf{A}_0 = \mathbf{I}$ to transform (3.53) into the equivalent form

$$\mathbf{A}_0(\mathbf{I}-\mathbf{K}'-i\eta(\mathbf{T}-\mathbf{T}_0))\phi - i\eta\phi = -2\mathbf{A}_0 g, \tag{3.54}$$

an equation which we have to consider in the space $C(\partial D)$. By Theorem 2.31 the difference $\mathbf{T}-\mathbf{T}_0$ is compact, and hence the combination $\mathbf{A}_0(\mathbf{I}-\mathbf{K}'-i\eta(\mathbf{T}-\mathbf{T}_0))$ is compact and the Riesz theory is applicable to (3.54). Since we have already shown uniqueness, we can now conclude existence from Corollary 1.20.

Numerical implementations of the combined double- and single-layer potential approach have been described by Greenspan and Werner [1], Kussmaul [1], Bolomey and Tabbara [1], Ruland [1], Giroire [1], and Meyer et al. [1], [2]. We would like to point out that in the numerical procedure for the exterior Neumann problem one can either use the regularized form (3.54) of the integral equation or directly discretize the singular integral equation (3.53) by making use of (2.62). An analysis of the appropriate choice of the coupling parameter η in order to minimize the condition number of the integral operators has been carried out by Kress and Spassov [1].

Another method leading to uniquely solvable integral equations for the exterior boundary-value problems was proposed by Jones [1] who suggested adding a series of radiating waves of the form

$$\chi(x,y) := ik \sum_{n=0}^{\infty} \sum_{m=-n}^{n} a_{nm} h_n^{(1)}(k|x|) Y_n^m\left(\frac{x}{|x|}\right) h_n^{(1)}(k|y|) \overline{Y_n^m}\left(\frac{y}{|y|}\right) \tag{3.55}$$

to the fundamental solution $\Phi(x,y)$. Here $h_n^{(1)}$ denote the spherical Hankel function of the first kind and of order n and Y_n^m denote the spherical harmonics introduced in (3.19). In the following analysis of Jones' method we assume D to be a connected domain containing the origin and choose a ball B of radius R and center at the origin such that $\bar{B} \subset D$. On the coefficients a_{nm} we impose the condition that the series (3.55) is uniformly convergent in x and y in any region $|x|, |y| \geqslant R+\varepsilon$, $\varepsilon > 0$, and that the series can be two times differentiated term by term with respect to any of the variables with the resulting series being uniformly convergent.

Replacing the fundamental solution $\Phi(x, y)$ by the modification

$$\Psi(x, y) = \Phi(x, y) + \chi(x, y),$$

we see that the modified double-layer potential

$$u(x) = \int_{\partial D} \frac{\partial \Psi(x, y)}{\partial \nu(y)} \psi(y)\, ds(y), \qquad x \in \mathbb{R}^3 \setminus \partial D \setminus \bar{B}, \tag{3.56}$$

with continuous density ψ is a solution of the exterior Dirichlet problem (3.17) provided ψ is a solution of the integral equation

$$\psi(x) + 2\int_{\partial D} \frac{\partial \Psi(x, y)}{\partial \nu(y)} \psi(y)\, ds(y) = 2f(x), \qquad x \in \partial D. \tag{3.57}$$

By our assumptions on the series (3.55) the kernel $\partial \chi(x, y)/\partial \nu(y)$ is continuous on $\partial D \times \partial D$, and hence by Theorem 2.6 the modified operator $\tilde{\mathbf{K}}$: $C(\partial D) \to C(\partial D)$ defined by

$$(\tilde{\mathbf{K}}\psi)(x) := 2\int_{\partial D} \frac{\partial \Psi(x, y)}{\partial \nu(y)} \psi(y)\, ds(y), \qquad x \in \partial D$$

is compact.

Theorem 3.35. The modified double-layer integral equation (3.57) for the exterior Dirichlet problem is uniquely solvable for all positive wave numbers $k > 0$ provided that either

$$|2a_{nm} + 1| < 1 \tag{3.58}$$

for all $n = 0, 1, 2, \ldots,\ m = -n, \ldots, n$, or

$$|2a_{nm} + 1| > 1 \tag{3.59}$$

for all $n = 0, 1, 2, \ldots,\ m = -n, \ldots, n$.

Proof. We have to show that the homogeneous integral equation only has the trivial solution. Let $\psi \in C(\partial D)$ be a solution of the homogeneous equation $\psi + \tilde{\mathbf{K}}\psi = 0$. Then u, as defined by (3.56), solves the homogeneous exterior Dirichlet problem. Hence, from the uniqueness of solutions to the exterior Dirichlet problem, $u = 0$ in $\mathbb{R}^3 \setminus D$ and from the jump relations of Theorem 2.21 (which remain valid for the modified double-layer potential) we have $\partial u_- / \partial \nu = 0$ on ∂D.

From the expansion (cf. Erdélyi et al. [1])

$$\frac{e^{ik|x-y|}}{4\pi|x-y|} = ik \sum_{n=0}^{\infty} \sum_{m=-n}^{n} j_n(k|x|)\, Y_n^m\!\left(\frac{x}{|x|}\right) h_n^{(1)}(k|y|)\, \overline{Y_n^m}\!\left(\frac{y}{|y|}\right), \qquad |x| < |y|, \tag{3.60}$$

and (3.55) we see that there exist constants α_{nm} such that u can be expanded in the form

$$u(x)=\sum_{n=0}^{\infty}\sum_{m=-n}^{n}\alpha_{nm}\{j_n(k|x|)+a_{nm}h_n^{(1)}(k|x|)\}Y_n^m\left(\frac{x}{|x|}\right). \qquad (3.61)$$

This expansion and its term by term derivatives are uniformly convergent in some domain $R_1 \leqslant |x| \leqslant R_2$, where $R+\varepsilon<R_1<R_2$. Using this expansion, the orthogonality of the spherical harmonics, and the Wronskian relation for the spherical Bessel functions, we now obtain from the second Green's theorem (3.5) that

$$0=\int_{\partial D}\left(u_-\frac{\partial \bar{u}_-}{\partial \nu}-\bar{u}_-\frac{\partial u_-}{\partial \nu}\right)ds=\int_{|x|=R_1}\left(u\frac{\partial \bar{u}}{\partial \nu}-\bar{u}\frac{\partial u}{\partial \nu}\right)ds$$

$$=\frac{i}{2k}\sum_{n=0}^{\infty}\sum_{m=-n}^{n}|\alpha_{nm}|^2(1-|1+2a_{nm}|^2). \qquad (3.62)$$

From this and the conditions (3.58) or (3.59), we can now conclude that $\alpha_{nm}=0$ for $n=0,1,2,\dots$, $m=-n,\dots,n$. Hence $u=0$ in $R_1\leqslant|x|\leqslant R_2$ and therefore $u=0$ in $D\setminus B$ by the analyticity of u (Theorem 3.5). The jump relation of Corollary 2.14 now implies that $\psi=0$.

In the same manner the modified single-layer potential

$$u(x)=\int_{\partial D}\Psi(x,y)\phi(y)\,ds(y), \qquad x\in\mathbb{R}^3\setminus\partial D\setminus\bar{B}, \qquad (3.63)$$

with continuous density ϕ solves the exterior Neumann problem (3.18) provided ϕ is a solution of the integral equation

$$\phi(x)-2\int_{\partial D}\frac{\partial\Psi(x,y)}{\partial\nu(x)}\phi(y)\,ds(y)=-2g(x), \qquad x\in\partial D. \qquad (3.64)$$

Noting that the operator $\tilde{\mathbf{K}}'$: $C(\partial D)\to C(\partial D)$ defined by

$$(\tilde{\mathbf{K}}'\phi)(x):=2\int_{\partial D}\frac{\partial\Psi(x,y)}{\partial\nu(x)}\phi(y)\,ds(y), \qquad x\in\partial D,$$

is compact and repeating essentially the same argument as used in the proof of Theorem 3.35 yields the following theorem.

Theorem 3.36. The modified single-layer integral equation (3.64) for the exterior Neumann problem is uniquely solvable for all positive wave numbers $k>0$ provided that either (3.58) or (3.59) is satisfied.

Note that in contrast to the combined single- and double-layer approach for the Neumann problem the modified single-layer approach only involves compact operators and avoids regularization techniques.

In our proof of Theorem 3.35 we have followed Ursell [2], who clarified parts of Jones' work, and Kleinman and Roach [2], who suggested various criteria for choosing the coefficients a_{nm}.

From (3.62) we see that for the validity of Theorems 3.35 and 3.36 it is crucial that all coefficients a_{nm} in the modification (3.55) are chosen to be different from zero. If we have only a finite number of nonvanishing coefficients, say $a_{nm}=0$ for $n>N$ for some $N\in\mathbb{N}$, then from (3.62) we get $\alpha_{nm}=0$ for all $n\leqslant N$ and hence by (3.61) we can conclude that u can be extended into the whole interior domain D and represents a solution to the homogeneous interior Neumann problem. Therefore, if we let $\tilde{\mathbf{K}}_N$ denote the operator $\tilde{\mathbf{K}}$ with $a_{nm}=0$ for $n>N$, the operator $\mathbf{I}+\tilde{\mathbf{K}}_N$ can have a nontrivial nullspace only at the interior Neumann eigenvalues. Using the fact that the operator $\mathbf{I}+\tilde{\mathbf{K}}$ with all coefficients different from zero has a trivial nullspace for all positive wave numbers, by a continuity argument employing the Neumann series of the operator $(\mathbf{I}+\tilde{\mathbf{K}})^{-1}(\tilde{\mathbf{K}}_N-\tilde{\mathbf{K}})$, we observe that in order to make the modified integral equation uniquely solvable for a fixed finite range of positive wave numbers it suffices to require only that a sufficiently large number of the coefficients a_{nm} be different from zero. In a more detailed analysis Jones [1] actually proved that in order to remove the first N interior eigenvalues it suffices to have the first N coefficients in the series (3.55) different from zero.

We shall conclude by showing that a special choice of the coefficients a_{nm} allows us to identify the fundamental solution Ψ with the Green's function for a ball with impedance boundary condition (cf. Ursell [1]). In particular, let the a_{nm} be defined by

$$a_{nm}:=\frac{-\alpha_n}{\beta_n}, \tag{3.65}$$

$n=0,1,2,\ldots,\ m=-n,\ldots,n$, where

$$\alpha_n:=kj_n'(kR)+i\eta j_n(kR),$$

$$\beta_n:=kh_n^{(1)\prime}(kR)+i\eta h_n^{(1)}(kR),$$

$n=0,1,2,\ldots,$ for some constant $\eta>0$. Using the Wronskian relation for the spherical Bessel functions, we see that

$$|\beta_n|^2=k^2|h_n^{(1)\prime}(kR)|^2+\eta^2|h_n^{(1)}(kR)|^2+\frac{2\eta}{kR^2}$$

$$|\beta_n-2\alpha_n|^2=k^2|h_n^{(1)\prime}(kR)|^2+\eta^2|h_n^{(1)}(kR)|^2-\frac{2\eta}{kR^2}$$

from which we conclude that $\beta_n \neq 0$ for all $n = 0, 1, 2, \ldots$, and that condition (3.58) is satisfied. The choice (3.65) of the coefficients a_{nm} now implies that for $|x| = R$ and $|y| > R$ the function

$$\Psi(x, y) = \Phi(x, y) + \chi(x, y)$$

satisfies the boundary condition

$$\frac{\partial \Psi(x, y)}{\partial \nu(x)} + i\eta\Psi(x, y) = 0, \tag{3.66}$$

that is, Ψ is the Green's function for the exterior of the ball B of radius R with an impedance boundary condition. We note that the unique solvability of the exterior boundary integral equations using a fundamental solution satisfying (3.66) can be shown in a manner that is considerably simpler than the above approach due to Jones (cf. Ursell [1]).

3.7 THE IMPEDANCE BOUNDARY-VALUE PROBLEM

We shall now consider the impedance boundary-value problem for the Helmholtz equation.

Exterior Impedance Boundary-Value Problem

Find a function $u \in \mathfrak{R}(\mathbb{R}^3 \setminus \bar{D})$ satisfying the Helmholtz equation in $\mathbb{R}^3 \setminus \bar{D}$, the Sommerfeld radiation condition at infinity, and the boundary condition

$$\frac{\partial u}{\partial \nu} + \lambda u = g \quad \text{on} \quad \partial D \tag{3.67}$$

where g and λ are given continuous functions defined on ∂D.

Theorem 3.37. The exterior impedance boundary-value problem has at most one solution provided

$$\operatorname{Im}(\bar{k}\lambda) \geq 0 \quad \text{on} \quad \partial D. \tag{3.68}$$

Proof. Let u satisfy the homogeneous boundary condition $(\partial u / \partial \nu) + \lambda u = 0$ on ∂D. Then

$$\operatorname{Im}\left(k \int_{\partial D} u \frac{\partial \bar{u}}{\partial \nu} \, ds \right) = \operatorname{Im}\left(\bar{k} \int_{\partial D} \lambda |u|^2 \, ds \right)$$

and the statement follows from Theorem 3.12.

Analogous to the exterior Neumann problem (i.e., the special case when $\lambda = 0$), the single-layer potential

$$u(x) = \int_{\partial D} \Phi(x, y)\phi(y)\, ds(y), \qquad x \in \mathbb{R}^3 \backslash \partial D,$$

is a solution of the exterior impedance problem (3.67) provided the continuous density ϕ is a solution to the integral equation

$$\phi - \mathbf{K}'\phi - \lambda \mathbf{S}\phi = -2g. \tag{3.69}$$

It is easily seen (cf. Theorem 3.22) that the homogeneous form of equation (3.69) has nontrivial solutions if and only if the wave number k is an interior Dirichlet eigenvalue. In fact, we have $N(\mathbf{I}-\mathbf{K}'-\lambda\mathbf{S}) = N(\mathbf{I}-\mathbf{K}') = V$, that is, $N(\mathbf{I}-\mathbf{K}'-\lambda\mathbf{S})$ is independent of λ. Therefore we shall not discuss the integral equation (3.69) any further but instead shall proceed to uniquely solvable integral equations via the combined single- and double-layer approach and the modified single-layer approach described in the previous section for the Neumann problem.

The combined single- and double-layer potential

$$u(x) = \int_{\partial D} \left\{ \Phi(x, y) + i\eta \frac{\partial \Phi(x, y)}{\partial \nu(y)} \right\} \phi(y)\, ds(y), \qquad x \in \mathbb{R}^3 \backslash \partial D \tag{3.70}$$

where $\eta \neq 0$ is chosen as in (3.50) solves the exterior impedance problem (3.67) provided the density $\phi \in \mathfrak{N}(\partial D)$ is a solution of the singular integral equation

$$(1 - i\eta\lambda)\phi - (\mathbf{K}' + i\eta\mathbf{T} + i\eta\lambda\mathbf{K} + \lambda\mathbf{S})\phi = -2g. \tag{3.71}$$

Using the same arguments as for the exterior Dirichlet problem (cf. the proof of Theorem 3.33), it can be shown that the homogeneous form of equation (3.71) only has the trivial solution $\phi = 0$ in $\mathfrak{N}(\partial D)$. Existence follows as in the proof of Theorem 3.34 by regularizing (3.71) into the equivalent form

$$\mathbf{A}_0[(1 - i\eta\lambda)\mathbf{I} - (\mathbf{K}' + i\eta(\mathbf{T} - \mathbf{T}_0) + i\eta\lambda\mathbf{K} + \lambda\mathbf{S})]\phi - i\eta\phi = -2\mathbf{A}_0 g.$$

Therefore we have the following theorem.

Theorem 3.38. The combined single- and double-layer integral equation (3.71) for the exterior impedance boundary-value problem is uniquely solvable for all wave numbers $\operatorname{Im} k \geqslant 0$ and all impedances satisfying (3.68).

An integral equation which we do not have to regularize is obtained by seeking for a solution in the form of a modified single-layer potential

$$u(x) = \int_{\partial D} \Psi(x, y)\phi(y)\, ds(y) \tag{3.72}$$

that solves the impedance boundary-value problem (3.67) provided the continuous density ϕ is a solution of the integral equation

$$\phi(x)-2\int_{\partial D}\left\{\frac{\partial\Psi(x,y)}{\partial\nu(x)}+\lambda(x)\Psi(x,y)\right\}\phi(y)\,ds(y)=-2g(x),$$

$$x\in\partial D. \quad (3.73)$$

Proceeding as in the case of the Neumann problem, we can prove the following result.

Theorem 3.39. The modified single-layer integral equation (3.73) for the exterior impedance boundary-value problem is uniquely solvable for all positive wave numbers $k>0$ provided that $\operatorname{Im}\lambda\geqslant 0$ on ∂D and either (3.58) or (3.59) is satisfied.

By using the integral equation (3.73), it can easily be shown that the solution to the impedance boundary value problem converges to the solution of the Neumann problem as $\lambda\to 0$. The corresponding and mathematically more challenging problem of the singular perturbation of the impedance problem into the Dirichlet problem as $\lambda\to\infty$ has been studied by Kirsch [6].

3.8 THE TRANSMISSION BOUNDARY-VALUE PROBLEM

The mathematical description of the diffraction of time-harmonic acoustic waves by an obstacle D with acoustic constants different from those of the surrounding medium $\mathbb{R}^3\setminus\bar{D}$ leads to a transmission problem of the following form.

Transmission Problem

Find two functions $u\in C^2(\mathbb{R}^3\setminus\bar{D})\cap C(\mathbb{R}^3\setminus D)$ and $u_0\in C^2(D)\cap C(\bar{D})$ satisfying the Helmholtz equations

$$\Delta u+k^2u=0 \quad\text{in}\quad \mathbb{R}^3\setminus\bar{D}, \qquad \Delta u_0+k_0^2u_0=0 \quad\text{in}\quad D, \qquad (3.74)$$

the radiation condition

$$\left(\frac{x}{|x|},\operatorname{grad}u(x)\right)-iku(x)=o\left(\frac{1}{|x|}\right), \qquad |x|\to\infty,$$

uniformly for all directions $x/|x|$ and the transmission conditions

$$\left.\begin{aligned}\mu u-\mu_0u_0&=f\\ \frac{\partial u}{\partial\nu}-\frac{\partial u_0}{\partial\nu}&=g\end{aligned}\right\}\quad\text{on}\quad \partial D. \qquad (3.75)$$

Here k and k_0, μ and μ_0 are given complex numbers and f and g are given continuous functions defined on ∂D. The second boundary condition in (3.75) has to be understood in the sense described in Theorem 2.21.

We note that the transmission conditions (3.75) are quite general in the sense that they always can be renormalized to read

$$u - u_0 = \tilde{f}, \qquad \lambda \frac{\partial u}{\partial \nu} - \lambda_0 \frac{\partial u_0}{\partial \nu} = \tilde{g} \quad \text{on} \quad \partial D$$

for appropriate functions $\tilde{f}$ and $\tilde{g}$ and complex numbers λ and λ_0. From a physical point of view, an appropriate choice of the constants μ and μ_0 guarantees the continuity of the pressure and the normal velocity of the acoustic wave across the boundary ∂D.

For the sake of simplicity, we assume all the constants k and k_0, μ and μ_0 to be positive. However, our results can easily be extended to the case of complex values of these parameters (cf. Kress and Roach [2]).

Theorem 3.40. The transmission problem has at most one solution.

Proof. Let u and u_0 satisfy the homogeneous transmission conditions $\mu u - \mu_0 u_0 = 0$, $(\partial u/\partial \nu) - (\partial u_0/\partial \nu) = 0$ on ∂D. Then from the second Green's theorem (3.5), we have

$$\operatorname{Im}\left(k \int_{\partial D} u \frac{\partial \bar{u}}{\partial \nu}\, ds \right) = \operatorname{Im}\left(k_0^2 k \frac{\mu_0}{\mu} \int_D |u_0|^2\, dx \right)$$

and from Theorem 3.12 we obtain $u = 0$ in $\mathbb{R}^3 \setminus D$. The homogeneous boundary data now imply that $u_0 = 0$, $\partial u_0/\partial \nu = 0$ on ∂D, and hence from the representation Theorem 3.1 we have that $u_0 = 0$ in D.

In order to prove the existence of a solution to the transmission problem we seek the solution in the form of combined double- and single-layer potentials

$$u(x) = \int_{\partial D} \left\{ \frac{\partial \Phi(x,y)}{\partial \nu(y)} \psi(y) + \mu \Phi(x,y) \phi(y) \right\} ds(y), \qquad x \in \mathbb{R}^3 \setminus \bar{D},$$

$$u_0(x) = \int_{\partial D} \left\{ \frac{\partial \Phi_0(x,y)}{\partial \nu(y)} \psi(y) + \mu_0 \Phi_0(x,y) \phi(y) \right\} ds(y), \qquad x \in D, \tag{3.76}$$

with continuous densities ψ and ϕ and Φ_0 denoting the fundamental solution Φ with k replaced by k_0. Using Theorems 2.13, 2.19, and 2.21, it can be seen that (3.76) defines a solution of the transmission problem (3.75) if ψ and ϕ are a solution of the system of integral equations

$$(\mu + \mu_0)\psi + (\mu \mathbf{K} - \mu_0 \mathbf{K}_0)\psi + (\mu^2 \mathbf{S} - \mu_0^2 \mathbf{S}_0)\phi = 2f,$$

$$(\mu + \mu_0)\phi - (\mathbf{T} - \mathbf{T}_0)\psi - (\mu \mathbf{K}' - \mu_0 \mathbf{K}_0')\phi = -2g. \tag{3.77}$$

On the product space $C(\partial D)\times C(\partial D)$ equipped with the norm $\left\|\begin{pmatrix}\psi\\ \phi\end{pmatrix}\right\|_\infty :=$ $\max(\|\psi\|_\infty, \|\phi\|_\infty)$, we introduce the operator $\mathbf{A}$ defined by

$$\mathbf{A} := \begin{pmatrix} -(\mu\mathbf{K}-\mu_0\mathbf{K}_0) & -(\mu^2\mathbf{S}-\mu_0^2\mathbf{S}_0) \\ (\mathbf{T}-\mathbf{T}_0) & (\mu\mathbf{K}'-\mu_0\mathbf{K}'_0) \end{pmatrix}.$$

$\mathbf{A}$ is obviously compact since by Theorems 2.30 and 2.31 all its components are compact. We can now write the system (3.77) in the abbreviated form

$$(\mu+\mu_0)\chi - \mathbf{A}\chi = 2h \tag{3.77'}$$

where $\chi = \begin{pmatrix}\psi\\ \phi\end{pmatrix}$ and $h = \begin{pmatrix}f\\ -g\end{pmatrix}$.

Theorem 3.41. The transmission problem has a unique solution.

Proof. The proof is accomplished by showing that (3.77) is uniquely solvable. Let $\chi = \begin{pmatrix}\psi\\ \phi\end{pmatrix}$ be a solution to the homogeneous equation $(\mu+\mu_0)\chi - \mathbf{A}\chi = 0$. Then the potentials u and u_0 given by (3.76) solve the homogeneous transmission problem. Therefore, by the uniqueness Theorem 3.40, we have $u=0$ in $\mathbb{R}^3\setminus D$ and $u_0 = 0$ in D. Now define

$$v(x) := \frac{1}{\mu_0}\int_{\partial D}\left\{\frac{\partial\Phi_0(x,y)}{\partial\nu(y)}\psi(y) + \mu_0\Phi_0(x,y)\phi(y)\right\} ds(y), \qquad x\in\mathbb{R}^3\setminus\bar{D}$$

$$v_0(x) := -\frac{1}{\mu}\int_{\partial D}\left\{\frac{\partial\Phi(x,y)}{\partial\nu(y)}\psi(y) + \mu\Phi(x,y)\phi(y)\right\} ds(y), \qquad x\in D.$$

Then by the jump relations for single- and double-layer potentials we have

$$\begin{aligned} &\mu_0 v - u_0 = \psi, && u + \mu v_0 = \psi \\ &\frac{\partial v}{\partial\nu} - \frac{1}{\mu_0}\frac{\partial u_0}{\partial\nu} = -\phi, && \frac{1}{\mu}\frac{\partial u}{\partial\nu} + \frac{\partial v_0}{\partial\nu} = -\phi \end{aligned} \qquad \text{on } \partial D. \tag{3.78}$$

Hence v and v_0 solve the homogeneous transmission problem

$$\Delta v + k_0^2 v = 0 \quad \text{in } \mathbb{R}^3\setminus\bar{D}, \qquad \Delta v_0 + k^2 v_0 = 0 \quad \text{in } D$$

with transmission conditions

$$\mu_0 v - \mu v_0 = 0, \qquad \frac{\partial v}{\partial\nu} - \frac{\partial v_0}{\partial\nu} = 0 \quad \text{on } \partial D.$$

From Theorem 3.40 we now see that $v=0$ in $\mathbb{R}^3 \setminus D$ and $v_0=0$ in D. Hence from (3.78) we can conclude that $\psi=\phi=0$.

3.9 INTEGRAL EQUATIONS BASED ON THE REPRESENTATION THEOREMS

Up to now, we have transformed the boundary-value problems to integral equations by seeking the solution in the form of surface potentials. However, it is also possible to obtain integral equations based on the representation Theorems 3.1 and 3.3. These equations will turn out to be adjoint to those derived by the surface potential approach. For brevity, we shall confine ourselves to the exterior boundary-value problems. For a detailed analysis see Kleinman and Roach [1].

Under the assumptions of Theorem 3.3, in particular $u \in \mathfrak{R}(\mathbb{R}^3 \setminus \bar{D})$, we can represent radiating solutions u to the Helmholtz equation in the form

$$u(x)=\int_{\partial D}\left\{u(y)\frac{\partial \Phi(x,y)}{\partial \nu(y)}-\frac{\partial u}{\partial \nu}(y)\Phi(x,y)\right\}ds(y), \qquad x \in \mathbb{R}^3 \setminus \bar{D}.$$

Letting x tend to the boundary and using Theorem 2.13, we see that

$$-u+\mathbf{K}u-\mathbf{S}\frac{\partial u}{\partial \nu}=0 \quad \text{on} \quad \partial D. \tag{3.79}$$

Taking the normal derivative on the boundary and using Theorem 2.19 shows that

$$-\frac{\partial u}{\partial \nu}+\mathbf{T}u-\mathbf{K}'\frac{\partial u}{\partial \nu}=0 \quad \text{on} \quad \partial D. \tag{3.80}$$

Observe that $u|_{\partial D} \in \mathfrak{N}(\partial D)$ since by assumption $u \in \mathfrak{R}(\mathbb{R}^3 \setminus \bar{D})$.

Now let u be the solution to the exterior Dirichlet problem with boundary values $f \in \mathfrak{N}(\partial D)$. Then, by Theorem 3.29, we have that $u \in \mathfrak{R}(\mathbb{R}^3 \setminus \bar{D})$ and using $u=f$ on ∂D and (3.80), we obtain the integral equation

$$\phi+\mathbf{K}'\phi=\mathbf{T}f \tag{3.81}$$

of the second kind for the unknown normal derivative $\phi := \partial u/\partial \nu$. Since we have already established the existence of a solution to the exterior Dirichlet problem, the existence of a solution to the integral equation (3.81) is immediate. Hence we need only be concerned with the question of uniqueness. Equation (3.81) is obviously the adjoint equation of equation (3.26) obtained from the double-layer potential approach. Hence by Theorem 3.17 and the Fredholm alternative, equation (3.81) is uniquely solvable if and only if the wave number k is not an interior Neumann eigenvalue. Having solved the integral equation for ϕ, the solution of the boundary-value problem is given through the representation Theorem 3.3.

Now let u be a solution to the exterior Neumann problem. Then using $\partial u/\partial \nu = g$ on ∂D and (3.79), we obtain the integral equation

$$\psi - \mathbf{K}\psi = -\mathbf{S}g \tag{3.82}$$

of the second kind for the unknown boundary values $\psi := u$ on ∂D. Since (3.82) is the adjoint of equation (3.29) obtained from the single-layer approach, by Theorem 3.22 and Fredholm's alternative, equation (3.82) is uniquely solvable if and only if the wave number k is not an interior Dirichlet eigenvalue.

It is also possible to derive integral equations of the first kind. For the Dirichlet problem we obtain from (3.79) the equation

$$\mathbf{S}\phi = -f + \mathbf{K}f \tag{3.83}$$

for the unknown normal derivative $\phi := \partial u/\partial \nu$. The existence of a solution to the integral equation (3.83) again follows from the existence of a solution to the exterior Dirichlet problem and uniqueness is settled by Theorem 3.30 with unique solvability if and only if k is not an interior Dirichlet eigenvalue.

As suggested by Burton and Miller [1] we can linearly combine the equations (3.81) and (3.83) to obtain the integral equation

$$\phi + \mathbf{K}'\phi - i\eta\mathbf{S}\phi = \mathbf{T}f - i\eta(\mathbf{K}f - f) \tag{3.84}$$

of the second kind. This combined Green's formula integral equation is the adjoint of equation (3.51) obtained by the combined double- and single-layer approach. Therefore, if $\eta \neq 0$ and $\eta \operatorname{Re} k \geqslant 0$, the following theorem is a consequence of Fredholm's alternative.

Theorem 3.42. The combined integral equation (3.84) for the exterior Dirichlet problem is uniquely solvable for all wave numbers satisfying $\operatorname{Im} k \geqslant 0$.

For the Neumann problem we find from (3.80) the equation

$$\mathbf{T}\psi = g + \mathbf{K}'g \tag{3.85}$$

of the first kind for the unknown boundary values $\psi := u$ on ∂D. This equation is uniquely solvable if and only if the wave number k is not an interior Neumann eigenvalue.

Combining equations (3.82) and (3.85) we get the equation

$$\psi - \mathbf{K}\psi - i\eta\mathbf{T}\psi = -\mathbf{S}g - i\eta(g + \mathbf{K}'g) \tag{3.86}$$

of the second kind. This combined Green's formula integral equation again is the adjoint of equation (3.53) obtained via the combined single- and double-layer approach. Because the operator $\mathbf{T}$ is not compact, a straightforward application of Fredholm's alternative is not possible and uniqueness for (3.86) has to be dealt with separately. Let $\psi \in \mathfrak{N}(\partial D)$ be a solution of the homogeneous equation $\psi - \mathbf{K}\psi - i\eta\mathbf{T}\psi = 0$ and define the double-layer potential v with density ψ. Then $i\eta(\partial v_-/\partial\nu) + v_- = 0$ on ∂D and applying Green's theorem as in the proof of Theorem 3.33, we can conclude that $v_- = \partial v_-/\partial\nu = 0$. Hence by Theorem 3.1 we have that $v = 0$ in D and by Theorem 2.21 v solves the homogeneous exterior Neumann problem. Hence $v = 0$ in $\mathbf{R}^3 \setminus D$ and from Corollary 2.14 we can now conclude that $\psi = 0$.

Theorem 3.43. The combined integral equation (3.86) for the exterior Neumann problem is uniquely solvable for all wave numbers satisfying $\operatorname{Im} k \geqslant 0$.

We can also use Jones' modification to derive integral equations that are uniquely solvable for all positive wave numbers. Theorem 3.3 remains valid after replacing the fundamental solution $\Phi(x, y)$ by $\Psi(x, y)$ as defined by (3.55). Thus, analogous to (3.81) and (3.82), we obtain the modified integral equations

$$\phi + \tilde{\mathbf{K}}'\phi = \tilde{\mathbf{T}}f \tag{3.87}$$

for the Dirichlet problem and

$$\psi - \tilde{\mathbf{K}}\psi = -\tilde{\mathbf{S}}g \tag{3.88}$$

for the Neumann problem where the modified operators $\tilde{\mathbf{S}}$ and $\tilde{\mathbf{T}}$ are defined in an obvious way. Since the equations (3.87) and (3.88) are the adjoints of (3.57) and (3.64), respectively, we have the following theorem.

Theorem 3.44. The Jones' modification (3.87) and (3.88) of the integral equations for the exterior Dirichlet and Neumann problems are uniquely solvable for all positive wave numbers $k > 0$ provided either (3.58) or (3.59) is satisfied.

In a similar way, integral equations based on equations (3.79) and (3.80) can be obtained for the impedance and the transmission problem and these are the adjoints of (3.69), (3.71), (3.73), and (3.77).

Closely related to the derivation of integral equations based on the representation Theorem 3.3 is a procedure known as the null-field method. We shall now briefly describe this method. For any radiating solution $u \in \mathfrak{R}(\mathbf{R}^3 \setminus \bar{D})$ of the Helmholtz equation we have

$$\int_{\partial D} \left\{ u(y) \frac{\partial \Phi(x, y)}{\partial \nu(y)} - \frac{\partial u}{\partial \nu}(y) \Phi(x, y) \right\} ds(y) = 0, \qquad x \in D. \tag{3.89}$$

We assume the origin to be contained in D. Then, using the expansion (3.60) and the orthonormality of the spherical harmonics Y_n^m over the unit sphere, we find that

$$\int_{\partial D}\left\{u(y)\frac{\partial v_n^m}{\partial \nu}(y)-\frac{\partial u}{\partial \nu}(y)v_n^m(y)\right\}ds(y)=0 \tag{3.90}$$

for $n=0,1,2,\ldots,\ m=-n,\ldots,n$ where, analogous to (3.19), the v_n^m are given by

$$v_n^m(y):=h_n^{(1)}(kr)Y_n^m(\theta,\phi). \tag{3.91}$$

Hence, for the exterior Dirichlet problem the unknown normal derivative $\phi:=\partial u/\partial \nu$ on ∂D is a solution of the moment problem

$$\int_{\partial D}\phi v_n^m\,ds=f_n^m \tag{3.92}$$

$n=0,1,2,\ldots,\ m=-n,\ldots,n$ where

$$f_n^m:=\int_{\partial D}f\frac{\partial v_n^m}{\partial \nu}\,ds.$$

The corresponding equations for the Neumann problem are

$$\int_{\partial D}\psi\frac{\partial v_n^m}{\partial \nu}\,ds=g_n^m \tag{3.93}$$

$n=0,1,2,\ldots,\ m=-n,\ldots,n$ where

$$g_n^m:=\int_{\partial D}gv_n^m\,ds.$$

Equations of this type were first derived by Waterman [1] for electromagnetic scattering and later for acoustic scattering problems (Waterman [2]). A possible approach for the numerical solution of the null-field equations is to choose a complete set of functions $\{w_n^m\}$ in $L^2(\partial D)$ and seek finite approximations of the form

$$\phi\approx\sum_{n=0}^{N}\sum_{m=-n}^{n}\alpha_{nm}w_n^m.$$

Such an approach leads to a linear system of equations for the coefficients α_{nm}. For further details we refer the reader to Waterman [2].

Remarkably, as shown by Martin [1] and later in a simplified manner by Colton and Kress [3], the null-field equations are uniquely solvable for all wave numbers, that is, no nonuniqueness difficulties occur at interior eigenvalues.

Theorem 3.45. The null-field equations for the exterior Dirichlet and Neumann problems are uniquely solvable for all wave numbers satisfying $\operatorname{Im} k \geqslant 0$.

Proof. Since existence to the boundary-value problems and hence to the null-field equations is already established, we need only to establish uniqueness. Our proof of this is based on Colton and Kress [3].

Let ϕ be a solution to the homogeneous null-field equations for the exterior Dirichlet problem, that is,

$$\int_{\partial D} \phi v_n^m \, ds = 0$$

$n = 0, 1, 2, \ldots,\ m = -n, \ldots, n$. Multiplying this equation by u_n^m as defined by (3.19) and reversing the steps leading from (3.89) to (3.92) shows that the single-layer potential

$$w(x) = \int_{\partial D} \phi(y) \Phi(x, y) \, ds(y), \qquad x \in \mathbb{R}^3 \setminus \partial D,$$

is identically zero in some ball contained in D and hence by Theorem 3.5 we have $w = 0$ in D. By Theorem 2.12 we now see that w is a solution to the homogeneous exterior Dirichlet problem. Hence $w = 0$ in $\mathbb{R}^3 \setminus D$ and from Corollary 2.20 we see that $\phi = 0$ on ∂D.

The uniqueness for the null-field equations to the exterior Neumann problem follows in a similar manner.

We note in closing that uniquely solvable null-field equations can also be obtained for the impedance and transmission problems (Colton and Kress [3]).

3.10 THE TWO-DIMENSIONAL CASE

It is occasionally important to look at boundary-value problems for the Helmholtz equation in $\mathbb{R}^2$ that describe acoustic scattering from infinitely long cylindrical bodies. Therefore, without going into any details, we would like to point out that all the results of this chapter remain valid in two dimensions after the appropriate modification of the fundamental solution and of the Sommerfeld radiation condition. We now quickly list these modifications.

The fundamental solution (2.30) has to be replaced by

$$\Phi(x, y) = \frac{i}{4} H_0^{(1)}(k|x - y|) \tag{3.94}$$

where $H_0^{(1)}$ denotes the Hankel function of the first kind of order zero. Since (3.94) has a singularity at $x = y$ of the form of the fundamental solution

$$\Phi_0(x, y) = \frac{1}{2\pi} \log \frac{1}{|x - y|} \tag{3.95}$$

of the Laplace equation in two dimensions, it can be verified that the properties of acoustic single- and double-layer potentials derived in Chapter 2 remain valid in $\mathbb{R}^2$.

The Sommerfeld radiation condition (3.7) has to be replaced by

$$\left(\frac{x}{|x|}, \operatorname{grad} u(x)\right) - iku(x) = o\left(\frac{1}{|x|^{1/2}}\right), \qquad |x| \to \infty, \tag{3.96}$$

uniformly for all directions $x/|x|$. Then the two-dimensional version of Theorem 3.3, Lemma 3.11, and the corresponding uniqueness results remain valid. Finally, we note that from the asymptotic formula

$$H_0^{(1)}(r) = \sqrt{\frac{2}{\pi r}}\, e^{i(r-\pi/4)}\left(1 + O\left(\frac{1}{r}\right)\right)$$

it can be seen that the fundamental solution (3.94) satisfies the radiation condition (3.96).

4

BOUNDARY-VALUE PROBLEMS FOR THE TIME-HARMONIC MAXWELL EQUATIONS AND THE VECTOR HELMHOLTZ EQUATION

The aim of this chapter is to extend the results of the previous chapter to the case of the time-harmonic Maxwell equations. In particular, we shall use the method of integral equations to establish the existence of solutions to the classical boundary-value problems in electromagnetic scattering theory. However, there are significant differences in the analysis needed to study the electromagnetic scattering problems and the acoustic scattering problems treated in the previous chapter. These differences lead to the need to examine the regularity properties of single-layer potentials in the space $C^{0,\alpha}$ as well as the need to develop new regularization techniques for treating the singular integral equations that arise in the course of our analysis. As in the case of acoustic wave propagation, a central difficulty is that the classical approach for solving electromagnetic boundary-value problems by the method of integral equations leads to equations that are noninvertible at interior eigenvalues.

We begin our analysis by deriving representation theorems and introducing the Silver–Müller radiation condition for the time-harmonic Maxwell equations. Because the elimination of either the magnetic or electric field from Maxwell's equations leads to the vector Helmholtz equation for the remaining field, we include a treatment of this equation in our discussion. This has the further advantage of clarifying the close relationship between acoustic and electromagnetic scattering theory. Having introduced the vector Helmholtz equation, we then proceed to consider boundary-value problems for the scattering of electromagnetic waves by a perfect conductor formulated either as a boundary-value problem for Maxwell's equations or for the vector Helmholtz equation. Included in our discussion are results on uniqueness, the existence of solutions via the classical approach leading to the previously mentioned problems of interior eigenvalues, and the recent approach of Kress [5], [6] and Knauff and Kress [1] that avoids these problems. After a brief consideration of the impedance boundary-value problem, we conclude by deriving integral equations for the solutions of the above boundary-value problems by means of the representation theorems. This approach leads to integral equations that are the adjoints of the ones obtained via the layer approach.

4.1 TIME-HARMONIC ELECTROMAGNETIC SCATTERING

We consider electromagnetic wave propagation in a homogeneous isotropic medium in $\mathbb{R}^3$ with electric permittivity ε, magnetic permeability μ, and electric conductivity σ. The electromagnetic wave with frequency $\omega > 0$ will be described by the electric and magnetic field

$$\mathcal{E}(x,t) = \left(\varepsilon + \frac{i\sigma}{\omega}\right)^{-1/2} E(x)e^{-i\omega t}$$

$$\mathcal{H}(x,t) = \mu^{-1/2} H(x)e^{-i\omega t}.$$

From the time dependent form of Maxwell's equations

$$\operatorname{curl} \mathcal{E} + \mu \frac{\partial \mathcal{H}}{\partial t} = 0, \qquad \operatorname{curl} \mathcal{H} - \varepsilon \frac{\partial \mathcal{E}}{\partial t} = \sigma \mathcal{E} \tag{4.1}$$

we conclude that the space dependent parts E and H satisfy the time-harmonic form of Maxwell's equations

$$\operatorname{curl} E - ikH = 0, \qquad \operatorname{curl} H + ikE = 0 \tag{4.2}$$

where the wave number k is given by $k^2 = (\varepsilon + (i\sigma/\omega))\mu\omega^2$. We choose the sign of k such that

$$\operatorname{Im} k \geqslant 0. \tag{4.3}$$

Therefore the mathematical description of the scattering of time-harmonic waves by an obstacle D leads to boundary-value problems for the reduced Maxwell equations. In particular, consider the scattering of a given incoming electromagnetic wave E^i, H^i by a perfectly conducting body D. Then for the total wave $E^{\text{tot}} = E^i + E^s$, $H^{\text{tot}} = H^i + H^s$, where E^s, H^s denotes the scattered wave, the tangential component of the electric field must vanish on the conducting surface ∂D, that is, $[\nu, E^{\text{tot}}] = 0$. The scattering by a body D that is not perfectly conducting but that does not allow the electromagnetic wave to penetrate deeply into the body leads to an impedance boundary condition of the form

$$[\nu,[\nu, H^{\text{tot}}]]-\psi[\nu, E^{\text{tot}}]=0$$

where ψ denotes the (possibly nonconstant) electromagnetic impedance of the obstacle D.

4.2 REPRESENTATION THEOREMS AND RADIATION CONDITIONS

We begin our analysis by establishing a representation theorem due to Stratton and Chu [1] that shows that any solution to the time-harmonic Maxwell equations can be represented as the electromagnetic field generated by a combination of surface distributions of electric and magnetic dipoles.

Theorem 4.1. Let $E, H \in C^1(D)\cap C(\bar{D})$ be a solution to Maxwell's equations

$$\operatorname{curl} E - ikH = 0, \qquad \operatorname{curl} H + ikE = 0 \quad \text{in} \quad D.$$

Then

$$\operatorname{curl}\int_{\partial D}[\nu(y), E(y)]\Phi(x,y)\,ds(y)$$

$$-\frac{1}{ik}\operatorname{curl}\operatorname{curl}\int_{\partial D}[\nu(y), H(y)]\Phi(x,y)\,ds(y) = \begin{cases} -E(x), & x\in D,\\ 0, & x\in\mathbb{R}^3\setminus\bar{D},\end{cases}$$

and

$$\operatorname{curl}\int_{\partial D}[\nu(y), H(y)]\Phi(x,y)\,ds(y)$$

$$+\frac{1}{ik}\operatorname{curl}\operatorname{curl}\int_{\partial D}[\nu(y), E(y)]\Phi(x,y)\,ds(y) = \begin{cases} -H(x), & x\in D,\\ 0, & x\in\mathbb{R}^3\setminus\bar{D}.\end{cases}$$

Proof. We shall use the notations introduced in the proof of Theorem 3.1 and choose an arbitrary fixed point $x\in D$ and an arbitrary fixed unit vector $e\in\mathbb{R}^3$. Using Maxwell's equations for E and H and the relation $\operatorname{curl}\operatorname{curl}\operatorname{curl} e\Phi = k^2\operatorname{curl} e\Phi$, we compute

$$\operatorname{div}\left\{[E,\operatorname{curl} e\Phi]-\frac{1}{ik}[H,\operatorname{curl}\operatorname{curl} e\Phi]\right\}=0 \quad \text{in} \quad D\setminus\{x\}.$$

Hence, from Gauss' theorem, we find

$$\int_{\partial D+\Omega_{x,r}}\Big\{\big(\nu(y),E(y),\operatorname{curl}_y e\Phi(x,y)\big) - \frac{1}{ik}\big(\nu(y),H(y),\operatorname{curl}_y\operatorname{curl}_y e\Phi(x,y)\big)\Big\}\,ds(y)=0. \tag{4.4}$$

With the help of Stokes' theorem and the second Maxwell equation, we see that

$$\int_{\Omega_{x,r}}\big(\nu(y),H(y),\operatorname{grad}_y\operatorname{div}_y e\Phi(x,y)\big)\,ds(y) = -ik\int_{\Omega_{x,r}}\big(\nu(y),E(y)\big)\operatorname{div}_y e\Phi(x,y)\,ds(y).$$

Then, since on $\Omega_{x,r}$ we have $\Phi(x,y)=O(1/r)$ and

$$\operatorname{div} e\Phi(x,y)=\frac{(\nu(y),e)}{4\pi r^2}+O\left(\frac{1}{r}\right) \qquad \operatorname{curl} e\Phi(x,y)=\frac{[\nu(y),e]}{4\pi r^2}+O\left(\frac{1}{r}\right),$$

we use $\operatorname{curl}\operatorname{curl} e\Phi = k^2 e\Phi + \operatorname{grad}\operatorname{div} e\Phi$ to obtain by straightforward calculation that

$$\lim_{r\to 0}\int_{\Omega_{x,r}}\Big\{\big(\nu(y),E(y),\operatorname{curl}_y e\Phi(x,y)\big) - \frac{1}{ik}\big(\nu(y),H(y),\operatorname{curl}_y\operatorname{curl}_y e\Phi(x,y)\big)\Big\}\,ds(y)=\big(e,E(x)\big).$$

Finally, by observing the symmetry relation (2.52) it is easily verified that

$$\big(\nu(y),E(y),\operatorname{curl}_y e\Phi(x,y)\big)=\big(e,\operatorname{curl}_x[\nu(y),E(y)]\Phi(x,y)\big)$$

and

$$\big(\nu(y),H(y),\operatorname{curl}_y\operatorname{curl}_y e\Phi(x,y)\big)=\big(e,\operatorname{curl}_x\operatorname{curl}_x[\nu(y),H(y)]\Phi(x,y)\big).$$

Hence we can now conclude from (4.4) that

$$\left(e,\int_{\partial D}\Big\{\operatorname{curl}_x[\nu(y),E(y)]\Phi(x,y) - \frac{1}{ik}\operatorname{curl}_x\operatorname{curl}_x[\nu(y),H(y)]\Phi(x,y)\Big\}\,ds(y)+E(x)\right)=0.$$

Since e is arbitrary, we have established Theorem 4.1 for $x\in D$.

If $x \in \mathbb{R}^3 \setminus \bar{D}$, the proof follows in a similar manner from the identity

$$\int_{\partial D}\Big\{(\nu(y), E(y), \operatorname{curl}_y e\Phi(x, y))$$

$$-\frac{1}{ik}(\nu(y), H(y), \operatorname{curl}_y \operatorname{curl}_y e\Phi(x, y))\Big\}\, ds(y) = 0.$$

The representation of H is now easily obtained by using $H = (1/ik)\operatorname{curl} E$.

Analogous to Theorem 3.5, we now have the following theorem.

Theorem 4.2. Any continuously differentiable solution to Maxwell's equations possesses analytic cartesian components.

In particular, the cartesian components of solutions to Maxwell's equations are automatically two times continuously differentiable. Therefore we can employ the vector identity

$$\operatorname{curl}\operatorname{curl} E = -\Delta E + \operatorname{grad}\operatorname{div} E$$

to prove the following result.

Theorem 4.3. Let E, H be a solution to Maxwell's equations. Then E and H are divergence free and satisfy the vector Helmholtz equation

$$\Delta E + k^2 E = 0, \qquad \Delta H + k^2 H = 0.$$

Conversely, let E (or H) be a solution to the vector Helmholtz equation satisfying $\operatorname{div} E = 0$ (or $\operatorname{div} H = 0$). Then E and $H := (1/ik)\operatorname{curl} E$ (or H and $E := (-1/ik)\operatorname{curl} H$) satisfy Maxwell's equations.

Let $a \in \mathbb{R}^3$ be a constant vector. Then

$$\begin{aligned} E_m(x) &:= \operatorname{curl}_x a\Phi(x, y) \\ H_m(x) &:= \frac{1}{ik}\operatorname{curl} E_m(x), \qquad x \in \mathbb{R}^3 \setminus \{y\} \end{aligned} \tag{4.5}$$

represent the electromagnetic field generated by a magnetic dipole located at the point $y \in \mathbb{R}^3$ and solve Maxwell's equations. Similarly,

$$\begin{aligned} H_e(x) &:= \operatorname{curl}_x a\Phi(x, y) \\ E_e(x) &:= -\frac{1}{ik}\operatorname{curl}_x H_e(x), \qquad x \in \mathbb{R}^3 \setminus \{y\} \end{aligned} \tag{4.6}$$

represent the electromagnetic field generated by an electric dipole. Theorem 4.1 obviously gives a representation of any solution of Maxwell's equations in terms of electric and magnetic dipoles distributed on the boundary surface and in this sense the fields (4.5) and (4.6) may be considered as fundamental solutions to Maxwell's equations.

By straightforward calculations, it can be seen that

$$E_m(x) = ik\Phi(x,y)\left[\frac{x}{|x|}, a\right] + O\left(\frac{1}{|x|^2}\right), \qquad |x| \to \infty,$$

$$H_m(x) = -ik\Phi(x,y)\left\{a - \left(a, \frac{x}{|x|}\right)\frac{x}{|x|}\right\} + O\left(\frac{1}{|x|^2}\right), \qquad |x| \to \infty,$$

uniformly for all directions $x/|x|$ and uniformly for all y contained in any bounded set of $\mathbb{R}^3$. From this, and the property $E_e = -H_m$, $H_e = E_m$, we conclude the following.

Theorem 4.4. Both the electromagnetic field E_m, H_m of a magnetic dipole and the electromagnetic field E_e, H_e of an electric dipole satisfy the Silver–Müller radiation conditions (Müller [5], Silver [1])

$$\left[H, \frac{x}{|x|}\right] - E = o\left(\frac{1}{|x|}\right), \qquad |x| \to \infty,$$

and

$$\left[E, \frac{x}{|x|}\right] + H = o\left(\frac{1}{|x|}\right), \qquad |x| \to \infty,$$

uniformly for all directions $x/|x|$.

As we shall see, it suffices to impose only one of these radiation conditions to completely characterize the behavior of solutions to Maxwell's equations at infinity.

Theorem 4.5. Let $E, H \in C^1(\mathbb{R}^3 \setminus \bar{D}) \cap C(\mathbb{R}^3 \setminus D)$ solve Maxwell's equations

$$\operatorname{curl} E - ikH = 0, \qquad \operatorname{curl} H + ikE = 0 \quad \text{in } \mathbb{R}^3 \setminus \bar{D}$$

and one of the Silver–Müller radiation conditions

$$\left[H, \frac{x}{|x|}\right] - E = o\left(\frac{1}{|x|}\right), \qquad |x| \to \infty, \tag{4.7}$$

or

$$\left[E, \frac{x}{|x|}\right] + H = o\left(\frac{1}{|x|}\right), \qquad |x| \to \infty, \tag{4.8}$$

uniformly for all directions $x/|x|$. Then

$$\operatorname{curl}\int_{\partial D}[\nu(y), E(y)]\Phi(x,y)\,ds(y)$$

$$-\frac{1}{ik}\operatorname{curl}\operatorname{curl}\int_{\partial D}[\nu(y), H(y)]\Phi(x,y)\,ds(y) = \begin{cases} 0, & x \in D, \\ E(x) & x \in \mathbb{R}^3\setminus\bar{D}, \end{cases}$$

and

$$\operatorname{curl}\int_{\partial D}[\nu(y), H(y)]\Phi(x,y)\,ds(y)$$

$$+\frac{1}{ik}\operatorname{curl}\operatorname{curl}\int_{\partial D}[\nu(y), E(y)]\Phi(x,y)\,ds(y) = \begin{cases} 0, & x \in D, \\ H(x), & x \in \mathbb{R}^3\setminus\bar{D}. \end{cases}$$

Proof. We shall again use the notations introduced in the proofs of Theorems 3.1 and 3.3. Proceeding as in Theorem 4.1, we see that the proof is established if we can show that

$$\int_{|y|=R}\Big\{\big(\nu(y), E(y), \operatorname{curl}_y e\Phi(x,y)\big)$$

$$-\frac{1}{ik}\big(\nu(y), H(y), \operatorname{curl}_y\operatorname{curl}_y e\Phi(x,y)\big)\Big\}\,ds(y) \to 0, \qquad R\to\infty.$$

To establish this we first verify that (4.7) implies that

$$\int_{|y|=R}|E|^2\,ds = O(1), \qquad R\to\infty. \tag{4.9}$$

We observe that from (4.7) it follows that

$$0 = \lim_{R\to\infty}\int_{|y|=R}|[H,\nu]-E|^2\,ds$$

$$= \lim_{R\to\infty}\int_{|y|=R}\{|[H,\nu]|^2 + |E|^2 - 2\operatorname{Re}(\nu,\bar{E},H)\}\,ds.$$

We now apply Gauss' theorem to obtain

$$\int_{|y|=R}(\nu,\bar{E},H)\,ds = \int_{\partial D}(\nu,\bar{E},H)\,ds + i\int_{D_R}(k|E|^2 - \bar{k}|H|^2)\,dy.$$

Inserting the real part of this equation into the previous equation, we find that

$$\lim_{R\to\infty}\left\{\int_{|y|=R}\{\|[H,\nu]\|^2+|E|^2\}\,ds+2\,\mathrm{Im}(k)\int_{D_R}\{|E|^2+|H|^2\}\,dy\right\}$$

$$=2\,\mathrm{Re}\int_{\partial D}(\nu,\bar{E},H)\,ds. \tag{4.10}$$

From (4.10) and the fact that $\mathrm{Im}\,k\geqslant 0$, it follows that (4.9) is true.

From Stokes' theorem and the identity $\mathrm{curl}\,\mathrm{curl}\,e\Phi=-\Delta e\Phi+\mathrm{grad}\,\mathrm{div}\,e\Phi$, we now see that

$$\int_{|y|=R}\Big\{\big(\nu(y),E(y),\mathrm{curl}_y e\Phi(x,y)\big)$$

$$-\frac{1}{ik}\big(\nu(y),H(y),\mathrm{curl}_y\mathrm{curl}_y e\Phi(x,y)\big)\Big\}\,ds(y)$$

$$=\int_{|y|=R}\{(\nu(y),E(y),\mathrm{curl}_y e\Phi(x,y))+(\nu(y),E(y))\mathrm{div}_y e\Phi(x,y)$$

$$+ik(\nu(y),H(y),e)\Phi(x,y)\}\,ds(y)$$

$$=\int_{|y|=R}\left(E(y),\left[\mathrm{curl}_y e\Phi(x,y),\frac{y}{|y|}\right]+\frac{y}{|y|}\mathrm{div}_y e\Phi(x,y)-ike\Phi(x,y)\right)ds(y)$$

$$-ik\int_{|y|=R}\left(e,\left[H(y),\frac{y}{|y|}\right]-E(y)\right)\Phi(x,y)\,ds(y)$$

$$=:I_1+I_2.$$

By straightforward calculations we see that

$$\left[\mathrm{curl}_y e\Phi(x,y),\frac{y}{|y|}\right]+\frac{y}{|y|}\mathrm{div}_y e\Phi(x,y)-ike\Phi(x,y)=O\left(\frac{1}{R^2}\right),\qquad |y|=R,$$

for the fundamental solution Φ, and using (4.9) and Schwarz's inequality, we can now deduce that $I_1=O(1/R)$, $R\to\infty$ The radiation condition (4.7) and $\Phi(x,y)=O(1/R)$, $|y|=R$, imply that $I_2=o(1)$, $R\to\infty$. Hence

$$\lim_{R\to\infty}\int_{|y|=R}\Big\{\big(\nu(y),E(y),\mathrm{curl}_y e\Phi(x,y)\big)$$

$$-\frac{1}{ik}\big(\nu(y),H(y),\mathrm{curl}_y\mathrm{curl}_y e\Phi(x,y)\big)\Big\}\,ds(y)=0,$$

which completes the proof in case condition (4.7) is satisfied.

Finally, let (4.8) be satisfied. Then $\tilde{E} := -H$ and $\tilde{H} := E$ solve Maxwell's equations and satisfy $[\tilde{H},(x/|x|)] - \tilde{E} = o(1/|x|)$, $|x| \to \infty$. Hence verifying the representation under the radiation condition (4.8) can be reduced to the previous case of the radiation condition (4.7).

Combining Theorems 4.4 and 4.5, we now have the following corollary.

Corollary 4.6. Any solution to Maxwell's equations satisfying the radiation condition

$$\left[H, \frac{x}{|x|}\right] - E = o\left(\frac{1}{|x|}\right), \qquad |x| \to \infty,$$

uniformly for all directions $x/|x|$ also satisfies

$$\left[E, \frac{x}{|x|}\right] + H = o\left(\frac{1}{|x|}\right), \qquad |x| \to \infty,$$

uniformly for all directions and vice versa.

Since straightforward calculations show that the cartesian components of the fundamental solutions (4.5) and (4.6) satisfy the Sommerfeld radiation condition uniformly for all y contained in any bounded set of $\mathbb{R}^3$, from Theorem 4.5 we can also conclude the following result.

Corollary 4.7. The cartesian components of any solution to Maxwell's equations satisfying the Silver–Müller radiation conditions also satisfy the Sommerfeld radiation condition for the scalar Helmholtz equation.

As we shall see later in Corollary 4.14, the converse of Corollary 4.7 is also true.

Theorem 4.8. Let $E, H \in C^1(\mathbb{R}^3 \setminus \bar{D})$ be a solution to Maxwell's equations satisfying one of the Silver–Müller radiation conditions. Then the expansions

$$E(x) = \frac{e^{ikr}}{r} \sum_{m=0}^{\infty} \frac{E_m(\theta, \phi)}{r^m}$$

and

$$H(x) = \frac{e^{ikr}}{r} \sum_{m=0}^{\infty} \frac{H_m(\theta, \phi)}{r^m}$$

are valid in the sense of Theorem 3.6.

Proof. This is an immediate consequence of Theorem 4.3, Corollary 4.7, and Theorem 3.6.

Corollary 4.9. Every solution E, H to Maxwell's equations satisfying one of the Silver–Müller radiation conditions has the asymptotic form

$$E(x) = \frac{e^{ikr}}{r} E_0(\theta, \phi) + O\left(\frac{1}{r^2}\right)$$

$$H(x) = \frac{e^{ikr}}{r} H_0(\theta, \phi) + O\left(\frac{1}{r^2}\right)$$

with the property

$$H_0 = [e_r, E_0]$$

and

$$(e_r, E_0) = (e_r, H_0) = 0$$

where e_r denotes the unit vector in the radial direction. The field $E_0\colon \Omega \to \mathbf{C}^3$ is called the far-field pattern or radiation pattern of E, H.

Proof. This follows from Theorem 4.8. The properties of E_0 and H_0 follow from the radiation conditions.

Corollary 4.10. Let $E, H \in C^1(\mathbb{R}^3 \setminus \bar{D})$ be a solution to Maxwell's equations satisfying one of the Silver–Müller radiation conditions for which the far-field pattern vanishes identically. Then $E = H = 0$ in $\mathbb{R}^3 \setminus \bar{D}$.

Proof. This is a consequence of Corollary 4.7 and Corollary 3.9.

We conclude this section by considering representation theorems and radiation conditions for the vector Helmholtz equation. To simplify notations, for any domain G with boundary ∂G of class C^2 we introduce the linear space of vector fields

$$\mathcal{F}(G) := \{E\colon \bar{G} \to \mathbf{C}^3 \mid E \in C^2(G) \cap C(\bar{G}), \operatorname{div} E, \operatorname{curl} E \in C(\bar{G})\}.$$

The assumptions $E, F \in \mathcal{F}(G)$ suffice for the application of the first vector Green's theorem

$$\int_G \{(E, \Delta F) + (\operatorname{curl} E, \operatorname{curl} F) + \operatorname{div} E \operatorname{div} F\}\, dx$$
$$= \int_{\partial G} \{(\nu, E, \operatorname{curl} F) + (\nu, E) \operatorname{div} F\}\, ds \quad (4.11)$$

and the second vector Green's theorem

$$\int_G \{(E, \Delta F) - (F, \Delta E)\}\, dx$$
$$= \int_{\partial G} \{(\nu, E, \operatorname{curl} F) + (\nu, E) \operatorname{div} F - (\nu, F, \operatorname{curl} E) - (\nu, F) \operatorname{div} E\}\, ds$$
$$(4.12)$$

in a bounded domain G. Both of these Green's theorems follow easily from Gauss' theorem.

Theorem 4.11. Let $E \in \mathfrak{F}(D)$ be a solution of the vector Helmholtz equation

$$\Delta E + k^2 E = 0 \quad \text{in} \quad D.$$

Then

$$\operatorname{curl}\int_{\partial D}[\nu(y), E(y)]\Phi(x,y)\,ds(y) - \operatorname{grad}\int_{\partial D}(\nu(y), E(y))\Phi(x,y)\,ds(y)$$

$$-\int_{\partial D}\{[\operatorname{curl} E(y), \nu(y)] + \nu(y)\operatorname{div} E(y)\}\Phi(x,y)\,ds(y)$$

$$= \begin{cases} -E(x), & x \in D, \\ 0, & x \in \mathbb{R}^3 \setminus \bar{D}. \end{cases}$$

Proof. In the setting of the proof of Theorem 4.1, we use the second vector Green's theorem to find

$$\int_{\partial D + \Omega_{x,r}} \{(\nu(y), E(y), \operatorname{curl}_y e\Phi(x,y)) + (\nu(y), E(y))\operatorname{div}_y e\Phi(x,y)$$

$$-[(\nu(y), e, \operatorname{curl} E(y)) + (\nu(y), e)\operatorname{div} E(y)]\Phi(x,y)\}\,ds(y) = 0. \tag{4.13}$$

From this we can easily derive the representation formula by the same type of calculations as carried out in Theorem 4.1.

Let $a \in \mathbb{R}^3$ be a constant vector and define vector fields satisfying the vector Helmholtz equation by

$$\begin{aligned} E_1(x) &:= \operatorname{curl}_x a\Phi(x,y) \\ E_2(x) &:= a\Phi(x,y) \\ E_3(x) &:= \operatorname{grad}_x \Phi(x,y), \qquad x \in \mathbb{R}^3 \setminus \{y\}. \end{aligned} \tag{4.14}$$

Then, Theorem 4.11 gives a representation of any solution of the vector Helmholtz equation in terms of the fields E_1, E_2, and E_3. By straightforward calculations (see Knauff and Kress [1]) it can be seen that the behavior at infinity of these fields is characterized by the following theorem.

Theorem 4.12. The fields E_1, E_2, E_3 satisfy the radiation condition

$$\left[\operatorname{curl} E, \frac{x}{|x|}\right] + \frac{x}{|x|}\operatorname{div} E - ikE = o\left(\frac{1}{|x|}\right)$$

uniformly for all directions $x/|x|$ and uniformly for all y contained in any bounded set of $\mathbb{R}^3$.

Theorem 4.13. Let $E \in \mathfrak{F}(\mathbb{R}^3 \setminus \bar{D})$ be a solution to the vector Helmholtz equation

$$\Delta E + k^2 E = 0 \quad \text{in} \quad \mathbb{R}^3 \setminus \bar{D}$$

satisfying the radiation condition

$$\left[\operatorname{curl} E, \frac{x}{|x|}\right] + \frac{x}{|x|} \operatorname{div} E - ikE = o\left(\frac{1}{|x|}\right) \tag{4.15}$$

uniformly for all directions $x/|x|$. Then

$$\operatorname{curl} \int_{\partial D} [\nu(y), E(y)] \Phi(x, y)\, ds(y) - \operatorname{grad} \int_{\partial D} (\nu(y), E(y)) \Phi(x, y)\, ds(y)$$

$$- \int_{\partial D} \{[\operatorname{curl} E(y), \nu(y)] + \nu(y) \operatorname{div} E(y)\} \Phi(x, y)\, ds(y)$$

$$= \begin{cases} 0, & x \in D, \\ E(x), & x \in \mathbb{R}^3 \setminus \bar{D}. \end{cases}$$

Proof. Proceeding as in Theorem 4.5, it has to be shown that

$$\int_{|y| = R} \{(\nu(y), E(y), \operatorname{curl}_y e \Phi(x, y)) + (\nu(y), E(y)) \operatorname{div}_y e \Phi(x, y)$$

$$- [(\nu(y), e, \operatorname{curl} E(y)) + (\nu(y), e) \operatorname{div} E(y)] \Phi(x, y)\}\, ds(y) \to 0, \qquad R \to \infty.$$

We first show that

$$\int_{|y| = R} |E|^2\, ds = O(1), \qquad R \to \infty. \tag{4.16}$$

Using the radiation condition and the first vector Green's theorem, we derive

$$\begin{aligned} &\lim_{R \to \infty} \int_{|y| = R} \{|[\operatorname{curl} E, \nu] + \nu \operatorname{div} E|^2 + |k|^2 |E|^2\}\, ds \\ &+ 2 \operatorname{Im}(k) \int_{D_R} \{|k|^2 |E|^2 + |\operatorname{curl} E|^2 + |\operatorname{div} E|^2\}\, dy \\ &= -2 \operatorname{Im}\left(k \int_{\partial D} \{(\nu, E, \operatorname{curl} \bar{E}) + (\nu, E) \operatorname{div} \bar{E}\}\right) ds \end{aligned} \tag{4.17}$$

from which (4.16) follows.

We now rearrange

$$\int_{|y|=R}\{(\nu(y), E(y), \operatorname{curl}_y e\Phi(x,y)) + (\nu(y), E(y))\operatorname{div}_y e\Phi(x,y)$$

$$- [(\nu(y), e, \operatorname{curl} E(y)) + (\nu(y), e)\operatorname{div} E(y)]\Phi(x,y)\}\, ds(y)$$

$$= \int_{|y|=R}\left(E(y), \left[\operatorname{curl}_y e\Phi(x,y), \frac{y}{|y|}\right] + \frac{y}{|y|}\operatorname{div}_y e\Phi(x,y) - ike\Phi(x,y)\right) ds(y)$$

$$- \int_{|y|=R}\left(e, \left[\operatorname{curl} E(y), \frac{y}{|y|}\right] + \frac{y}{|y|}\operatorname{div} E(y) - ikE(y)\right)\Phi(x,y)\, ds(y)$$

and complete the proof analogously to Theorem 4.5.

Corollary 4.14. Let E be a solution to the vector Helmholtz equation satisfying the radiation condition (4.15). Then the cartesian components of E satisfy the Sommerfeld radiation condition (3.7) and vice versa.

Proof. This can be seen either by straightforward calculations using the representation formulas of Theorems 3.3 and 4.13 or deduced from the identities

$$(\nu(y), E(y), \operatorname{curl}_y e\Phi(x,y)) + (\nu(y), E(y))\operatorname{div}_y e\Phi(x,y)$$

$$= (e, E(y))\frac{\partial\Phi(x,y)}{\partial\nu(y)} + (\nu(y), \operatorname{curl}_y[e, E(y)])\Phi(x,y)$$

$$+ (\nu(y), \operatorname{curl}_y \Phi(x,y)[E(y), e])$$

and

$$(\nu(y), e, \operatorname{curl}_y E(y)) + (\nu(y), e)\operatorname{div}_y E(y)$$

$$= \frac{\partial}{\partial\nu(y)}(e, E(y)) + (\nu(y), \operatorname{curl}_y[e, E(y)]).$$

From these relations it follows from Stokes theorem that

$$\int_{\partial D}\{(\nu(y), E(y), \operatorname{curl}_y e\Phi(x,y)) + (\nu(y), E(y))\operatorname{div}_y e\Phi(x,y)$$

$$- [(\nu(y), e, \operatorname{curl}_y E(y)) + (\nu(y), e)\operatorname{div}_y E(y)]\Phi(x,y)\}\, ds(y)$$

$$= \int_{\partial D}\left\{(e, E(y))\frac{\partial\Phi(x,y)}{\partial\nu(y)} - \Phi(x,y)\frac{\partial}{\partial\nu(y)}(e, E(y))\right\} ds(y),$$

which in view of (4.13) demonstrates that the representation formulas for the vector Helmholtz equation stemming from Theorems 3.1 and 3.3 and Theorems 4.11 and 4.13 can be transformed into each other.

Corollary 4.14 combined with Theorem 4.3 shows that the converse of Corollary 4.7 is also valid.

4.3 THE BOUNDARY-VALUE PROBLEMS FOR A PERFECT CONDUCTOR: UNIQUENESS THEOREMS

We shall consider the following interior and exterior boundary-value problems for Maxwell's equations:

Interior Maxwell Boundary-Value Problem

Find two vector fields $E, H \in C^1(D) \cap C(\bar{D})$ satisfying Maxwell's equations in D and the boundary condition

$$[\nu, E] = c \quad \text{on} \quad \partial D \tag{4.18}$$

where $c \in C^{0,\alpha}(\partial D)$ is a given tangential field with the additional property that its surface divergence $\operatorname{Div} c$ exists in the sense of the limit integral definition and is of class $C^{0,\alpha}(\partial D)$, that is, $c \in \mathbb{S}^{0,\alpha}(\partial D)$.

Exterior Maxwell Boundary-Value Problem

Find two vector fields $E, H \in (C^1(\mathbb{R}^3 \setminus \bar{D}) \cap C(\mathbb{R}^3 \setminus D))$ satisfying Maxwell's equations in $\mathbb{R}^3 \setminus \bar{D}$, the Silver–Müller radiation condition (4.7) and (4.8), and the boundary condition

$$[\nu, E] = c \quad \text{on} \quad \partial D \tag{4.19}$$

where $c \in \mathbb{S}^{0,\alpha}(\partial D)$ is a given tangential field.

From the vector formula (2.75), we observe that the condition on the given tangential field c to possess a continuous surface divergence is necessary for the existence of a solution to the boundary-value problems. As we shall see later, Hölder continuity of the boundary data is required for our boundary integral equation treatment of the boundary-value problems.

Recalling Theorem 4.3 we eliminate the magnetic field and obtain from the boundary-value problem for Maxwell's equations a boundary-value problem for the vector Helmholtz equation $\Delta E + k^2 E = 0$ with boundary conditions of the form

$$[\nu, E] = c, \qquad \operatorname{div} E = 0 \quad \text{on} \quad \partial D. \tag{4.20}$$

Similarly, eliminating the electric field and using (2.75) we obtain a boundary-value problem for $\Delta H + k^2 H = 0$ with boundary conditions

$$[[\operatorname{curl} H, \nu], \nu] = ik[c, \nu], \qquad (\nu, H) = \frac{i}{k}\operatorname{Div} c \quad \text{on} \quad \partial D. \tag{4.21}$$

Hence, in addition to the boundary-value problems for Maxwell's equations, we shall also consider the following slightly more general boundary-value problems for the vector Helmholtz equation.

Interior Electric Boundary-Value Problem

Find a vector field $E \in \mathfrak{F}(D)$ (i.e., a vector field $E \in C^2(D) \cap C(\bar{D})$ with $\operatorname{div} E, \operatorname{curl} E \in C(\bar{D})$) satisfying the vector Helmholtz equation in D and the boundary condition

$$[\nu, E] = c, \qquad \operatorname{div} E = \gamma \quad \text{on} \quad \partial D \tag{4.22}$$

where $\gamma \in C^{0,\alpha}(\partial D)$ is a given function and $c \in \mathfrak{T}^{0,\alpha}(\partial D)$ is a given tangential field.

Exterior Electric Boundary-Value Problem

Find a vector field $E \in \mathfrak{F}(\mathbb{R}^3 \setminus \bar{D})$ satisfying the vector Helmholtz equation in $\mathbb{R}^3 \setminus \bar{D}$, the radiation condition (4.15), and the boundary condition

$$[\nu, E] = c, \qquad \operatorname{div} E = \gamma \quad \text{on} \quad \partial D \tag{4.23}$$

where γ and c are given as in the interior problem.

Interior Magnetic Boundary-Value Problem

Find a vector field $H \in \mathfrak{F}(D)$ satisfying the vector Helmholtz equation in D and the boundary condition

$$[[\operatorname{curl} H, \nu], \nu] = d, \qquad (\nu, H) = \delta \quad \text{on} \quad \partial D \tag{4.24}$$

where $\delta \in C^{0,\alpha}(\partial D)$ is a given function and $d \in C^{0,\alpha}(\partial D)$ is a given tangential field.

Exterior Magnetic Boundary-Value Problem

Find a vector field $H \in \mathfrak{F}(\mathbb{R}^3 \setminus \bar{D})$ satisfying the vector Helmholtz equation in $\mathbb{R}^3 \setminus \bar{D}$, the radiation condition (4.15), and the boundary condition

$$[[\operatorname{curl} H, \nu], \nu] = d, \qquad (\nu, H) = \delta \quad \text{on} \quad \partial D \tag{4.25}$$

where δ and d are given as in the interior problem.

From the relation $\Delta \operatorname{div} E = \operatorname{div} \Delta E$ we observe that for any solution E of the vector Helmholtz equation $\operatorname{div} E$ solves the scalar Helmholtz equation. In addition, for the exterior problems, $\operatorname{div} E$ satisfies the Sommerfeld radiation condition (3.7) provided E satisfies the radiation condition (4.15). This can be seen from taking the divergence in the representation Theorem 4.13 to obtain

$$\operatorname{div} E(x) = \int_{\partial D} \Big\{ k^2(\nu(y), E(y))\Phi(x,y) - \big(\operatorname{grad}_y \Phi(x,y), \nu(y), \operatorname{curl} E(y)\big) + \operatorname{div} E(y) \frac{\partial \Phi(x,y)}{\partial \nu(y)} \Big\}\, ds(y), \qquad x \in \mathbb{R}^3 \setminus \bar{D}.$$

Hence, from the special boundary condition $\operatorname{div} E = 0$ on ∂D for the electric boundary-value problem, we can conclude in the case of the interior problem for $\operatorname{Im} k > 0$ and in the case of the exterior problem for $\operatorname{Im} k \geqslant 0$ that $\operatorname{div} E = 0$ in D or $\mathbb{R}^3 \setminus \bar{D}$, respectively. Hence, in view of Theorem 4.3, the Maxwell boundary-value problem and the special case of the electric boundary-value problem where $\gamma = 0$ are equivalent. Similarly, the Maxwell boundary-value problem is equivalent to the magnetic boundary-value problem where $\operatorname{Div}[d, \nu] + k^2\delta = 0$. In this case, under the assumption that $\operatorname{curl}\operatorname{curl} H \in C(\bar{D})$ or $C(\mathbb{R}^3 \setminus D)$, we see from (2.75),

$$\begin{aligned} \frac{\partial}{\partial \nu} \operatorname{div} H &= (\nu, \operatorname{curl}\operatorname{curl} H) + (\nu, \Delta H) \\ &= -\operatorname{Div}[\nu, \operatorname{curl} H] - k^2(\nu, H), \end{aligned}$$

and the boundary conditions on H that $\operatorname{div} H$ satisfies the homogeneous Neumann condition $(\partial/\partial\nu)\operatorname{div} H = 0$ on ∂D. The required regularity for H follows from the fact that $H \in C^{0,\alpha}(\bar{D})$ or $C^{0,\alpha}(\mathbb{R}^3 \setminus D)$ which is a consequence of the analysis in the next section and the following lemma.

Lemma 4.15. Let $A \in C^2(D) \cap C^{0,\alpha}(\bar{D})$ be a solution of the vector Helmholtz equation with the property that $\operatorname{div} A \in C(\bar{D})$ and that the surface divergence $\operatorname{Div}[\nu, \operatorname{curl} A]$ exists and is of class $C^{0,\alpha}(\partial D)$, that is, $[\nu, \operatorname{curl} A] \in \mathfrak{T}^{0,\alpha}(\partial D)$. Then $\operatorname{div} A \in C^{1,\alpha}(\bar{D})$ and $\operatorname{curl}\operatorname{curl} A \in C^{0,\alpha}(\bar{D})$.

Proof. Taking the divergence in the representation Theorem 4.11, we find that

$$\operatorname{div} A(x) = \int_{\partial D} \Big\{ -k^2(\nu(y), A(y))\Phi(x,y) + \big(\operatorname{grad}_y \Phi(x,y), \nu(y), \operatorname{curl} A(y)\big) - \operatorname{div} A(y) \frac{\partial \Phi(x,y)}{\partial \nu(y)} \Big\}\, ds(y), \qquad x \in D.$$

From this, using Gauss' theorem (2.73), it follows that

$$\operatorname{div} A(x) = \int_{\partial D} \left\{ \phi(y)\Phi(x,y) - \operatorname{div} A(y) \frac{\partial \Phi(x,y)}{\partial \nu(y)} \right\} ds(y), \qquad x \in D, \tag{4.26}$$

where we have set $\phi := -k^2(\nu, A) - \operatorname{Div}[\nu, \operatorname{curl} A]$. Now letting x tend to the boundary and using Theorem 2.13, we obtain the integral equation

$$\operatorname{div} A(x) + 2\int_{\partial D} \frac{\partial \Phi(x,y)}{\partial \nu(y)} \operatorname{div} A(y)\, ds(y) = 2\int_{\partial D} \Phi(x,y)\phi(y)\, ds(y),$$

$$x \in \partial D,$$

which we can rewrite in the abbreviated form

$$\psi + \mathbf{K}\psi = \mathbf{S}\phi$$

for $\psi := \operatorname{div} A$ on ∂D. Since by assumption we have $\phi \in C^{0,\alpha}(\partial D)$, from Theorem 2.31 we have $\mathbf{S}\phi \in C^{1,\alpha}(\partial D)$, and thus from Theorem 2.30 we conclude that $\psi \in C^{1,\alpha}(\partial D)$, that is, $\operatorname{div} A \in C^{1,\alpha}(\partial D)$. Then, using Theorems 2.17 and 2.23, we finally conclude from (4.26) that $\operatorname{div} A \in C^{1,\alpha}(\bar{D})$. The statement on curl curl A follows from the identity $\operatorname{curl}\operatorname{curl} A = k^2 A + \operatorname{grad}\operatorname{div} A$.

Lemma 4.15 is, of course, also valid for A defined in an exterior domain.

Theorem 4.16. Let $\operatorname{Im} k > 0$. Then the interior Maxwell boundary-value problem, and the interior electric and the interior magnetic boundary-value problems have at most one solution.

Proof. For any solution $E \in \mathfrak{F}(D)$ of the vector Helmholtz equation, we have from the first Green's theorem (4.11) that

$$\int_{\partial D} \{(\nu, \bar{E}, \operatorname{curl} E) + (\nu, \bar{E}) \operatorname{div} E\}\, ds = \int_D \{|\operatorname{curl} E|^2 + |\operatorname{div} E|^2 - k^2 |E|^2\}\, dx.$$

Because the left-hand side of this equation vanishes if E satisfies either the homogeneous electric or magnetic boundary condition, splitting the right-hand side into real and imaginary parts and using $\operatorname{Im} k > 0$ shows that

$$\int_D |E|^2\, dx = 0.$$

Hence, $E = 0$ in D.

For k real, we in general do not have uniqueness for the interior problems. As is easily seen, the fields

$$E(x) = \operatorname{curl} x u(x), \qquad H(x) = \frac{1}{ik} \operatorname{curl} E(x)$$

satisfy Maxwell's equations provided u is a solution to the Helmholtz equation. In particular, if we choose u to be a nontrivial solution to the homogeneous Dirichlet problem for the unit ball as described by (3.19), we see that the homogeneous boundary condition $[\nu, E]=0$ on the boundary is satisfied. In general, as in the case of the Dirichlet and Neumann problem, it can be shown that for any domain D there exists for each of the interior Maxwell, interior electric, and interior magnetic problems a countable set of positive wave numbers k, called eigenvalues, accumulating only at infinity for which the homogeneous problem has nontrivial solutions (see Müller and Niemeyer [1]).

If we denote the set of interior eigenvalues of the Dirichlet problem by $\mathcal{D}$, of the Maxwell problem by $\mathcal{M}$, and of the electric boundary-value problem by $\mathcal{E}$, we have the relation $\mathcal{E}=\mathcal{M}\cup\mathcal{D}$. To see this, we note that by the elimination process described in Theorem 4.3 it is obvious that $\mathcal{M}\subset\mathcal{E}$. To show $\mathcal{D}\subset\mathcal{E}$ we note that for any solution u of the homogeneous Dirichlet problem, $\operatorname{grad} u$ solves the homogeneous electric problem. Hence $\mathcal{M}\cup\mathcal{D}\subset\mathcal{E}$. Conversely, if E is a solution of the homogeneous electric problem, then $\operatorname{div} E$ solves the homogeneous Dirichlet problem. Hence, either $\operatorname{div} E=0$ and by Theorem 4.3 the field E leads to a nontrivial solution of the homogeneous Maxwell problem or $\operatorname{div} E$ is a nontrivial solution of the homogeneous Dirichlet problem. Hence $\mathcal{E}\subset\mathcal{M}\cup\mathcal{D}$. Similarly, we have $\mathcal{H}=\mathcal{M}\cup\mathcal{N}$ where $\mathcal{H}$ denotes the interior magnetic eigenvalues and $\mathcal{N}$ the interior Neumann eigenvalues.

Uniqueness results for the exterior electromagnetic boundary-value problems are based on the following result.

Theorem 4.17. Let $E\in\mathcal{F}(\mathbb{R}^3\setminus\bar{D})$ be a solution to the vector Helmholtz equation satisfying the radiation condition (4.15) and

$$\operatorname{Im}\left(k\int_{\partial D}\{(\nu, E, \operatorname{curl}\bar{E})+(\nu, E)\operatorname{div}\bar{E}\}\,ds\right)\geqslant 0. \tag{4.27}$$

Then $E=0$ in $\mathbb{R}^3\setminus D$.

Proof. If $\operatorname{Im} k>0$, then from (4.17) we see that

$$\int_{D_R}|E|^2\,dx\to 0, \qquad R\to\infty.$$

Hence $E=0$ in $\mathbb{R}^3\setminus D$. If $\operatorname{Im} k=0$ then from (4.17) we see that

$$\int_{|x|=R}|E|^2\,dx\to 0, \qquad R\to\infty.$$

Then, using Corollary 4.14 and applying Lemma 3.11 to the cartesian components of E, we find that $E=0$ in $\mathbb{R}^3\setminus D$.

Theorem 4.18. The exterior Maxwell boundary-value problem and the exterior electric and magnetic boundary-value problem have no more than one solution.

Proof. This follows from Theorem 4.17.

4.4 EXISTENCE OF SOLUTIONS TO THE ELECTROMAGNETIC BOUNDARY-VALUE PROBLEMS BY INTEGRAL EQUATIONS OF THE SECOND KIND

We shall now reduce the electromagnetic boundary-value problems to integral equations of the second kind that, as opposed to the acoustic boundary-value problems, we must discuss in spaces of Hölder continuous functions rather than merely continuous functions.

Theorem 4.19. The electromagnetic field of a surface distribution of magnetic dipoles

$$\begin{aligned} E(x) &= \operatorname{curl} \int_{\partial D} \Phi(x, y) a(y)\, ds(y), \\ & \qquad\qquad\qquad\qquad\qquad\qquad x \in \mathbb{R}^3 \setminus \partial D \\ H(x) &= \frac{1}{ik} \operatorname{curl} E(x), \end{aligned} \tag{4.28}$$

with tangential density $a \in C^{0,\alpha}(\partial D)$, $0 < \alpha < 1$, solves the interior Maxwell problem in D provided a is a solution of the integral equation

$$a(x) - 2\int_{\partial D} [\nu(x), \operatorname{curl}_x\{\Phi(x, y) a(y)\}]\, ds(y) = -2c(x), \qquad x \in \partial D. \tag{4.29}$$

It solves the exterior Maxwell problem in $\mathbb{R}^3 \setminus \bar{D}$ provided a is a solution of the integral equation

$$a(x) + 2\int_{\partial D} [\nu(x), \operatorname{curl}_x\{\Phi(x, y) a(y)\}]\, ds(y) = 2c(x), \qquad x \in \partial D. \tag{4.30}$$

Proof. From Theorems 4.3 and 4.4 we see that E, H satisfy the Maxwell equations in $\mathbb{R}^3 \setminus \partial D$ and the radiation conditions (4.7) and (4.8). By Theorem 2.24 we have $E \in C^{0,\alpha}(\bar{D})$ and $E \in C^{0,\alpha}(\mathbb{R}^3 \setminus D)$ and the boundary conditions (4.18) or (4.19) are fulfilled if a is a solution to the integral equation (4.29) or (4.30), respectively. Furthermore, since E satisfies $[\nu, E] = c$ on ∂D where

$\operatorname{Div} c \in C^{0,\alpha}(\partial D)$, we see from Lemma 4.15 as applied to

$$A(x) := \int_{\partial D} \Phi(x, y) a(y)\, ds(y), \qquad x \in \mathbb{R}^3 \setminus \partial D,$$

that $H \in C^{0,\alpha}(\bar{D})$ or $H \in C^{0,\alpha}(\mathbb{R}^3 \setminus D)$ for the interior or exterior problem, respectively.

Since for the interior problem we now have $\operatorname{div} A \in C^{1,\alpha}(\bar{D})$, we see from Corollary 2.25 that $\operatorname{div} A$ restricted to $\mathbb{R}^3 \setminus \bar{D}$ has boundary values in class $C^{1,\alpha}(\partial D)$. Hence by Theorem 3.27 we have $\operatorname{div} A \in C^{1,\alpha}(\mathbb{R}^3 \setminus D)$ and again using the identity $\operatorname{curl}\operatorname{curl} A = -\Delta A + \operatorname{grad}\operatorname{div} A$, we can conclude that $H \in C^{0,\alpha}(\mathbb{R}^3 \setminus D)$. Similarly, for the exterior problem we also have $H \in C^{0,\alpha}(\bar{D})$. We can now apply (2.75) to the jump relation $a = [\nu, E_+] - [\nu, E_-]$ of Corollary 2.25 to obtain $\operatorname{Div} a \in C^{0,\alpha}(\partial D)$. Thus we have established the following result.

Corollary 4.20. Any solution a to the integral equation (4.29) or (4.30) automatically belongs to $\mathfrak{T}^{0,\alpha}(\partial D)$ if $c \in \mathfrak{T}^{0,\alpha}(\partial D)$.

In an analogous manner, we can prove the following theorems for the electric and magnetic boundary-value problems.

Theorem 4.21. The vector field

$$E(x) = \operatorname{curl} \int_{\partial D} \Phi(x, y) a(y)\, ds(y) - \int_{\partial D} \Phi(x, y) \lambda(y) \nu(y)\, ds(y),$$

$$x \in \mathbb{R}^3 \setminus \partial D, \quad (4.31)$$

with tangential density a and scalar function λ of class $C^{0,\alpha}(\partial D)$ solves the interior electric boundary-value problem provided a and λ solve the system of integral equations

$$a(x) - 2\int_{\partial D} [\nu(x), \operatorname{curl}_x\{\Phi(x, y) a(y)\}]\, ds(y)$$

$$+ 2\int_{\partial D} \Phi(x, y)[\nu(x), \nu(y)]\lambda(y)\, ds(y) = -2c(x), \qquad (4.32)$$

$$\lambda(x) - 2\int_{\partial D} \frac{\partial \Phi(x, y)}{\partial \nu(y)} \lambda(y)\, ds(y) = -2\gamma(x), \qquad x \in \partial D.$$

It solves the exterior electric boundary value problem provided a and λ solve the system of integral equations

$$a(x)+2\int_{\partial D}[\nu(x),\operatorname{curl}_x\{\Phi(x,y)a(y)\}]\,ds(y)$$

$$-2\int_{\partial D}\Phi(x,y)[\nu(x),\nu(y)]\lambda(y)\,ds(y)=2c(x), \tag{4.33}$$

$$\lambda(x)+2\int_{\partial D}\frac{\partial\Phi(x,y)}{\partial\nu(y)}\lambda(y)\,ds(y)=2\gamma(x), \qquad x\in\partial D.$$

Theorem 4.22. The vector field

$$H(x)=\int_{\partial D}\Phi(x,y)[\nu(y),b(y)]\,ds(y)+\operatorname{grad}\int_{\partial D}\Phi(x,y)\mu(y)\,ds(y),$$

$$x\in\mathbb{R}^3\setminus\partial D \tag{4.34}$$

with tangential density b and scalar function μ of class $C^{0,\alpha}(\partial D)$ solves the interior magnetic boundary-value problem provided b and μ solve the system of integral equations

$$b(x)+2\int_{\partial D}[\nu(x),[\nu(x),\operatorname{curl}_x\{\Phi(x,y)[\nu(y),b(y)]\}]]\,ds(y)=2d(x),$$

$$\mu(x)+2\int_{\partial D}\Phi(x,y)(\nu(x),\nu(y),b(y))\,ds(y) \tag{4.35}$$

$$+2\int_{\partial D}\frac{\partial\Phi(x,y)}{\partial\nu(x)}\mu(y)\,ds(y)=2\delta(x), \qquad x\in\partial D.$$

It solves the exterior magnetic boundary-value problem provided b and μ solve the system of integral equations

$$b(x)-2\int_{\partial D}[\nu(x),[\nu(x),\operatorname{curl}_x\{\Phi(x,y)[\nu(y),b(y)]\}]]\,ds(y)=-2d(x),$$

$$\mu(x)-2\int_{\partial D}\Phi(x,y)(\nu(x),\nu(y),b(y))\,ds(y) \tag{4.36}$$

$$-2\int_{\partial D}\frac{\partial\Phi(x,y)}{\partial\nu(x)}\mu(y)\,ds(y)=-2\delta(x), \qquad x\in\partial D.$$

The proof of the existence of solutions to these integral equations based on the first and second parts of Fredholm's alternative was first given by Müller

[1], [3] and Weyl [2]. In the following we shall describe a slightly modified version of their approach.

Recalling the definition (2.82) of the compact operators $\mathbf{M}, \mathbf{M}'$: $\mathfrak{T}^{0,\alpha}(\partial D) \to \mathfrak{T}^{0,\alpha}(\partial D)$, we can write the integral equations (4.29) and (4.30) in the short form

$$a - \mathbf{M}a = -2c \tag{4.29'}$$

and

$$a + \mathbf{M}a = 2c. \tag{4.30'}$$

Note that equation (4.29′) for the interior Maxwell problem is equivalent to the equation

$$b + \mathbf{M}'b = -2[\nu, c] \tag{4.29''}$$

for the density $b := [\nu, a]$. Hence the integral equations of the interior and exterior Maxwell problems are adjoint.

As we shall see, the nullspace of the operator $\mathbf{I}+\mathbf{M}$ corresponds to solutions of the homogeneous interior Maxwell problem. Therefore we introduce the linear space

$$\mathfrak{M} := \{[\nu, H]|_{\partial D} \,|\, E, H \in C^1(D) \cap C(\bar{D}), \operatorname{curl} E - ikH = 0,$$
$$\operatorname{curl} H + ikE = 0 \quad \text{in} \quad D, \qquad [\nu, E] = 0 \quad \text{on} \quad \partial D\}.$$

If k is not an interior Maxwell eigenvalue, then obviously $\mathfrak{M} = \{0\}$. Note that the pair $H, -E$ satisfies Maxwell's equations if and only if the pair E, H does.

Note also that because of Theorem 2.32 the nullspace of $\mathbf{I}+\mathbf{M}$ in the spaces $C(\partial D)$ and $C^{0,\alpha}(\partial D)$ is the same.

Theorem 4.23. $N(\mathbf{I}+\mathbf{M}) = \mathfrak{M}$.

Proof. Let $a \in N(\mathbf{I}+\mathbf{M})$ and define an electromagnetic field E, H by (4.28). Then $[\nu, E_+] = 0$ on ∂D and from the uniqueness Theorem 4.18 it follows that $E = 0$ in $\mathbb{R}^3 \setminus D$. From the conclusions on the regularity of H leading to Corollary 4.20, we see that $ikH = \operatorname{curl}\operatorname{curl} A = k^2 A + \operatorname{grad}\operatorname{div} A$ has continuous tangential components across the boundary ∂D. Thus $[\nu, H_-] = 0$ on ∂D, that is, $H, -E$ form a nontrivial solution to the interior Maxwell problem. Finally, from Corollary 2.25, we find $a = [\nu, E_+] - [\nu, E]$. Hence $a \in \mathfrak{M}$.

Conversely, let $a \in \mathfrak{M}$, that is, $a = [\nu, H]|_{\partial D}$ where E, H is a solution of the homogeneous interior Maxwell problem. Then from the representation Theorem 4.1 we have

$$\operatorname{curl} \int_{\partial D} \Phi(x, y) a(y)\, ds(y) = 0, \qquad x \in \mathbb{R}^3 \setminus \bar{D}.$$

Passing to the limit $x \to \partial D$ and using Theorem 2.26, we have $a + \mathbf{M}a = 0$, that is, $a \in N(\mathbf{I}+\mathbf{M})$.

By Fredholm's alternative Theorem 1.30 we have $\dim N(\mathbf{I}+\mathbf{M}) = \dim N(\mathbf{I}+\mathbf{M}') = m_M$ where $m_M = 0$ if k is not an interior Maxwell eigenvalue and $m_M \in \mathbb{N}$ if k is an eigenvalue. For the second case, we have the following theorem.

Theorem 4.24. Let $b_1,\dots,b_m$ be a basis for $N(\mathbf{I}+\mathbf{M}')$ and define

$$\begin{aligned} E_j(x) &:= \operatorname{curl}\int_{\partial D}\Phi(x,y)\left[b_j(y),\nu(y)\right] ds(y), \\ H_j(x) &:= \frac{1}{ik}\operatorname{curl} E_j(x), \qquad x \in \mathbb{R}^3\setminus\partial D, \end{aligned} \tag{4.37}$$

$j=1,\dots,m_M$. Then

$$b_j = \left[\nu,\left[\nu, E_{j+}\right]\right] \quad \text{on} \quad \partial D \tag{4.38}$$

$j=1,\dots,m_M$, and the tangential fields

$$a_j := \left[\nu, \bar{H}_{j+}\right] \tag{4.39}$$

$j=1,\dots,m_M$, form a basis of $N(\mathbf{I}+\mathbf{M})$. The matrix

$$\langle a_j, b_l\rangle = \int_{\partial D}\left(\nu, E_{l+}, \bar{H}_{j+}\right) ds, \qquad j,l=1,\dots,m_M,$$

is regular and hence by Theorem 1.31 the Riesz number is one.

Proof. Since $b_j + \mathbf{M}'b_j = 0$ we have $[\nu, E_{j-}] = 0$ on ∂D. Using the jump relations of Corollary 2.25, we therefore see that $[b_j, \nu] = [\nu, E_{j+}]$ on ∂D. The pair E_j, H_j is a solution of the homogeneous interior problem. Since interior Maxwell eigenvalues are real, the pair $\bar{E}_j, -\bar{H}_j$ also solves the homogeneous problem. As in the proof of the previous theorem, we have the continuity result $[\nu, H_{j+}] = [\nu, H_{j-}]$ and therefore by Theorem 4.23 $[\nu, \bar{H}_{j+}] \in \mathfrak{M} = N(\mathbf{I}+\mathbf{M})$.

Assume α_j, $j=1,\dots,m_M$, satisfies

$$\sum_{j=1}^{m_M} \alpha_j \langle a_j, b_l\rangle = 0, \qquad l=1,\dots,m_M,$$

and define

$$E := \sum_{j=1}^{m_M} \bar{\alpha}_j E_j, \qquad H := \sum_{j=1}^{m_M} \bar{\alpha}_j H_j.$$

Then

$$\int_{\partial D} (\nu, \bar{H}_+, E_+)\, ds = 0$$

that is,

$$\int_{\partial D} \{(\nu, E_+, \operatorname{curl} \bar{E}_+) + (\nu, E_+) \operatorname{div} \bar{E}_+\}\, ds = 0$$

and from Theorem 4.17 we conclude $E = H = 0$ in $\mathbb{R}^3 \setminus D$. In particular, $[\nu, E_+] = 0$ on ∂D and therefore $\sum_{j=1}^{m_M} \bar{\alpha}_j b_j = 0$. Hence $\alpha_j = 0, j = 1, \ldots, m_M$, and the proof is completed as in Theorem 3.18.

Remark 4.25. Since the interior Maxwell eigenvalues are real, we can choose the basis of the solutions to the homogeneous interior Maxwell problem in such a way that the electric fields are purely imaginary and the magnet fields are real. Therefore we can select the basis $b_1, \ldots, b_{m_M}$ of $N(\mathbf{I} + \mathbf{M}')$ in Theorem 4.24 such that the $a_1, \ldots, a_{m_M}$ are real valued.

We are now able to obtain existence results on the interior and exterior Maxwell problems.

Theorem 4.26. The interior Maxwell problem is solvable if and only if

$$\int_{\partial D} (c, H)\, ds = 0 \tag{4.40}$$

for all solutions E, H to the homogeneous interior Maxwell problem.

Proof. The proof proceeds as in Theorem 3.20 with Fredholm's alternative applied to the integral equation (4.29″) and use being made of Theorem 4.23.

The necessity of the solvability condition (4.40) follows from the second vector Green's theorem (4.12) applied to a solution of the inhomogeneous and a solution of the homogeneous problem.

Theorem 4.27. The exterior Maxwell problem is uniquely solvable.

Proof. This is proved in a manner analogous to Theorem 3.21 using the integral equation (4.30). The necessary modification of (4.28) in the case when k is an interior Maxwell eigenvalue is

$$\begin{aligned} E(x) &= \operatorname{curl} \int_{\partial D} \Phi(x, y) a(y)\, ds(y) + \sum_{j=1}^{m_M} \alpha_j H_j(x), \\ H(x) &= \frac{1}{ik} \operatorname{curl} E(x), \qquad x \in \mathbb{R}^3 \setminus \partial D, \end{aligned} \tag{4.41}$$

where the H_j, $j=1,\ldots,m_M$, are defined by (4.37). Using (4.39) and Remark 4.25, we observe that (4.41) is a solution of the exterior Maxwell problem if a and the coefficients α_j, $j=1,\ldots,m_M$, are chosen such that

$$a+\mathbf{M}a=2c-2\sum_{j=1}^{m_M}\alpha_j a_j. \tag{4.42}$$

The proof is now completed as in Theorem 3.21 by using Theorem 4.24.

To discuss the system of integral equations given in Theorems 4.21 and 4.22, we introduce the product space $X^{0,\alpha}(\partial D):=\mathfrak{T}^{0,\alpha}(\partial D)\times C^{0,\alpha}(\partial D)$ endowed with the product norm

$$\left\|\begin{pmatrix} a\\ \lambda\end{pmatrix}\right\|_{0,\alpha}:=\max(\|a\|_{0,\alpha},\|\lambda\|_{0,\alpha}).$$

In an obvious notation, we define operators $\mathbf{L},\mathbf{L}'$: $X^{0,\alpha}(\partial D)\to X^{0,\alpha}(\partial D)$ of the form

$$\mathbf{L}=\begin{pmatrix}\mathbf{L}_{11} & \mathbf{L}_{12}\\ \mathbf{L}_{21} & \mathbf{L}_{22}\end{pmatrix}\qquad \mathbf{L}':=\begin{pmatrix}\mathbf{L}'_{11} & \mathbf{L}'_{12}\\ \mathbf{L}'_{21} & \mathbf{L}'_{22}\end{pmatrix}$$

where

$$\begin{aligned}
&\mathbf{L}_{11}(a):=\mathbf{M}(a), && \mathbf{L}'_{11}(b):=\mathbf{M}'(\mathrm{b}),\\
&\mathbf{L}_{12}(\lambda):=-[\nu,\mathbf{S}(\lambda\nu)], && \mathbf{L}'_{21}(b):=(\nu,\mathbf{S}[\nu,b]),\\
&\mathbf{L}_{21}(a):=0, && \mathbf{L}'_{12}(\mu):=0,\\
&\mathbf{L}_{22}(\lambda):=\mathbf{K}(\lambda), && \mathbf{L}'_{22}(\mu):=\mathbf{K}'(\mu).
\end{aligned} \tag{4.43}$$

From Theorems 2.30, 2.31, and 2.32 we see that the operators $\mathbf{L}$ and $\mathbf{L}'$ are compact since their components are. We introduce a bilinear form $\langle\cdot,\cdot\rangle$: $X^{0,\alpha}(\partial D)\times X^{0,\alpha}(\partial D)\to\mathbf{C}$ by

$$\left\langle\begin{pmatrix} a\\ \lambda\end{pmatrix},\begin{pmatrix} b\\ \mu\end{pmatrix}\right\rangle:=\int_{\partial D}(a,b)\,ds+\int_{\partial D}\lambda\mu\,ds. \tag{4.44}$$

Then the operators $\mathbf{L}$ and $\mathbf{L}'$ are adjoint with respect to this dual system.

We can now rewrite the systems of integral equations of Theorems 4.21 and 4.22 in the abbreviated operator form

$$\begin{pmatrix} a\\ \lambda\end{pmatrix}-\mathbf{L}\begin{pmatrix} a\\ \lambda\end{pmatrix}=-2\begin{pmatrix} c\\ \gamma\end{pmatrix} \tag{4.32'}$$

and

$$\binom{a}{\lambda}+\mathbf{L}\binom{a}{\lambda}=2\binom{c}{\gamma} \tag{4.33'}$$

for the interior and exterior electric problem and

$$\binom{b}{\mu}+\mathbf{L}'\binom{b}{\mu}=2\binom{d}{\delta} \tag{4.35'}$$

and

$$\binom{b}{\mu}-\mathbf{L}'\binom{b}{\mu}=-2\binom{d}{\delta} \tag{4.36'}$$

for the interior and exterior magnetic problem.

We introduce the linear space

$$\mathcal{H}:=\left\{\left(\begin{array}{c}[\nu, H]|_{\partial D}\\ \operatorname{div} H|_{\partial D}\end{array}\right)\middle|\begin{array}{lll}H\in\mathfrak{F}(D), & \Delta H+k^2H=0 & \text{in}\quad D\\ [[\operatorname{curl} H,\nu],\nu]=0, & (\nu, H)=0 & \text{on}\quad \partial D\end{array}\right\}.$$

Theorem 4.28. $N(\mathbf{I}+\mathbf{L})=\mathcal{H}$.

Proof. Let $\binom{a}{\lambda}\in N(\mathbf{I}+\mathbf{L})$ and define a field E by (4.31). Then $[\nu, E_+]=0$ and $\operatorname{div} E_+=0$ on ∂D and from the uniqueness Theorem 4.18 it follows that $E=0$ in $\mathbb{R}^3\setminus D$. By the jump relations of Corollary 2.25, we now obtain $(\nu, E_-)=0$ on ∂D and as in the proof of Theorem 4.23 it can be shown that $\operatorname{curl} E$ has continuous tangential components across the boundary ∂D. Thus $[[\operatorname{curl} E_-,\nu],\nu]=0$ on ∂D, that is, E is a solution to the homogeneous interior magnetic problem. Finally, from Corollary 2.25 we see that $a=-[\nu, E_-]$ and $\lambda=-\operatorname{div} E_-$. Hence $\binom{a}{\lambda}\in\mathcal{H}$.

Conversely, let $\binom{a}{\lambda}\in\mathcal{H}$, that is, $a=[\nu, H]|_{\partial D}$ and $\lambda=\operatorname{div} H|_{\partial D}$ where H is a solution to the homogeneous interior magnetic problem. Then from the representation Theorem 4.11 we have

$$\operatorname{curl}\int_{\partial D}\Phi(x, y)a(y)\,ds(y)-\int_{\partial D}\Phi(x, y)\lambda(y)\nu(y)\,ds(y)=0,\qquad x\in\mathbb{R}^3\setminus\bar{D}.$$

Passing to the limit $x\to\partial D$ and taking the tangential component we find with the aid of Theorem 2.26 that $a+\mathbf{L}_{11}a+\mathbf{L}_{12}\lambda=0$. Taking the divergence and letting $x\to\partial D$, we obtain with the aid of Theorem 2.13 that $\lambda+\mathbf{L}_{22}\lambda=0$. Hence $\binom{a}{\lambda}\in N(\mathbf{I}+\mathbf{L})$. Note that as in the case of Theorem 4.23 the nullspaces are the same in the space $C(\partial D)$ and $C^{0,\alpha}(\partial D)$.

By Fredholm's alternative we have $\dim N(\mathbf{I}+\mathbf{L})=\dim N(\mathbf{I}+\mathbf{L}')=m_H$ where $m_H=0$ if k is not an interior magnetic eigenvalue and where $m_H\in\mathbb{N}$ if k is an eigenvalue. For the second case we have the following theorem.

Theorem 4.29. Let $\binom{b_1}{\mu_1},\dots,\binom{b_{m_H}}{\mu_{m_H}}$ be a basis for $N(\mathbf{I}+\mathbf{L}')$ and define

$$H_j(x):=\int_{\partial D}\Phi(x,y)[\nu(y),b_j(y)]\,ds(y)+\operatorname{grad}\int_{\partial D}\Phi(x,y)\mu_j(y)\,ds(y),\qquad x\in\mathbb{R}^3\setminus\partial D,\quad(4.45)$$

$j=1,\dots,m_H$. Then

$$b_j=-[[\operatorname{curl}H_{j+},\nu],\nu],\qquad \mu_j=-(\nu,H_{j+})\quad\text{on}\quad\partial D,\qquad(4.46)$$

$j=1,\dots,m_H$, and

$$a_j:=[\nu,\bar{H}_{j+}],\qquad \lambda_j:=\operatorname{div}\bar{H}_{j+}\qquad(4.47)$$

form a basis for $N(\mathbf{I}+\mathbf{L})$. The matrix

$$\left\langle\binom{a_j}{\lambda_j},\binom{b_l}{\mu_l}\right\rangle=\int_{\partial D}\{(\nu,\bar{H}_{j+},\operatorname{curl}H_{l+})-\operatorname{div}\bar{H}_{j+}(\nu,H_{l+})\}\,ds,\qquad j,l=1,\dots,m_H,$$

is regular and hence by Theorem 1.31 the Riesz number is one.

Proof. Since $\binom{b_j}{\mu_j}+\mathbf{L}'\binom{b_j}{\mu_j}=0$ we clearly have $[[\operatorname{curl}H_{j-},\nu],\nu]=0$ and $(\nu,H_{j-})=0$ on ∂D, that is H_j (and also $\bar{H}_j$) is a solution of the homogeneous interior magnetic problem. Then (4.46) follows from the jump relations of Corollaries 2.20 and 2.25. Since by the same jump relations we also have $[\nu,H_{j+}]=[\nu,H_{j-}]$ and $\operatorname{div}H_{j+}=\operatorname{div}H_{j-}$ on ∂D, $\binom{a_j}{\lambda_j}\in\mathcal{H}=N(\mathbf{I}+\mathbf{L})$ by Theorem 4.28.

Assume α_j, $j=1,\dots,m_H$, satisfies

$$\sum_{j=1}^{m_H}\alpha_j\left\langle\binom{a_j}{\lambda_j},\binom{b_l}{\mu_l}\right\rangle=0,\qquad l=1,\dots,m_H,$$

and define

$$H:=\sum_{j=1}^{m_H}\bar{\alpha}_jH_j.$$

Then

$$\int_{\partial D}\{(\nu,\bar{H}_+,\operatorname{curl}H_+)-(\nu,H_+)\operatorname{div}\bar{H}_+\}\,ds=0$$

and therefore

$$\operatorname{Im}\int_{\partial D}\{(\nu, H_+, \operatorname{curl}\overline{H}_+)+(\nu, H_+)\operatorname{div}\overline{H}_+\}\, ds = 0.$$

From Theorem 4.17 we can now conclude that $H=0$ in $\mathbb{R}^3\setminus D$. The proof is now completed as in Theorem 3.18.

Theorem 4.30. The interior magnetic boundary-value problem is solvable if and only if

$$\int_{\partial D}\{(\nu, H, d)+\delta \operatorname{div} H\}\, ds = 0 \tag{4.48}$$

is satisfied for all solutions H to the homogeneous interior magnetic problem.

Proof. This is proved in an analogous manner to Theorem 3.20 with Fredholm's alternative applied to the integral equation (4.35) using Theorem 4.28.

The necessity of the solvability condition (4.48) follows from the second vector Green's theorem (4.12) applied to a solution of the inhomogeneous problem and a solution of the homogeneous problem.

Theorem 4.31. The exterior electric boundary-value problem is uniquely solvable.

Proof. The proof follows that of Theorem 3.21 using the integral equation (4.33). The modification of (4.31) in the case of an interior magnetic eigenvalue is

$$\begin{aligned} E(x) &= \operatorname{curl}\int_{\partial D}\Phi(x, y)a(y)\, ds(y)-\int_{\partial D}\Phi(x, y)\lambda(y)\nu(y)\, ds(y) \\ &\quad+\sum_{j=1}^{m_H}\alpha_j H_j(x), \qquad x\in\mathbb{R}^3\setminus\partial D \end{aligned} \tag{4.49}$$

where the H_j, $j=1,\dots,m_M$, are defined by (4.45). This leads to the integral equation

$$\begin{pmatrix} a \\ \lambda\end{pmatrix}+\mathbf{L}\begin{pmatrix} a \\ \lambda\end{pmatrix}=2\begin{pmatrix} c \\ \gamma\end{pmatrix}-2\sum_{j=1}^{m_H}\alpha_j\begin{pmatrix} a_j \\ \lambda_j\end{pmatrix}.$$

Here we have again assumed the basis of $N(\mathbf{I}+\mathbf{L})$ given by (4.47) to be real valued.

Finally, without giving the proofs, we state the corresponding theorems for the interior electric and exterior magnetic boundary-value problems. We introduce the linear space

$$\mathfrak{E}:=\left\{\begin{pmatrix}[[\operatorname{curl} E,\nu],\nu]|_{\partial D}\\ (\nu,E)|_{\partial D}\end{pmatrix}\middle|\begin{array}{ll}E\in\mathfrak{F}(D), & \Delta E+k^2E=0 \text{ in } D\\ [\nu,E]=0, & \operatorname{div}E=0 \text{ on } \partial D\end{array}\right\}.$$

Theorem 4.32. $N(\mathbf{I}-\mathbf{L}')=\mathfrak{E}$.

Theorem 4.33. Let $\binom{c_1}{\gamma_1},\ldots,\binom{c_{m_E}}{\gamma_{m_E}}$ be a basis of $N(\mathbf{I}-\mathbf{L})$ and define

$$E_j(x):=\operatorname{curl}\int_{\partial D}\Phi(x,y)c_j(y)\,ds(y)-\int_{\partial D}\Phi(x,y)\gamma_j(y)\nu(y)\,ds(y),$$

$$x\in\mathbb{R}^3\backslash\partial D, \quad (4.50)$$

$j=1,\ldots,m_E$. Then

$$c_j=[\nu,E_j]_+,\qquad \gamma_j=\operatorname{div}E_{j+}\quad\text{on}\quad \partial D, \quad (4.51)$$

$j=1,\ldots,m_E$, and

$$d_j:=[\operatorname{curl}\bar{E}_{j+},\nu],\nu],\qquad \delta_j:=(\nu,\bar{E}_{j+}) \quad (4.52)$$

form a basis of $N(\mathbf{I}-\mathbf{L}')$. The matrix

$$\left\langle\binom{d_j}{\delta_j},\binom{c_l}{\gamma_l}\right\rangle=\int_{\partial D}\{-(\nu,E_{l+},\operatorname{curl}\bar{E}_{j-})+(\nu,\bar{E}_{j-})\operatorname{div}E_{l+}\}\,ds,$$

$$j,l=1,\ldots,m_E,$$

is regular and hence by Theorem 1.31 the Riesz number is one.

Theorem 4.34. The interior electric boundary-value problem is solvable if and only if

$$\int_{\partial D}\{(c,\operatorname{curl}E)-(\nu,E)\gamma\}\,ds=0 \quad (4.53)$$

is satisfied for all solutions E to the homogeneous interior electric problem.

Theorem 4.35. The exterior magnetic boundary-value problem is uniquely solvable.

4.5 BOUNDARY INTEGRAL EQUATIONS OF THE FIRST KIND

As in the case of boundary-value problems in acoustic scattering, it is also possible to use integral equations of the first kind in the study of electromagnetic scattering.

Theorem 4.36. The electromagnetic field of a surface distribution of electric dipoles

$$E(x)=\operatorname{curl}\operatorname{curl}\int_{\partial D}\Phi(x,y)[\nu(y),b(y)]\,ds(y),$$
$$H(x)=\frac{1}{ik}\operatorname{curl}E(x),\qquad x\in\mathbf{R}^3\setminus\partial D, \tag{4.54}$$

with tangential density $b\in\mathfrak{S}_{\perp}^{0,\alpha}(\partial D)$, $0<\alpha<1$ (i.e., the surface divergence $\operatorname{Div}[\nu, b]$ exists and is of class $C^{0,\alpha}(\partial D)$) is a solution of the interior and the exterior Maxwell problems provided b is a solution of the singular integral equation

$$\left[\nu(x),\operatorname{curl}\operatorname{curl}\int_{\partial D}\Phi(x,y)[\nu(y),b(y)]\,ds(y)\right]=c(x),\qquad x\in\partial D. \tag{4.55}$$

Recalling the definition (2.85) of the operator $\mathbf{N}$: $\mathfrak{S}_{\perp}(\partial D)\to\mathfrak{T}(\partial D)$, we rewrite (4.55) in the short form

$$\mathbf{N}b=2c \tag{4.55$'$}$$

and establish the following theorem.

Theorem 4.37. For any inhomogeneity $c\in\mathfrak{S}^{0\alpha}(\partial D)$, the integral equation (4.55) of the first kind for the Maxwell boundary-value problem has a unique solution provided k is not an interior Maxwell eigenvalue. If k is an eigenvalue, then the integral equation is solvable if and only if c satisfies condition (4.40) and in this case the solution is not unique.

Proof. We proceed along the lines of Theorem 3.30. Existence is shown by solving the interior and exterior Maxwell problems using the integral equations of Theorem 4.19 and then using the representation Theorems 4.1 and 4.5. The solution of equation (4.55) is given by

$$b=\frac{2}{k^2}\left[\mathbf{N}\left[(\mathbf{I}-\mathbf{M})^{-1}(\mathbf{I}+\mathbf{M})^{-1}c,\nu\right],\nu\right] \tag{4.56}$$

that in particular shows that the inverse $\mathbf{N}^{-1}$ of $\mathbf{N}$ is not bounded.

For the electric and magnetic boundary-value problems, we obtain the following results by exchanging the approaches of Theorems 4.21 and 4.22.

Theorem 4.38. The vector field

$$E(x)=\int_{\partial D}\Phi(x,y)[\nu(y),b(y)]\,ds(y)+\operatorname{grad}\int_{\partial D}\Phi(x,y)\mu(y)\,ds(y),$$
$$x\in\mathbf{R}^3\setminus\partial D, \tag{4.57}$$

with tangential field b and scalar function μ of class $C^{0,\alpha}(\partial D)$ is a solution of the interior and exterior electric boundary-value problem provided b and μ satisfy the system of singular integral equations

$$\begin{aligned}
&\int_{\partial D} \Phi(x,y)[\nu(x),[\nu(y),b(y)]]\,ds(y)\\
&\qquad+\int_{\partial D}[\nu(x),\operatorname{grad}_x\{\Phi(x,y)\mu(y)\}]\,ds(y)=c(x),\\
&\int_{\partial D}\operatorname{div}_x\{\Phi(x,y)[\nu(y),b(y)]\}\,ds(y)\\
&\qquad-k^2\int_{\partial D}\Phi(x,y)\mu(y)\,ds(y)=\gamma(x),\qquad x\in\partial D.
\end{aligned}\tag{4.58}$$

Theorem 4.39. The vector field

$$H(x)=\operatorname{curl}\int_{\partial D}\Phi(x,y)a(y)\,ds(y)-\int_{\partial D}\Phi(x,y)\lambda(y)\nu(y)\,ds(y),\qquad x\in\mathbb{R}^3\setminus\partial D,\tag{4.59}$$

with tangential field $a\in\mathfrak{T}^{0,\alpha}(\partial D)$ and scalar function $\mu\in C^{0,\alpha}(\partial D)$ is a solution of the interior and exterior magnetic boundary-value problems provided a and λ satisfy the system of singular integral equations

$$\begin{aligned}
&\left[\nu(x),\left[\nu(x),\operatorname{curl}\operatorname{curl}\int_{\partial D}\Phi(x,y)a(y)\,ds(y)\right]\right]\\
&\qquad-\int_{\partial D}[\nu(x),[\nu(x),\operatorname{curl}_x\{\Phi(x,y)\lambda(y)\nu(y)\}]]\,ds(y)=d(x)\\
&\int_{\partial D}(\nu(x),\operatorname{curl}_x\{\Phi(x,y)a(y)\})\,ds(y)\\
&\qquad-\int_{\partial D}\Phi(x,y)\lambda(y)(\nu(x),\nu(y))\,ds(y)=\delta(x),\qquad x\in\partial D.
\end{aligned}\tag{4.60}$$

We introduce operators $\mathbf{R}$: $X^{0,\alpha}(\partial D)\to X^{0,\alpha}(\partial D)$ and $\mathbf{Q}$: $\mathfrak{T}^{0,\alpha}(\partial D)\times C^{0,\alpha}(\partial \mathrm{D})\to X^{0,\alpha}(\partial D)$ of the form

$$\mathbf{R}=\begin{pmatrix}\mathbf{R}_{11} & \mathbf{R}_{12}\\ \mathbf{R}_{21} & \mathbf{R}_{22}\end{pmatrix},\qquad \mathbf{Q}=\begin{pmatrix}\mathbf{Q}_{11} & \mathbf{Q}_{12}\\ \mathbf{Q}_{21} & \mathbf{Q}_{22}\end{pmatrix}$$

where

$$\mathbf{R}_{11}(b) := [\nu, \mathbf{S}[\nu, b]],$$

$$\mathbf{R}_{22}(\mu) := -k^2\mathbf{S}(\mu), \tag{4.61}$$

$$(\mathbf{R}_{12}\mu)(x) := 2\int_{\partial D}[\nu(x), \operatorname{grad}_x\{\Phi(x,y)\mu(y)\}]\,ds(y), \qquad x \in \partial D,$$

$$(\mathbf{R}_{21}b)(x) := 2\int_{\partial D}\operatorname{div}_x\{\Phi(x,y)[\nu(y), b(y)]\}\,ds(y), \qquad x \in \partial D,$$

$$\mathbf{Q}_{11}(a) := [\nu, \mathbf{N}[a, \nu]],$$

$$\mathbf{Q}_{22}(\lambda) := -(\nu, \mathbf{S}(\lambda\nu)), \tag{4.62}$$

$$(\mathbf{Q}_{12}\lambda)(x) := -2\int_{\partial D}[\nu(x), [\nu(x), \operatorname{curl}_x\{\Phi(x,y)\lambda(y)\nu(y)\}]]\,ds(y),$$

$$x \in \partial D,$$

$$(\mathbf{Q}_{21}a)(x) := 2\int_{\partial D}(\nu(x), \operatorname{curl}_x\{\Phi(x,y)a(y)\})\,ds(y), \quad x \in \partial D.$$

Out of these eight operators, $\mathbf{R}_{11}$, $\mathbf{R}_{22}$, and $\mathbf{Q}_{22}$ are clearly compact and the operator $\mathbf{Q}_{11}$ is unbounded. For the remaining operators $\mathbf{R}_{12}$, $\mathbf{R}_{21}$, $\mathbf{Q}_{12}$, and $\mathbf{Q}_{21}$, the integrals have to be understood in the sense of Cauchy's principal value. These four operators are bounded by Theorems 2.17 and 2.24, but they are not compact. We now write the systems of integral equations (4.58) and (4.60) in the short form

$$\mathbf{R}\begin{pmatrix} b \\ \mu \end{pmatrix} = 2\begin{pmatrix} c \\ \gamma \end{pmatrix} \tag{4.58$'$}$$

for the electric boundary-value problem and

$$\mathbf{Q}\begin{pmatrix} a \\ \lambda \end{pmatrix} = 2\begin{pmatrix} d \\ \delta \end{pmatrix} \tag{4.60$'$}$$

for the magnetic boundary-value problem and note that both are singular. We also note that the adjoints $\mathbf{R}'$ and $\mathbf{Q}'$ of $\mathbf{R}$ and $\mathbf{Q}$ with respect to the dual system (4.44) are given by $\mathbf{R}' = \mathbf{R}$ and $\mathbf{Q}' = \mathbf{Q}$.

Analogous to Theorem 4.37 we have the following theorems.

Theorem 4.40. For any inhomogeneity $\binom{c}{\gamma} \in \mathfrak{T}^{0,\alpha}(\partial D) \times C^{0,\alpha}(\partial D)$ the integral equation (4.58) of the first kind for the electric boundary-value problem has a unique solution provided k is not an interior electric eigenvalue. If k is an eigenvalue, then the integral equation is solvable if and only if c and γ satisfy condition (4.53).

Theorem 4.41. For any inhomogeneity $\binom{d}{\delta} \in \mathfrak{T}^{0,\alpha}(\partial D) \times C^{0,\alpha}(\partial D)$ the integral equation (4.60) of the first kind for the magnetic boundary-value problem has a unique solution provided k is not an interior magnetic eigenvalue. If k is an eigenvalue, then the integral equation is solvable if and only if d and δ satisfy condition (4.48).

If k is not an interior eigenvalue of the electric or magnetic boundary-value problem, the inverse $\mathbf{R}^{-1}: \mathfrak{S}^{0,\alpha}(\partial D) \times C^{0,\alpha}(\partial D) \to X^{0,\alpha}(\partial D)$ is explicitly given by

$$\mathbf{R}^{-1} = -\mathbf{Q}(\mathbf{I}-\mathbf{L})^{-1}(\mathbf{I}+\mathbf{L})^{-1} \tag{4.63}$$

and the inverse $\mathbf{Q}^{-1}: X^{0,\alpha}(\partial D) \to \mathfrak{S}^{0,\alpha}(\partial D) \times C^{0,\alpha}(\partial D)$ is given by

$$\mathbf{Q}^{-1} = -\mathbf{R}(\mathbf{I}-\mathbf{L}')^{-1}(\mathbf{I}+\mathbf{L}')^{-1}. \tag{4.64}$$

4.6. MODIFIED INTEGRAL EQUATIONS

For the same reasons as in acoustic scattering, it is desirable to develop modifications of the integral equations for the exterior electromagnetic boundary-value problems that are uniquely solvable for all wave numbers. Hence in the following analysis we shall describe the analogue of the combined double- and single-layer potential approach in the electromagnetic situation. As opposed to the Dirichlet and Neumann problems, the integral equations for all three electromagnetic problems become singular and thus the regularization procedures are different from those used for the scalar problem.

Knauff and Kress [1] suggested seeking the solution to the exterior electric boundary-value problem in the combination of the forms (4.31) and (4.57), namely,

$$\begin{aligned} E(x) = {} & \operatorname{curl}\int_{\partial D} \Phi(x,y)a(y)\,ds(y) - \int_{\partial D} \Phi(x,y)\lambda(y)\nu(y)\,ds(y) \\ & + i\eta\left\{\int_{\partial D} \Phi(x,y)[\nu(y),a(y)]\,ds(y) + \operatorname{grad}\int_{\partial D} \Phi(x,y)\lambda(y)\,ds(y)\right\}, \\ & \qquad x \in \mathbf{R}^3 \setminus \partial D, \end{aligned} \tag{4.65}$$

where $\eta \neq 0$ is an arbitrary real number such that

$$\eta \operatorname{Re} k \geq 0. \tag{4.66}$$

Then (4.65) is a solution to the exterior electric boundary-value problem if the densities $a \in \mathfrak{T}^{0,\alpha}(\partial D)$ and $\lambda \in C^{0,\alpha}(\partial D)$ are solutions of the integral equation

$$\begin{pmatrix} a \\ \lambda \end{pmatrix} + \mathbf{L}\begin{pmatrix} a \\ \lambda \end{pmatrix} + i\eta \mathbf{R}\begin{pmatrix} a \\ \lambda \end{pmatrix} = 2\begin{pmatrix} c \\ \gamma \end{pmatrix}. \tag{4.67}$$

Since the operator $\mathbf{R}$ is bounded but not compact, the integral equation (4.67) is singular and requires regularization before applying the Riesz theory. The following technique is a slight simplification of the methods used by Kress [6].

Theorem 4.42. The combined integral equation (4.67) for the exterior electric boundary-value problem is uniquely solvable for all wave numbers $\operatorname{Im} k \geqslant 0$.

Proof. We first show that if a solution exists to the integral equation it is unique. Let $\begin{pmatrix} a \\ \lambda \end{pmatrix}$ be a solution of the homogeneous form of equation (4.67). Then the field E defined by (4.65) solves the homogeneous exterior electric boundary value problem and thus $E = 0$ in $\mathbb{R}^3 \backslash D$. Then from the jump relations of Corollaries 2.18 and 2.25 and the regularity argument employed in the proof of Theorem 4.19, we have

$$\begin{aligned} -[\nu, E_-] &= a, & -[\nu, \operatorname{curl} E_-] &= i\eta[\nu, a] \\ -\operatorname{div} E_- &= \lambda, & -(\nu, E_-) &= -i\eta\lambda \end{aligned} \tag{4.68}$$

on ∂D. Hence the first vector Green's theorem (4.11) yields

$$i\eta \int_{\partial D} (|a|^2 + |\lambda|^2)\, ds = \int_D (|\operatorname{curl} E|^2 + |\operatorname{div} E|^2 - k^2|E|^2)\, dx.$$

The imaginary part of this equation reads

$$\eta \int_{\partial D} (|a|^2 + |\lambda|^2)\, ds = -2 \operatorname{Re} k \operatorname{Im} k \int_D |E|^2\, dx$$

from which we deduce $a = 0$, $\lambda = 0$ because of (4.66) and the fact that $\operatorname{Im} k \geqslant 0$.

For small η the existence of a solution to the inhomogeneous equation (4.67) follows by Corollary 1.20 since $\mathbf{L}$ is compact and $\mathbf{I} + i\eta\mathbf{R}$ has a bounded inverse given by a Neumann series provided $|\eta| < \|\mathbf{R}\|^{-1}$. For arbitrary η we regularize the equation (4.67) by actually showing that the solution to the exterior electric problem can always be represented in the form (4.65).

To achieve this, we consider the solution to the exterior electric boundary-value problem as the solution of a special case of the following problem: Find a vector field $F \in \mathfrak{F}(\mathbf{R}^3 \backslash \bar{D})$ satisfying the vector Helmholtz equation in

$\mathbb{R}^3 \setminus \bar{D}$, the radiation condition (4.15), and the boundary condition

$$[\nu, F] = c^*, \qquad \operatorname{div} F + i\eta(\nu, F) = \gamma^* \quad \text{on} \quad \partial D \tag{4.69}$$

where $c^* \in \mathfrak{S}^{0,\alpha}(\partial D)$ and $\gamma^* \in C^{0,\alpha}(\partial D)$ are given. In our special case, the boundary values are specified by $c^* := c$ and $\gamma^* := \gamma + i\eta(\nu, E)$ where E denotes the solution to the exterior electric problem. Note that the existence of such a solution is guaranteed by the approach (4.65) if the parameter η is chosen small enough.

In the subsequent analysis we shall show that the boundary-value problem described by (4.69) has a unique solution and that this solution can be uniquely expressed in the form

$$F(x) = \operatorname{curl} \int_{\partial D} \Phi(x, y) a(y)\, ds(y) - \beta \int_{\partial D} \Phi(x, y) \lambda(y) \nu(y)\, ds(y)$$

$$+ i\eta \left\{ \int_{\partial D} \Phi(x, y)[\nu(y), a(y)]\, ds(y) + \operatorname{grad} \int_{\partial D} \Phi(x, y) \lambda(y)\, ds(y) \right\},$$

$$x \in \mathbb{R}^3 \setminus \partial D, \tag{4.70}$$

with $a \in \mathfrak{T}^{0,\alpha}(\partial D)$ and $\lambda \in C^{0,\alpha}(\partial D)$ where β is a fixed, but arbitrary positive constant. If we now choose $\beta = 1$, the proof of Theorem 4.42 is complete.

We start by proving uniqueness for the boundary-value problem (4.69). If F satisfies the homogeneous conditions $[\nu, F] = 0$, $\operatorname{div} F + i\eta(\nu, F) = 0$ on ∂D, then

$$\operatorname{Im}\left(k \int_{\partial D} \{(\nu, F, \operatorname{curl} \bar{F}) + (\nu, F) \operatorname{div} \bar{F}\}\, ds \right) = \eta \operatorname{Re} k \int_{\partial D} |(\nu, F)|^2\, ds \geq 0$$

and from Theorem 4.17 we conclude that $F = 0$ in $\mathbb{R}^3 \setminus D$.

Straightforward combination of the Theorems 4.21, 4.22, 4.38, and 4.39 shows that (4.70) satisfies the boundary condition (4.69) if a and λ satisfy the integral equation

$$\mathbf{A}\begin{pmatrix} a \\ \lambda \end{pmatrix} + \mathbf{B}\begin{pmatrix} a \\ \lambda \end{pmatrix} = 2\begin{pmatrix} c^* \\ \gamma^* \end{pmatrix} \tag{4.71}$$

where

$$\mathbf{A} = \begin{pmatrix} \mathbf{I} & i\eta \mathbf{R}_{12} \\ 0 & (\beta + \eta^2)\mathbf{I} \end{pmatrix}$$

and $\mathbf{B}$ has the form

$$\mathbf{B} = \begin{pmatrix} \mathbf{B}_{11} & \mathbf{B}_{12} \\ \mathbf{B}_{21} & \mathbf{B}_{22} \end{pmatrix}$$

with $\mathbf{B}_{11}=\mathbf{L}_{11}+i\eta\mathbf{R}_{11}$, $\mathbf{B}_{12}=\beta\mathbf{L}_{12}$, $\mathbf{B}_{21}=i\eta(\mathbf{R}_{21}+\mathbf{Q}_{21})-\eta^2\mathbf{L}'_{21}$, and $\mathbf{B}_{22}=\beta\mathbf{L}_{22}+i\eta(\mathbf{R}_{22}+\beta\mathbf{Q}_{22})-\eta^2\mathbf{L}'_{22}$. Note that since $\mathbf{R}_{12}$ is bounded, $\mathbf{A}$ has a bounded inverse. From our previous analysis we see that the operators $\mathbf{B}_{11}$, $\mathbf{B}_{12}$, and $\mathbf{B}_{22}$ are compact. The operator $\mathbf{B}_{21}$ is also compact since $\mathbf{R}_{21}+\mathbf{Q}_{21}$ has the form

$$(\mathbf{R}_{21}+\mathbf{Q}_{21})(a)(x)=2\int_{\partial D}(\nu(x)-\nu(y),\operatorname{grad}_x\Phi(x,y),a(y))\,ds(y)$$

from which we can deduce compactness by Corollary 2.9. Hence $\mathbf{B}$ is compact and Corollary 1.20 can be applied to equation (4.71). Note that this result also remains valid for the limiting case $\beta=0$.

The uniqueness of solutions to the integral equation (4.71) follows by repeating the arguments used in establishing the uniqueness of solutions to equation (4.67). In particular, the formulas corresponding to (4.68) read

$$\begin{aligned}-[\nu,F_-]&=a, & -[\nu,\operatorname{curl}F_-]&=i\eta[\nu,a]\\ -\operatorname{div}F_-&=\beta\lambda, & -(\nu,F_-)&=-i\eta\lambda\end{aligned}\tag{4.72}$$

on ∂D and in the case $\beta>0$ the proof is completed as above. For later use we also want to verify that in the limiting case $\beta=0$ we still have uniqueness if k is not an interior Dirichlet eigenvalue. In this case we see from the previous argument that if $\begin{pmatrix}c^*\\ \gamma^*\end{pmatrix}=0$, then $a=0$ on ∂D. From (4.72) we see that $\operatorname{div}F_-=0$ on ∂D. Hence, by our assumption on k it follows that $\operatorname{div}F=0$ in D. Then from the transformation

$$\operatorname{div}F(x)=-i\eta k^2\int_{\partial D}\Phi(x,y)\lambda(y)\,ds(y),\qquad x\subset\mathbb{R}^3\setminus\partial D$$

we can conclude that $\lambda=0$ by the jump relation of Corollary 2.20.

The corresponding approach obtained by seeking a solution H in the form

$$\begin{aligned}H(x)=&\int_{\partial D}\Phi(x,y)[\nu(y),b(y)]\,ds(y)+\operatorname{grad}\int_{\partial D}\Phi(x,y)\mu(y)\,ds(y)\\ &-i\eta\left\{\operatorname{curl}\int_{\partial D}\Phi(x,y)b(y)\,ds(y)-\int_{\partial D}\Phi(x,y)\mu(y)\nu(y)\,ds(y)\right\},\\ &\qquad x\in\mathbb{R}^3\setminus\partial D,\end{aligned}\tag{4.73}$$

solves the exterior magnetic problem provided the densities $b\in\mathbb{S}^{0,\alpha}_{\perp}(\partial D)$ and $\mu\in C^{0,\alpha}(\partial D)$ satisfy the singular integral equation

$$\begin{pmatrix}b\\ \mu\end{pmatrix}-\mathbf{L}'\begin{pmatrix}b\\ \mu\end{pmatrix}+i\eta\mathbf{Q}\begin{pmatrix}b\\ \mu\end{pmatrix}=-2\begin{pmatrix}d\\ \delta\end{pmatrix}.\tag{4.74}$$

Theorem 4.43. The combined integral equation (4.74) for the exterior magnetic boundary-value problem is uniquely solvable for all wave numbers $\operatorname{Im} k \geqslant 0$.

Proof. By Theorem 4.35 we know that a unique solution to the exterior magnetic boundary-value problem exists. Then we can consider this field as the solution to an exterior electric boundary-value problem which by the preceding theorem we can uniquely represent in the form (4.65) with η replaced by $1/\eta$. But then we also have a representation in the form (4.73) with $b := ia/\eta$ and $\mu := i\lambda/\eta$. Uniqueness for the integral equation (4.74) follows by the same argument as for (4.67).

We wish to point out that it is also possible to obtain a direct existence proof for equation (4.74) without relying on Theorem 4.35. This can be done by regularizing the equation in a manner similar to the one we shall use for the exterior Maxwell problem.

We now observe that the combined magnetic and electric dipole approach

$$E(x) = \operatorname{curl}\int_{\partial D} \Phi(x,y)a(y)\,ds(y)$$

$$+ i\eta \operatorname{curl}\operatorname{curl}\int_{\partial D} \Phi(x,y)[\nu(y), a(y)]\,ds(y), \qquad (4.75)$$

$$H(x) = \frac{1}{ik}\operatorname{curl} E(x), \qquad x \in \mathbb{R}^3 \setminus \partial D,$$

solves the exterior Maxwell boundary-value problem provided the density $a \in \mathfrak{S}_{\perp}^{0,\alpha}(\partial D)$ is a solution of the singular integral equation

$$a + \mathbf{M}a + i\eta \mathbf{N}a = 2c. \qquad (4.76)$$

Theorem 4.44. The combined integral equation (4.76) for the exterior Maxwell problem is uniquely solvable for all wave numbers $\operatorname{Im} k \geqslant 0$.

Proof. Let $a \in \mathfrak{S}_{\perp}^{0,\alpha}(\partial D)$ be a solution of the homogeneous equation $a + \mathbf{M}a + i\eta\mathbf{N}a = 0$. Then the electromagnetic field defined by (4.75) solves the homogeneous exterior Maxwell problem and therefore $E = H = 0$ in $\mathbb{R}^3 \setminus D$. The jump relations of Corollary 2.25 and the transformation (2.86) yield

$$-[\nu, E_-] = a, \qquad -[\nu, \operatorname{curl} E_-] = i\eta k^2[\nu, a] \quad \text{on} \quad \partial D.$$

Hence from Gauss' theorem we have

$$i\eta k^2 \int_{\partial D} |a|^2\,ds = \int_{\partial D} (\nu, \bar{E}_-, \operatorname{curl} E_-)\,ds$$

$$= \int_D \{|\operatorname{curl} E|^2 - k^2|E|^2\}\,dx.$$

Multiplying by $\bar{k}^2$ and then taking the imaginary part, we get

$$\eta|k|^4\int_{\partial D}|a|^2\,ds = -2\,\mathrm{Re}\,k\,\mathrm{Im}\,k\int_D|\mathrm{curl}\,E|^2\,dx$$

from which we can conclude that $a=0$.

To prove existence we follow a regularization technique introduced by Kress [5]. First, we choose a value k_0 that is not an interior Dirichlet or Maxwell eigenvalue and indicate by a subscript that the operators are taken for the fundamental solution Φ with k set equal to k_0. In this case, by Theorem 4.27 we can solve the exterior Maxwell problem in the form

$$E(x)=\mathrm{curl}\int_{\partial D}\Phi_0(x,y)\tilde{a}(y)\,ds(y)$$

where $\tilde{a}$ is given by

$$\tilde{a}=2(I+\mathbf{M}_0)^{-1}c$$

with a bounded operator $(I+\mathbf{M}_0)^{-1}$. We can now consider

$$F(x):=\int_{\partial D}\Phi_0(x,y)\tilde{a}(y)\,ds(y),\qquad x\in\mathbb{R}^3\setminus\bar{D}$$

as the solution to a boundary-value problem of the form (4.69) where η is replaced by $1/\eta$ and the boundary values are specified by $c^*:=[\nu,F]$ and $\gamma^*:=\mathrm{div}\,F+(i/\eta)(\nu,F)$ or in terms of the bounded operators $\mathbf{S}_0$ and $\mathbf{R}_{21,0}$

$$c^*=\frac{1}{2}[\nu,\mathbf{S}_0\tilde{a}],\qquad \gamma^*=\frac{1}{2}\mathbf{R}_{21,0}[\tilde{a},\nu]+\frac{i}{2\eta}(\nu,\mathbf{S}_0\tilde{a}).$$

Following our previous treatment of the boundary conditions (4.69), we can represent F with the special parameter $\beta=0$ in the form

$$F(x)=\mathrm{curl}\int_{\partial D}\Phi(x,y)b(y)\,ds(y)$$
$$+\frac{i}{\eta}\left\{\int_{\partial D}\Phi(x,y)[\nu(y),b(y)]\,ds(y)+\mathrm{grad}\int_{\partial D}\Phi(x,y)\lambda(y)\,ds(y)\right\},$$
$$x\in\mathbb{R}^3\setminus\bar{D},$$

where the densities $b\in\mathbb{S}^{0,\alpha}(\partial D)$ and $\lambda\in C^{0,\alpha}(\partial D)$ are given by

$$\begin{pmatrix}b\\\lambda\end{pmatrix}=2(\mathbf{A}_0+\mathbf{B}_0)^{-1}\begin{pmatrix}c^*\\\gamma^*\end{pmatrix}$$

with $(\mathbf{A}_0+\mathbf{B}_0)^{-1}$ a bounded operator. Now setting $a:=(i/\eta)[\nu, b]$ we get a representation of the electric field $E=\operatorname{curl} F$ in the required form (4.75). Summarizing our construction of a, we see that we can write

$$a=2\mathbf{C}_0 c$$

where $\mathbf{C}_0$: $\mathfrak{T}^{0,\alpha}(\partial D)\rightarrow\mathfrak{T}^{0,\alpha}(\partial D)$ is bounded. (More precisely, we have $\mathbf{C}_0$: $\mathfrak{S}^{0,\alpha}(\partial D)\rightarrow\mathfrak{S}^{0,\alpha}_{\perp}(\partial D)$.) Since, of course, $a=2\mathbf{C}_0 c$ is a solution to

$$a+\mathbf{M}_0 a+i\eta\mathbf{N}_0=2c$$

the operator $\mathbf{C}_0$ is the inverse of $\mathbf{I}+\mathbf{M}_0+i\eta\mathbf{N}_0$.

Now, for an arbitrary wave number k we can transform equation (4.76) into the equivalent form

$$a+\mathbf{C}_0[(\mathbf{M}-\mathbf{M}_0)+i\eta(\mathbf{N}-\mathbf{N}_0)]a=2\mathbf{C}_0 c \tag{4.77}$$

where the operator $\mathbf{N}-\mathbf{N}_0$ is compact in $\mathfrak{T}^{0,\alpha}(\partial D)$ by Theorem 2.33. The proof can now be completed as in Theorem 3.34.

Numerical implementations of the modified integral equations of this section have been considered by Knauff [1], Knauff and Kress [2], and Mautz and Harrington [2].

It should also be possible to extend Jones' modifications to the electromagnetic problems. To our knowledge this has not yet been carried out.

4.7 THE IMPEDANCE BOUNDARY-VALUE PROBLEM

We shall now briefly consider the following exterior impedance boundary-value problem.

Exterior Impedance Boundary-Value Problem

Find two vector fields $E, H\in C^1(\mathbb{R}^3\setminus\bar{D})\cap C(\mathbb{R}^3\setminus D)$ satisfying Maxwell's equations in $\mathbb{R}^3\setminus\bar{D}$, the Silver–Müller radiation condition at infinity, and the boundary condition

$$[\nu,[\nu, H]-\psi[\nu, E]]=d \tag{4.78}$$

where $d\in C^{0,\alpha}(\partial D)$ is a given tangential field and $\psi\in C^{0,\alpha}(\partial D)$ a given function.

Theorem 4.45. The exterior impedance boundary-value problem has at most one solution provided

$$\operatorname{Re}\psi>0 \quad \text{on} \quad \partial D. \tag{4.79}$$

Proof. Let E, H satisfy the homogeneous boundary condition $[\nu,[\nu, H]-\psi[\nu, E]]=0$. Then by Theorem 4.3 the electric field E is a solution to the vector Helmholtz equation satisfying the radiation condition (4.15) and

$$[\nu,[\nu,\operatorname{curl} E]-ik\psi[\nu, E]]=0, \qquad \operatorname{div} E=0 \quad \text{on} \quad \partial D.$$

Hence,

$$\operatorname{Im}\left(k\int_{\partial D}\{(\nu, E,\operatorname{curl}\bar{E})+(\nu, E)\operatorname{div}\bar{E}\}\,ds\right)=|k|^2\int_{\partial D}\operatorname{Re}\psi|[\nu, E]|^2\,ds\geqslant 0$$

and the proof is completed by using Theorem 4.17.

We now try to find the solution to the exterior impedance problem in the form of the electromagnetic fields of a combination of magnetic and electric dipoles as in (4.75) where we assume the density a belongs to $\mathcal{S}^{0,\alpha}(\partial D)\cap\mathcal{S}_{\perp}^{0,\alpha}(\partial D)$. Then (4.75) solves the impedance problem if a is a solution of the singular integral equation

$$(\eta k+\psi)a+\psi\mathbf{M}a-\eta k\mathbf{M}'a+i\eta\psi\mathbf{N}a+\frac{i}{k}[\nu,\mathbf{N}[a,\nu]]=-2d. \qquad (4.80)$$

Theorem 4.46. The combined integral equation (4.80) for the exterior impedance boundary-value problem is uniquely solvable for all wave numbers $\operatorname{Im} k\geqslant 0$ and all impedances satisfying (4.79).

Proof. Uniqueness is shown as in the uniqueness proof for equation (4.76) in Theorem 4.44. Existence is obtained by a rather lengthy regularization of the singular integral equation, the details of which can be found in Colton and Kress [2].

4.8 INTEGRAL EQUATIONS BASED ON THE REPRESENTATION THEOREMS

We finally derive integral equations for the exterior problems based on the representation Theorems 4.5 and 4.13. Consider first the exterior Maxwell problem with boundary data in class $C^{0,\alpha}(\partial D)$. Letting x tend to the boundary in the representation Theorem 4.5, making use of Theorems 2.17, 2.26, and the transformation (2.86), we obtain the equations

$$[\nu, E]-\mathbf{M}[\nu, E]-\frac{1}{ik}\mathbf{N}[\nu,[\nu, H]]=0 \qquad (4.81)$$

and

$$[\nu, H]-\mathbf{M}[\nu, H]+\frac{1}{ik}\mathbf{N}[\nu,[\nu, E]]=0. \qquad (4.82)$$

Let E, H be the solution to the exterior Maxwell problem. Then from (4.82) we obtain the integral equation

$$b + \mathbf{M}'b = -\frac{1}{ik}[\nu, \mathbf{N}[\nu, c]] \tag{4.83}$$

of the second kind for the unknown tangential components $b := [\nu, [\nu, H]]$ of the magnetic field H. By our previous analysis the existence of a solution to the exterior Maxwell problem and, therefore, the existence of a solution to the integral equation (4.83) is already established. Since equation (4.83) is the adjoint of the integral equation (4.30) obtained from the magnetic dipole approach, by Theorem 4.23 and the Fredholm alternative, equation (4.83) is uniquely solvable if and only if k is not an interior Maxwell eigenvalue.

From (4.81) we obtain the singular integral equation

$$\mathbf{N}b = ik(c - \mathbf{M}c) \tag{4.84}$$

of the first kind. Existence is again already established, and by Theorem 4.37 the solution is unique if and only if k is not an interior Maxwell eigenvalue.

Because equations (4.83) and (4.84) are derived from the magnetic and electric fields in the representation Theorem 4.5, they are called the magnetic field equations and the electric field equations, respectively. They were first used by Maue [1].

We can add equations (4.83) and (4.84) to obtain the combined magnetic and electric field equation,

$$b + \mathbf{M}'b + i\eta\mathbf{N}b = -\frac{1}{ik}[\nu, \mathbf{N}[\nu, c]] - \eta k(c - \mathbf{M}c) \tag{4.85}$$

of the second kind that is the adjoint of equation (4.76) associated with the combined magnetic and electric dipole approach. Since the operator $\mathbf{N}$ is not compact, uniqueness for (4.85) has to be treated separately. Let $b \in \mathfrak{T}_{\perp}^{0,\alpha}(\partial D)$ be a solution to the homogeneous equation $b + \mathbf{M}'b + i\eta\mathbf{N}b = 0$ and define the electromagnetic field E, H of the electric dipole distribution $[\nu, b]$ as in Theorem 4.36. Then $[\nu, [\nu, H_-]] + \eta k[\nu, E_-] = 0$ on ∂D and proceeding as in the uniqueness proof of Theorem 4.44, we find that $E = H = 0$ in D. By the jump relations, E, H satisfy the homogeneous exterior Maxwell problem in $\mathbb{R}^3 \setminus D$. Hence $E = H = 0$ in $\mathbb{R}^3 \setminus D$ and finally $b = 0$ on ∂D. Therefore, we have the following theorem.

Theorem 4.47. The combined magnetic and electric field equation (4.85) for the exterior Maxwell problem is uniquely solvable for all wave numbers $\operatorname{Im} k \geqslant 0$.

The numerical implementation of equation (4.85) was considered by Mautz and Harrington [1].

Now let E be a solution to the vector Helmholtz equation satisfying the radiation condition (4.15). Under the assumption that $E \in C^2(\mathbb{R}^3 \setminus \bar{D}) \cap$

$C^{0,\alpha}(\mathbb{R}^3 \setminus D)$, $\operatorname{div} E, \operatorname{curl} E \in C^{0,\alpha}(\mathbb{R}^3 \setminus D)$, from the representation Theorem 4.13 and Theorems 2.17 and 2.26 we obtain the equations

$$\begin{pmatrix} [\nu, E] \\ \operatorname{div} E \end{pmatrix} - \mathbf{L}\begin{pmatrix} [\nu, E] \\ \operatorname{div} E \end{pmatrix} + \mathbf{R}\begin{pmatrix} [[\operatorname{curl} E, \nu], \nu] \\ (\nu, E) \end{pmatrix} = 0 \tag{4.86}$$

and

$$-\begin{pmatrix} [[\operatorname{curl} E, \nu], \nu] \\ (\nu, E) \end{pmatrix} + \mathbf{Q}\begin{pmatrix} [\nu, E] \\ \operatorname{div} E \end{pmatrix} - \mathbf{L}'\begin{pmatrix} [[\operatorname{curl} E, \nu], \nu] \\ (\nu, E) \end{pmatrix} = 0. \tag{4.87}$$

If E is the solution to the exterior electric boundary-value problem, then from (4.87) we obtain the integral equation

$$\begin{pmatrix} b \\ \mu \end{pmatrix} + \mathbf{L}'\begin{pmatrix} b \\ \mu \end{pmatrix} = \mathbf{Q}\begin{pmatrix} c \\ \gamma \end{pmatrix} \tag{4.88}$$

of the second kind for the unknown tangential components $b := [[\operatorname{curl} E, \nu], \nu]$ and the normal component $\mu := (\nu, E)$. This equation is adjoint to the equation (4.33) obtained by the layer approach and is uniquely solvable if and only if k is not an interior magnetic eigenvalue. From (4.86) we obtain the singular integral equation

$$\mathbf{R}\begin{pmatrix} b \\ \mu \end{pmatrix} = \mathbf{L}\begin{pmatrix} c \\ \gamma \end{pmatrix} - \begin{pmatrix} c \\ \gamma \end{pmatrix} \tag{4.89}$$

of the first kind which has a unique solution provided k is not an interior electric eigenvalue. A linear combination of the last two equations gives

$$\begin{pmatrix} b \\ \mu \end{pmatrix} + \mathbf{L}'\begin{pmatrix} b \\ \mu \end{pmatrix} + i\eta\mathbf{R}\begin{pmatrix} b \\ \mu \end{pmatrix} = \mathbf{Q}\begin{pmatrix} c \\ \gamma \end{pmatrix} + i\eta\left[\mathbf{L}\begin{pmatrix} c \\ \gamma \end{pmatrix} - \begin{pmatrix} c \\ \gamma \end{pmatrix}\right] \tag{4.90}$$

which is the adjoint of equation (4.67) associated with the combined layer approach. As demonstrated in Kress [6], this equation is uniquely solvable for all wave numbers $\operatorname{Im} k \geqslant 0$.

In the same manner, we can derive equations for the magnetic boundary-value problem that are the adjoints of the corresponding layer equations.

5

LOW FREQUENCY BEHAVIOR OF SOLUTIONS TO BOUNDARY-VALUE PROBLEMS IN SCATTERING THEORY

In this chapter, we shall study the behavior of solutions to boundary-value problems in acoustic and electromagnetic scattering theory as the frequency or wave number tends to zero. We first consider the case of acoustic waves and show that in the limiting case when the wave number is zero the corresponding integral equations can be solved by the method of successive approximations. This leads to an iterative procedure for solving the integral equations of acoustic scattering theory for small values of the wave number. We then consider the corresponding case of electromagnetic waves and show that for simply connected domains a similar regular perturbation argument can be made to establish an iterative procedure for solving the exterior Maxwell problem. However, in the case of multiply connected domains, the solutions of the exterior electromagnetic boundary-value problems are not unique in the potential theoretic limit. Hence, in this case the study of the limiting behavior of solutions, as the wave number tends to zero, leads to the investigation of a problem in singular perturbation theory. We shall study this problem in the final section of this chapter and establish necessary and sufficient conditions for the existence of the low frequency limit for solutions of the exterior electric boundary-value problem, the exterior magnetic boundary-value problem, and the exterior Maxwell problem.

5.1 ITERATIVE METHODS FOR SOLVING THE EXTERIOR DIRICHLET AND NEUMANN PROBLEMS

We begin by observing that the integral equation methods in Chapter 3 remain valid in the potential theoretic case $k=0$ if we replace the Sommerfeld radiation condition (3.7) by the condition

$$u(x)=o(1), \qquad |x|\to\infty, \tag{5.1}$$

with the limit holding uniformly with respect to all directions $x/|x|$. In particular, this assumption suffices to establish the exterior Green's representation Theorem 3.3 in the potential theoretic case (cf. Martensen [1]). From this representation we see that any harmonic function satisfying (5.1) automatically satisfies the stronger property

$$u(x)=O\left(\frac{1}{|x|}\right), \qquad \operatorname{grad} u(x)=O\left(\frac{1}{|x|^2}\right), \qquad |x|\to\infty, \tag{5.2}$$

uniformly for all directions.

In the following analysis, we shall distinguish the fundamental solutions of the Helmholtz equation and the Laplace equation by writing

$$\Phi_k(x,y)=\frac{e^{ik|x-y|}}{4\pi|x-y|}$$

and use subscripts to distinguish the operators in both cases.

To motivate the reader, we recall that the first existence results on the integral equations of potential theory were not based on the Riesz–Fredholm theory but were obtained by Neumann [1] with the aid of successive approximations. Although Neumann's convergence proofs were confined to the case of convex domains, they were later extended to arbitrary regions D with the aid of the following result on the spectrum of the integral operators $\mathbf{K}_0$ and $\mathbf{K}_0'$ (see (2.77) and (2.78)) due to Plemelj [1].

Theorem 5.1. In the potential theoretic case $k=0$ the integral operators $\mathbf{K}_0$ and $\mathbf{K}_0'$ have spectrum

$$\sigma(\mathbf{K}_0)=\sigma(\mathbf{K}_0')\subset[-1,1).$$

In particular, -1 is an eigenvalue with m linearly independent eigenfunctions where m denotes the number of components of D.

Proof. Let λ be an eigenvalue of $\mathbf{K}_0'$ with eigenfunction ϕ, that is, $\lambda\phi-\mathbf{K}\phi=0$. Define the single-layer potential u with density ϕ. Then from Theorems 2.12 and 2.19 we have

$$u_+=u_- \quad \text{on} \quad \partial D \tag{5.3}$$

and

$$\frac{\partial u_\pm}{\partial \nu} = \frac{1}{2}\mathbf{K}_0'\phi \mp \frac{1}{2}\phi = \frac{1}{2}(\lambda \mp 1)\phi \quad \text{on} \quad \partial D.$$

Hence

$$(\lambda + 1)\frac{\partial u_+}{\partial \nu} = (\lambda - 1)\frac{\partial u_-}{\partial \nu} \quad \text{on} \quad \partial D. \tag{5.4}$$

Applying the first Green's theorem (3.4) with the help of (5.2), we now see that

$$(1+\lambda)\int_{\mathbf{R}^3\setminus D} |\operatorname{grad} u|^2\, dx = (1-\lambda)\int_D |\operatorname{grad} u|^2\, dx. \tag{5.5}$$

Now define

$$I(u) := \int_D |\operatorname{grad} u|^2\, dx, \qquad \hat{I}(u) := \int_{\mathbf{R}^3\setminus D} |\operatorname{grad} u|^2\, dx.$$

Assume $I(u) = \hat{I}(u) = 0$. Then $\operatorname{grad} u = 0$ in $\mathbf{R}^3$ and from the jump relation of Corollary 2.20 we have $\phi = 0$, which is a contradiction to the fact that ϕ is an eigenfunction. Hence we can write

$$\lambda = \frac{I(u) - \hat{I}(u)}{I(u) + \hat{I}(u)}$$

which implies that $\lambda \in [-1, 1]$. Now assume $\lambda = 1$. Then from (5.5) we have $\operatorname{grad} u = 0$ in $\mathbf{R}^3\setminus D$ and therefore $u = 0$ in $\mathbf{R}^3\setminus D$ because of (5.1). Then from (5.3) we have $u_- = 0$ on ∂D and from the uniqueness of the interior Dirichlet problem for the Laplace equation we obtain $u = 0$. Therefore by Corollary 2.20 we have $\phi = 0$, that is, $\lambda = 1$ is not an eigenvalue. Hence $\sigma(\mathbf{K}_0') \subset [-1, 1)$ and by the Fredholm alternative the adjoint operator $\mathbf{K}_0$ has the same spectrum as $\mathbf{K}_0'$.

The fact that -1 is an eigenvalue of multiplicity m is a consequence of the following theorem.

Theorem 5.2. $N(\mathbf{I} + \mathbf{K}_0) = \{v|_{\partial D} \mid v \in C^1(D),\ \operatorname{grad} v = 0 \text{ in } D\}$.

Proof. This is the special case $k = 0$ of Theorem 3.17; alternatively, the result follows directly from (2.40).

We now consider the integral equation (3.29) for the exterior Neumann problem. The successive approximations

$$\phi_{n+1} = (1-\beta)\phi_n + \beta\mathbf{K}_0'\phi_n - 2\beta g, \qquad n = 0, 1, 2, \ldots \tag{5.6}$$

with arbitrary $\phi_0 \in C(\partial D)$ and where $\beta \in (0,1)$ converge in the Banach space

$C(\partial D)$ to the unique solution ϕ of the equation $\phi - \mathbf{K}'_0\phi = -2g$. This follows from Theorem 1.36 applied to the operator $\mathbf{A} = (1-\beta)\mathbf{I} + \beta\mathbf{K}'_0$ which by Theorem 5.1 has spectral radius less than one if $\beta \in (0,1)$. The optimal choice of the relaxation parameter β yielding the minimal spectral radius of $\mathbf{A}$ depends on the eigenvalues of $\mathbf{K}'_0$. If all the eigenvalues are negative (which is the case if ∂D is a sphere), then $\beta = \frac{2}{3}$ is optimal (Kleinman [1]).

Using the power series for the exponential function, we easily see that

$$\|\mathbf{K}_k - \mathbf{K}_0\| = O(k^2), \qquad \|\mathbf{K}'_k - \mathbf{K}'_0\| = O(k^2) \tag{5.7}$$

in both $C(\partial D)$ and $C^{0,\alpha}(\partial D)$. Hence, using Theorem 1.37, we have the following theorem due to Ahner and Kleinman [1] and Kleinman and Wendland [1].

Theorem 5.3. For any fixed $\beta \in (0,1)$ the successive approximations

$$\phi_{n+1} = (1-\beta)\phi_n + \beta\mathbf{K}'_k\phi_n - 2\beta g, \qquad n = 0,1,2,\ldots \tag{5.8}$$

with arbitrary $\phi_0 \in C(\partial D)$ converge in the Banach space $C(\partial D)$ to the unique solution ϕ of the equation $\phi - \mathbf{K}'_k\phi = -2g$ of the exterior Neumann problem for the Helmholtz equation provided k is sufficiently small.

Kleinman and Wendland also give estimates on the range of wave numbers k where (5.8) converges in the case of convex domains D and describe the discretization of the integral equation. In particular, they show that the discrete linear system can also be solved by successive approximations and that the rate of convergence does not depend on the discretization if the discretization error is small enough.

Similarly to (5.8), the adjoint equation (3.82) based on the representation Theorem 3.3 and the integral equation (3.25) for the interior Dirichlet problem can be solved by successive approximation for small values of k.

For the exterior Dirichlet problem, the integral equations (3.26) and (3.81) are not uniquely solvable since, for $k = 0$, -1 is an eigenvalue of $\mathbf{K}_0$ and $\mathbf{K}'_0$, that is, $k = 0$ is an interior Neumann eigenvalue. Therefore we cannot expect straightforward convergence of successive approximations for these equations. By eliminating the eigenvalue -1 through the use of the eigenfunctions for this eigenvalue, Ahner [1] developed modifications of the integral equations (3.26) and (3.81) that can be solved by successive approximations for sufficiently small values of k. A further method that avoids dealing with the eigenfunctions of $\mathbf{K}_0$ and $\mathbf{K}'_0$ was suggested by Colton and Kress [1]. In this approach, the fundamental solution Φ_k is replaced by a Green's function G_k defined in the exterior of a ball B of radius R with $\bar{B} \subset D$. For these modified integral operators, denoted by $\mathbf{K}_{k,R}$ and $\mathbf{K}'_{k,R}$, we have in the potential theoretic case that $\sigma(\mathbf{K}_{0,R}) = \sigma(\mathbf{K}'_{0,R}) \subset (-1,1)$. Then, as in the case of the exterior Neumann problem, the successive approximations will converge for sufficiently small values of k. The eigenvalues of $\mathbf{K}_{0,R}$ and $\mathbf{K}'_{0,R}$, and therefore the convergence rate of the iteration scheme, depend on the radius R of the ball B. By using an appropriate scalar product, it can be shown that the operator $\mathbf{K}_{0,R}$ can be symmetrized and hence its eigenvalues can be characterized by a variational principle (Kress and Roach [1]). From this it can be seen that the eigenvalues increase as R increases, that is, by using relaxation methods we can make the spectral radius of the operator smaller by choosing R larger. In particular, when D is the unit ball and B is a concentric ball with radius $R<1$, we have the eigenvalues

$$\sigma(\mathbf{K}_{0,R}) = \sigma(\mathbf{K}'_{0,R}) = \left\{ -\frac{1}{2m+1}(1-(2m+2)R^{2m+1}) | m=0,1,2,\ldots \right\} \cup \{0\}.$$

Iteration methods for the impedance boundary-value problem have been described by Ahner [2], and for the transmission problem by Kittappa and Kleinman [1].

5.2 ITERATIVE METHODS FOR ELECTROMAGNETIC PROBLEMS

In the potential theoretic case $k=0$, the time-harmonic Maxwell equations separate into the system

$$\operatorname{div} E = 0, \qquad \operatorname{curl} E = 0 \tag{5.9}$$

for the electric field E and the same system for the magnetic field H. Solutions to the system (5.9) are called harmonic vector fields. Again, our investigations on the integral equation methods for the Maxwell boundary-value problem remain valid after replacing the Silver–Müller radiation condition (4.7) and (4.8) by

$$E(x) = o(1), \qquad |x| \to \infty \tag{5.10}$$

where the limit holds uniformly for all directions $x/|x|$. The representation Theorems 4.1 and 4.5 take the form

$$-\operatorname{grad} \int_{\partial D} (\nu(y), E(y)) \Phi_0(x,y)\, ds(y)$$

$$+\operatorname{curl} \int_{\partial D} [\nu(y), E(y)] \Phi_0(x,y)\, ds(y) = \begin{cases} -E(x), & x \in D, \\ 0, & x \in \mathbb{R}^3 \setminus \bar{D}, \end{cases} \tag{5.11}$$

$$\operatorname{div} \int_{\partial D} [\nu(y), E(y)] \Phi_0(x,y)\, ds(y) = 0 \quad x \in \mathbb{R}^3,$$

for harmonic vector fields $E \in C^1(D) \cap C(\bar{D})$ and

$$-\operatorname{grad}\int_{\partial D}(\nu(y), E(y))\Phi_0(x, y)\, ds(y)$$

$$+\operatorname{curl}\int_{\partial D}[\nu(y), E(y)]\,\Phi_0(x, y)\, ds(y) = \begin{cases} 0, & x \in D \\ E(x), & x \in \mathbb{R}^3 \setminus \bar{D} \end{cases} \tag{5.12}$$

$$\operatorname{div}\int_{\partial D}[\nu(y), E(y)]\,\Phi_0(x, y)\, ds(y) = 0, \qquad x \in \mathbb{R}^3$$

for harmonic vector fields $E \in C^1(\mathbb{R}^3 \setminus \bar{D}) \cap C(\mathbb{R}^3 \setminus D)$ (Martensen [1]). From (5.12), we observe that harmonic fields satisfying condition (5.10) automatically satisfy

$$E(x) = O\left(\frac{1}{|x|^2}\right), \qquad |x| \to \infty, \tag{5.13}$$

uniformly for all directions. The fact that (5.11) and (5.12) are the limiting forms of Theorems 4.1 and 4.5 follows from the transformation (2.86) using (2.75) and eliminating the magnetic field H by Maxwell's equations.

In the case of the vector Laplace equation $\Delta E = 0$, we have to replace the radiation condition (4.15) by

$$E(x) = o(1), \qquad |x| \to \infty, \tag{5.14}$$

with the limit holding uniformly with respect to all directions $x/|x|$. Then the limiting form $k = 0$ of the representation Theorems 4.11 and 4.13 is valid (Kress [3]). In particular, from this fact we see that

$$E(x) = O\left(\frac{1}{|x|}\right), \qquad \operatorname{div} E(x), \operatorname{curl} E(x) = O\left(\frac{1}{|x|^2}\right), \qquad |x| \to \infty \tag{5.15}$$

uniformly for all directions.

We again use subscripts to distinguish the operators in the potential theoretic case. The spectrum of the potential theoretic operators $\mathbf{M}_0$ and $\mathbf{M}'_0$ (see (2.82) and (2.83)) has been studied by Müller and Niemeyer [1], Werner [2], Kress [1], [3], and Gray [1]. Analogous to Theorem 5.1, we have the following result.

Theorem 5.4. In the potential theoretic case $k = 0$, the integral operators $\mathbf{M}_0$ and $\mathbf{M}'_0$ have a spectrum

$$\sigma(\mathbf{M}_0) = \sigma(\mathbf{M}'_0) \subset [-1, 1].$$

If all the components of D are simply connected, then

$$\sigma(\mathbf{M}_0) = \sigma(\mathbf{M}'_0) \subset (-1,1);$$

otherwise 1 and -1 are eigenvalues with p linearly independent eigenfunctions where p denotes the topological genus of D.

Proof. From Theorem 2.32 we see that the spectrum of $\mathbf{M}_0$ is the same in $C(\partial D)$ and $C^{0,\alpha}(\partial D)$. Let λ be an eigenvalue of $\mathbf{M}_0$ with eigenfunction $a \in C^{0,\alpha}(\partial D)$, that is, $\lambda a - \mathbf{M}_0 a = 0$. Define a vector potential A by

$$A(x) := \int_{\partial D} \Phi_0(x,y) a(y)\, ds(y), \qquad x \in \mathbb{R}^3 \setminus \partial D. \tag{5.16}$$

Then from Theorems 2.12 and 2.24 we have

$$A_+ = A_-, \qquad \operatorname{div} A_+ = \operatorname{div} A_- \quad \text{on} \quad \partial D \tag{5.17}$$

and

$$[\nu, \operatorname{curl} A_\pm] = \tfrac{1}{2}\mathbf{M}_0 a \pm \tfrac{1}{2} a = \tfrac{1}{2}(\lambda \pm 1) a \quad \text{on} \quad \partial D.$$

Hence

$$(\lambda - 1)[\nu, \operatorname{curl} A_+] = (\lambda + 1)[\nu, \operatorname{curl} A_-] \text{ on } \partial D. \tag{5.18}$$

Now assume $\operatorname{div} A \neq 0$ on ∂D and apply the potential theoretic form of the representation Theorems 4.11 and 4.13 to the vector potential A. Multiply the formula of Theorem 4.13 by $(\lambda - 1)$ and of Theorem 4.11 by $(\lambda + 1)$, subtract, and take the divergence to get, with the help of (5.18), that

$$(\lambda - 1)\operatorname{div} A(x) = -2\int_{\partial D} \frac{\partial \Phi_0(x,y)}{\partial \nu(y)} \operatorname{div} A(y)\, ds(y), \qquad x \in \mathbb{R}^3 \setminus \bar{D}.$$

Letting x tend to the boundary and using Theorem 2.13 shows that

$$\lambda \operatorname{div} A + \mathbf{K}_0 \operatorname{div} A = 0 \quad \text{on} \quad \partial D, \tag{5.19}$$

that is, $-\lambda$ is an eigenvalue of $\mathbf{K}_0$ and therefore $\lambda \in (-1,1]$ by Theorem 5.1.

If $\operatorname{div} A = 0$ on ∂D, we can use the first vector Green's theorem to obtain from (5.17) and (5.18) that

$$(1-\lambda)\int_{\mathbb{R}^3 \setminus D} |\operatorname{curl} A|^2\, dx = (1+\lambda)\int_D |\operatorname{curl} A|^2\, dx$$

and from this we deduce that $\lambda \in [-1,1]$ as in Theorem 5.1.

Since $\mathbf{M}_0 a = \lambda a$ we have $\mathbf{M}'_0[a, \nu] = -\lambda[a, \nu]$ and hence by the Fredholm alternative the spectrum of $\mathbf{M}_0$ and $\mathbf{M}'_0$ is symmetric with respect to the origin. The statement on the eigenvalues -1 and 1 now follows from the following theorem.

Theorem 5.5. $N(\mathbf{I}+\mathbf{M}_0) = \{[\nu, E]|_{\partial D} | E \in C^1(D \cap C(\bar{D}),$

$$\operatorname{div} E = 0, \quad \operatorname{curl} E = 0 \text{ in } D, \quad (\nu, E) = 0 \text{ on } \partial D\}.$$

Proof. Let a be a solution of $a + \mathbf{M}_0 a = 0$ and define the vector potential A by (5.16) and set $E := \operatorname{curl} A$. By (5.18) we have $[\nu, E_+] = 0$ on ∂D and by (5.19) we see that $\operatorname{div} A|_{\partial D} \in N(\mathbf{I} - \mathbf{K}_0)$. Hence by Theorem 5.1 we have $\operatorname{div} A = 0$ on ∂D and from the uniqueness property for the Dirichlet problem for harmonic functions, it follows that $\operatorname{div} A = 0$ in $\mathbb{R}^3$. We now have $\operatorname{curl} E = -\Delta A + \operatorname{grad} \operatorname{div} A = 0$ and $\operatorname{div} E = \operatorname{div} \operatorname{curl} A = 0$ in $\mathbb{R}^3 \setminus \partial D$, that is, E is a harmonic field in D and $\mathbb{R}^3 \setminus \bar{D}$.

Using the representation formula (5.12) and $[\nu, E_+] = 0$ on ∂D, we observe that we can write $E = \operatorname{grad} u$ in $\mathbb{R}^3 \setminus D$ for some harmonic function $u \in C^2(\mathbb{R}^3 \setminus \bar{D}) \cap C^1(\mathbb{R}^3 \setminus D)$. From the boundary condition $[\nu, \operatorname{grad} u] = 0$ on ∂D, we see that $u = \text{const}$ on ∂D where the constant might be different on the components of ∂D. Then using the first Green's theorem and Stokes' theorem we have

$$\int_{\mathbb{R}^3 \setminus D} |E|^2 \, dx = -\int_{\partial D} \bar{u} \frac{\partial u}{\partial \nu} \, ds = -\int_{\partial D} \bar{u}(\nu, \operatorname{curl} A) \, ds = 0$$

from which we can conclude that $E = 0$ in $\mathbb{R}^3 \setminus D$. From the jump relations of Corollary 2.25, we now see that $(\nu, E_-) = 0$ and $a = -[\nu, E_-]$ on ∂D.

Conversely, let E be a harmonic vector field in D with vanishing normal components on the boundary ∂D and define $a := [\nu, E]$ on ∂D. Then from (5.11) we have

$$\operatorname{curl} \int_{\partial D} \Phi_0(x, y) a(y) \, ds(y) = 0, \qquad x \in \mathbb{R}^3 \setminus \bar{D}.$$

Letting x tend to the boundary and using Theorem 2.26 we see that $a + \mathbf{M}_0 a = 0$, that is, $a \in N(\mathbf{I} + \mathbf{M}_0)$.

Harmonic vector fields with vanishing normal components on the boundary are called Neumann vector fields (Martensen [1]). If all the components of D are simply connected, there exists only the trivial Neumann vector field $E = 0$. In this case, we can represent any curl free vector field in the form $E = \operatorname{grad} u$ with a harmonic function $u \in C^2(D) \cap C^1(\bar{D})$. Then from $(\nu, E) = 0$ on ∂D we have the homogeneous Neumann condition $\partial u / \partial \nu = 0$ on ∂D. Hence $\operatorname{grad} u = 0$ in D.

If D has multiply connected components, then, as shown by Martensen [1] and Werner [2], there exist p linearly independent Neumann vector fields in D.

We do not want to repeat this proof here, but instead we give an explicit example of the field generated by a static current on an infinite straight line. In cylindrical coordinates (ρ, θ, z) this field is given by

$$E(x) = \frac{e_\theta}{\rho}$$

where e_θ denotes the unit vector in the azimuthal direction. This vector field is a Neumann vector field in any body of revolution that does not contain points lying on the z-axis, that is, in any torus. In view of the representation formula (5.11), we observe that nontrivial Neumann fields cannot have identically vanishing tangential components. Hence the tangential components of linearly independent Neumann fields are linearly independent on the boundary. Therefore we have that $\dim N(\mathbf{I}+\mathbf{M}_0) = p$.

In the case $p = 0$ we have from Theorem 5.4 that the successive approximations

$$a_{n+1} = -\mathbf{M}_0 a_n + 2c, \qquad n = 0, 1, 2, \ldots \tag{5.20}$$

with arbitrary a_0 converge in the Banach space $C^{0,\alpha}(\partial D)$ to the unique solution a of the equation $a + \mathbf{M}_0 a = 2c$. If we now observe that

$$\|\mathbf{M}_k - \mathbf{M}_0\| = O(k^2) \tag{5.21}$$

both in $C(\partial D)$ and $C^{0,\alpha}(\partial D)$, then from Theorem 1.37 we arrive at the following convergence result for the exterior Maxwell problem (Gray [1]).

Theorem 5.6. If $\mathbb{R}^3 \setminus D$ is simply connected, then the successive approximations

$$a_{n+1} = -\mathbf{M}_k a_n + 2c, \qquad n = 0, 1, 2, \ldots \tag{5.22}$$

with arbitrary $a_0 \in C^{0,\alpha}(\partial D)$ converge in the Banach space $C^{0,\alpha}(\partial D)$ to the unique solution a of the equation $a + \mathbf{M}_k a = 2c$ of the exterior Maxwell problem (4.19) provided k is sufficiently small.

In the potential theoretic case $k = 0$, extensions of this iterative method have been given by Kress [2] using deflation methods. It is also possible to treat the exterior Maxwell problem in a manner similar to Ahner's [1] approach for the exterior Dirichlet problem, but with the disadvantage that one needs to know the eigenelements in the potential theoretic case, that is, the Neumann vector fields.

5.3 LOW WAVE NUMBER BEHAVIOR OF SOLUTIONS TO THE EXTERIOR ELECTROMAGNETIC BOUNDARY-VALUE PROBLEMS

For the exterior Dirichlet and Neumann problems, uniqueness also remains valid in the potential theoretic case $k = 0$. Hence, by using the uniquely solvable integral equations of Section 3.6, it is easily seen that the solutions to the exterior acoustic boundary-value problems depend continuously on the wave number k as $k \to 0$.

In contrast to the case of acoustic waves, the solutions of the exterior electromagnetic boundary-value problems are not in general unique in the potential theoretic limit. Hence, in the case of electromagnetic waves, the study of the limiting behavior of solutions as the wave number tends to zero leads to the investigation of a singular perturbation problem of the type described in Section 1.4.

We first show that the operators $\mathbf{K}_0$ and $\mathbf{K}'_0$ satisfy the assumptions of Theorem 1.32, and start with the analog of Theorem 3.18 for the case $k = 0$.

Theorem 5.7. Let $\phi_1, \ldots, \phi_m$ be a real basis for $N(\mathbf{I} + \mathbf{K}'_0)$ and define

$$u_j(x) := \int_{\partial D} \Phi_0(x, y)\phi_j(y)\, ds(y), \qquad x \in \mathbb{R}^3 \setminus \partial D, \tag{5.23}$$

$j = 1, \ldots, m$. Then

$$\phi_j = -\frac{\partial u_{j+}}{\partial \nu} \quad \text{on} \quad \partial D, \tag{5.24}$$

$j = 1, \ldots, m$, and the functions

$$\psi_j := -u_{j+} \quad \text{on} \quad \partial D \tag{5.25}$$

$j = 1, \ldots, m$, form a basis for $N(\mathbf{I} + \mathbf{K}_0)$. The matrix

$$\langle \psi_j, \phi_l \rangle = \int_{\partial D} u_{j+} \frac{\partial u_{l+}}{\partial \nu}\, ds, \qquad j, l = 1, \ldots, m,$$

is regular and hence by Theorem 1.31 the Riesz number is one.

Proof. The proof is the same as for Theorem 3.18 except that instead of the uniqueness Theorem 3.12 we apply the first Green's Theorem 3.4 and the asymptotic behavior (5.2) to conclude that if u is harmonic in $\mathbb{R}^3 \setminus D$ and

$$\int_{\partial D} u_+ \frac{\partial u_+}{\partial \nu}\, ds = 0$$

then $u = 0$ in $\mathbb{R}^3 \setminus D$.

A closer examination of the proof of Theorem 5.7 shows that $u_j = \text{const}$ in D, where the constant might be different for different components of D. Hence we have $u_{j+} = \text{const}$ on ∂D and therefore the vector fields

$$Y_j := -\operatorname{grad} u_j \tag{5.26}$$

$j=1,\dots,m$, are harmonic in $\mathbb{R}^3\setminus D$ and satisfy

$$[\nu, Y_j]=0 \quad \text{on} \quad \partial D. \tag{5.27}$$

Harmonic vector fields with vanishing tangential components on the boundary and satisfying condition (5.13) at infinity are called Dirichlet vector fields (Martensen [1]). The $Y_j, j=1,\dots,m$, defined by (5.26) form a basis of the linear space of Dirichlet fields in $\mathbb{R}^3\setminus D$ since

$$N(\mathbf{I}+\mathbf{K}_0')=\{(\nu, Y)|_{\partial D} \mid Y \text{ is a Dirichlet field in } \mathbb{R}^3\setminus D\}.$$

In view of (5.24) and (5.26) we only have to show that for any Dirichlet field Y in $\mathbb{R}^3\setminus D$, the normal component $\phi:=(\nu, Y)$ on ∂D satisfies $\phi+\mathbf{K}_0'\phi=0$. But this follows from the representation formula (5.12), which yields

$$\operatorname{grad}\int_{\partial D}\Phi_0(x,y)\phi(y)\,ds(y)=0, \qquad x\in D,$$

and then using Theorem 2.19 to take the normal component on the boundary of this expression.

From the power series for the exponential function, we see that $\mathbf{K}_k$ and $\mathbf{K}_k'$ satisfy the first condition of Theorem 1.32 with $s=2$. We use the bases $\phi_1,\dots,\phi_m$ for $N(\mathbf{I}+\mathbf{K}_0')$ and $\psi_1,\dots,\psi_m$ for $N(\mathbf{I}+\mathbf{K}_0)$ introduced in Theorem 5.7 and define

$$u_{j,k}(x):=\int_{\partial D}\Phi_k(x,y)\phi_j(y)\,ds(y), \qquad x\in\mathbb{R}^3\setminus\partial D,$$

$j=1,\dots,m$. Then, using $\partial u_{j-}/\partial\nu=0$ on ∂D, we can apply the second Green's theorem (3.5) to obtain

$$\begin{aligned}
\langle(\mathbf{I}+\mathbf{K}_k)\psi_j,\phi_l\rangle &= \langle\psi_j,(\mathbf{I}+\mathbf{K}_k')\phi_l\rangle\\
&= -2\int_{\partial D} u_{j-}\frac{\partial u_{l,k-}}{\partial\nu}\,ds\\
&= 2\int_{\partial D}\left\{u_{l,k-}\frac{\partial u_{j-}}{\partial\nu}-u_{j-}\frac{\partial u_{l,k-}}{\partial\nu}\right\}ds\\
&= 2\int_D\{u_{l,k}\Delta u_j-u_j\Delta u_{l,k}\}\,dx\\
&= 2k^2\int_D u_j u_{l,k}\,dx\\
&= 2k^2\int_D u_j u_l\,dx+O(k^3).
\end{aligned}$$

Since the matrix $\int_D u_j\, u_l\, dx$, $j, l = 1,\dots,m$, is positive definite, the third condition of Theorem 1.32 in the form of (1.12) is satisfied.

We now consider the operators $\mathbf{M}_0$ and $\mathbf{M}_0'$ and prove the analog of Theorem 4.24 for the case $k=0$.

Theorem 5.8. Let $b_1,\dots,b_p$ be a (real) basis of $N(\mathbf{I}+\mathbf{M}_0')$ and define

$$A_j(x) := \int_{\partial D} \Phi_0(x,y)\big[b_j(y), \nu(y)\big]\, ds(y), \qquad x \in \mathbb{R}^3 \setminus \partial D, \quad (5.28)$$

$j=1,\dots,p$. Let w_j, $j=1,\dots,p$, be solutions of the Neumann problem satisfying

$$\Delta w_j = 0 \quad \text{in} \quad D \qquad (5.29)$$

and

$$\frac{\partial w_j}{\partial \nu} = (\nu, A_j) \quad \text{on} \quad \partial D. \qquad (5.30)$$

The w_j are uniquely determined up to an additive constant. Then

$$b_j = \big[\nu, [\nu, \operatorname{curl} A_{j+}]\big] \quad \text{on} \quad \partial D, \qquad (5.31)$$

$j=1,\dots,p$, and the tangential fields

$$a_j := [\nu, A_{j+}] - [\nu, \operatorname{grad} w_j] \quad \text{on} \quad \partial D, \qquad (5.32)$$

$j=1,\dots,p$, form a basis of $N(\mathbf{I}+\mathbf{M}_0)$. The matrix

$$\langle a_j, b_l \rangle = \int_{\partial D} (\nu, \operatorname{curl} A_{l+}, A_{j+})\, ds, \qquad j,l=1,\dots,p, \qquad (5.33)$$

is regular and hence by Theorem 1.31 the Riesz number is one.

Proof. Since $b_j + \mathbf{M}_0' b_j = 0$ is equivalent to $-[\nu, b_j] + \mathbf{M}_0[\nu, b_j] = 0$, if we set $\lambda = 1$ in (5.19) we see that $\operatorname{div} A_j|_{\partial D} \in N(\mathbf{I}+\mathbf{K}_0)$. Hence by Theorem 5.2 we have $\operatorname{div} A_j = \text{const}$ on ∂D where the constant might be different on the components of ∂D. Note that $\operatorname{div} A_j \in C^1(\mathbb{R}^3 \setminus D)$ by Theorem 3.27. Using Stokes' theorem and the first Green's theorem, we see that

$$\begin{aligned}\int_{\mathbb{R}^3 \setminus D} |\operatorname{grad} \operatorname{div} A_j|^2\, dx &= -\int_{\partial D} \operatorname{div} A_j \frac{\partial}{\partial \nu} \operatorname{div} A_j\, ds \\ &= -\int_{\partial D} \operatorname{div} A_j (\nu, \operatorname{curl}\operatorname{curl} A_j)\, ds \\ &= 0.\end{aligned}$$

From this we conclude that $\operatorname{div} A_j = 0$ in $\mathbb{R}^3 \setminus D$ since $\operatorname{div} A_j(x) = O(1/|x|^2)$, $|x| \to \infty$. From the uniqueness of solutions to the interior Dirichlet problem for harmonic functions, we now see that

$$\operatorname{div} A_j = 0 \quad \text{in} \quad \mathbb{R}^3. \tag{5.34}$$

In particular, the Neumann problems for the w_j are solvable since by Gauss' theorem we have

$$\int_{\partial D} (\nu, A_j)\, ds = \int_D \operatorname{div} A_j \, dx = 0.$$

Using (5.34) we now have $\operatorname{curl}\operatorname{curl} A_j = -\Delta A_j + \operatorname{grad}\operatorname{div} A_j = 0$ and $\operatorname{div}\operatorname{curl} A_j = 0$ in $\mathbb{R}^3 \setminus \partial D$, that is, $\operatorname{curl} A_j$ is a harmonic field in $\mathbb{R}^3 \setminus \partial D$. Since $b_j + \mathbf{M}_0' b_j = 0$, we have the boundary condition $[\nu, \operatorname{curl} A_{j-}] = 0$ on ∂D and using this we can conclude as in Theorem 5.5 that

$$\operatorname{curl} A_j = 0 \quad \text{in} \quad D. \tag{5.35}$$

Hence, from the jump relation of Corollary 2.27 we see that (5.31) is true.

By construction, the fields

$$E_j := A_j - \operatorname{grad} w_j \quad \text{in} \quad D, \tag{5.36}$$

$j = 1, \ldots, p$, are Neumann vector fields. Hence by Theorem 5.5 we have $[\nu, E_j]|_{\partial D} \in N(\mathbf{I} + \mathbf{M}_0)$. Since $\operatorname{curl} A_j$ is harmonic, by (2.75) we have $\operatorname{Div}[\nu, \operatorname{curl} A_{j+}] = 0$ and therefore from Gauss' theorem (2.73) we see that

$$\int_{\partial D} (\nu, \operatorname{grad} w_j, b_l)\, ds = 0,$$

which establishes (5.33). Now, let α_j, $j = 1, \ldots, p$, be a solution of

$$\sum_{j=1}^{p} \alpha_j \langle a_j, b_l \rangle = 0, \qquad l = 1, \ldots, p$$

and define

$$A := \sum_{j=1}^{p} \alpha_j A_j.$$

Then

$$\int_{\partial D} (\nu, \operatorname{curl} A_+, A_+)\, ds = 0$$

and from the first vector Green's theorem (4.11) and (5.34) we conclude that $\operatorname{curl} A = 0$ in $\mathbb{R}^3 \setminus D$. In particular, $[\nu, \operatorname{curl} A_+] = 0$ on ∂D and therefore $\sum_{j=1}^{p} \alpha_j b_j = 0$. Hence $\alpha_j = 0$, $j = 1, \dots, p$, and the proof is concluded as in Theorem 3.18.

From (5.35) and the jump relations of Corollary 2.25, we see that the harmonic vector fields

$$Z_j := \operatorname{curl} A_j \quad \text{in} \quad \mathbb{R}^3 \setminus D, \tag{5.37}$$

$j = 1, \dots, m$, satisfy

$$(\nu, Z_j) = 0 \quad \text{on} \quad \partial D \tag{5.38}$$

and are therefore Neumann fields in $\mathbb{R}^3 \setminus D$. Actually, they form a basis for the linear space of Neumann fields in $\mathbb{R}^3 \setminus D$ satisfying condition (5.13) at infinity as is seen from the property

$$N(\mathbf{I} + \mathbf{M}'_0) = \{ Z|_{\partial D} \mid Z \text{ is a Neumann field in } \mathbb{R}^3 \setminus D \}.$$

In view of (5.31) and (5.37), we can establish this identity by showing that for any Neumann field Z in $\mathbb{R}^3 \setminus D$ the tangential component $b := Z$ on ∂D satisfies $b + \mathbf{M}'_0 b = 0$. But this follows from the representation formula (5.12) which yields

$$\operatorname{curl} \int_{\partial D} \Phi_0(x, y) [b(y), \nu(y)] \, ds(y) = 0, \qquad x \in D,$$

by taking the tangential component with the aid of Theorem 2.26.

The operators $\mathbf{M}_k$ and $\mathbf{M}'_k$ again satisfy the first of the conditions of Theorem 1.32 with $s = 2$. We now consider the bases $a_1, \dots, a_p$ of $N(\mathbf{I} + \mathbf{M}_0)$ and $b_1, \dots, b_p$ of $N(\mathbf{I} + \mathbf{M}'_0)$ introduced in Theorem 5.8 and define

$$A_{j,k}(x) := \int_{\partial D} \Phi_k(x, y) [b_j(y), \nu(y)] \, ds(y), \qquad x \in \mathbb{R}^3 \setminus \partial D,$$

$j = 1, \dots, p$. Using $(\nu, E_j) = 0$ on ∂D and $(\mathbf{I} + \mathbf{M}'_k) b_l = -2[\nu, [\nu, \operatorname{curl} A_{l,k-}]]$ on ∂D, we can apply the second vector Green's theorem (4.12) to obtain

$$\begin{aligned}
\langle (\mathbf{I} + \mathbf{M}_k) a_j, b_l \rangle &= \langle a_j, (\mathbf{I} + \mathbf{M}'_k) b_l \rangle \\
&= 2 \int_{\partial D} (\nu, E_{j-}, \operatorname{curl} A_{l,k-}) \, ds \\
&= -2k^2 \int_D (E_j, A_{l,k}) \, dx \\
&= -2k^2 \int_D (E_j, A_l) \, dx + O(k^3).
\end{aligned}$$

Since

$$\int_D (E_j, \operatorname{grad} w_l)\,dx = \int_{\partial D} w_l(\nu, E_j)\,ds = 0$$

the matrix $\int_D (E_j, A_l)\,dx = \int_D (E_j, E_l)\,dx$, $j, l = 1,\ldots,p$, is positive definite and therefore the third condition of Theorem 1.32 in the form of (1.12) is satisfied.

We now are in the position to state the main result of this section.

Theorem 5.9. The existence of constants $e_1,\ldots,e_m$ and functions c_0, γ_0 such that the limits

$$\begin{aligned} \|c - c_0\|_{\alpha, \partial D} &\to 0, & k &\to 0, \\ \|\operatorname{Div} c - \operatorname{Div} c_0\|_{\alpha, \partial D} &\to 0, & k &\to 0, \\ \|\gamma - \gamma_0\|_{\alpha, \partial D} &\to 0, & k &\to 0, \end{aligned} \tag{5.39}$$

and

$$\frac{1}{k^2}\int_{\partial D} \gamma(\nu, Y_j)\,ds \to e_j, \qquad k \to 0, \tag{5.40}$$

$j = 1,\ldots,m$, exist are necessary and sufficient for the convergence

$$\begin{aligned} \|E - E_0\|_{\alpha, \mathbb{R}^3\setminus D} &\to 0, & k &\to 0, \\ \|\operatorname{div} E - \operatorname{div} E_0\|_{\alpha, \mathbb{R}^3\setminus D} &\to 0, & k &\to 0, \\ \|\operatorname{curl} E - \operatorname{curl} E_0\|_{\alpha, \mathbb{R}^3\setminus D} &\to 0, & k &\to 0, \end{aligned} \tag{5.41}$$

of the solution E of the exterior electric boundary-value problem (4.23). The limiting field E_0 solves the exterior electric boundary-value problem for the vector Laplace equation

$$\Delta E_0 = 0 \quad \text{in} \quad \mathbb{R}^3\setminus D \tag{5.42}$$

satisfying the boundary condition

$$[\nu, E_0] = c_0, \qquad \operatorname{div} E_0 = \gamma_0 \quad \text{on} \quad \partial D \tag{5.43}$$

and at infinity

$$E_0(x) = o(1), \qquad |x| \to \infty, \tag{5.44}$$

uniformly for all directions $x/|x|$. It is uniquely determined by the additional properties

$$\int_{\mathbf{R}^3\setminus D}(E_0, Y_j)\,dx = e_j, \tag{5.45}$$

$j = 1,\dots,m$.

Proof. ***Necessity.*** The necessity of the conditions (5.39) follows from (5.41) with the help of (2.75). The necessity of (5.40) and the formula (5.45) follow by the second vector Green's theorem from

$$\int_{\partial D}\gamma(\nu, Y_j)\,ds = \int_{\partial D}\operatorname{div}E(\nu, Y_j)\,ds$$

$$= -\int_{\mathbf{R}^3\setminus D}(\Delta E, Y_j)\,dx = k^2\int_{\mathbf{R}^3\setminus D}(E, Y_j)\,dx.$$

Sufficiency. We use the integral equation (4.88) obtained from the representation Theorem 4.13. Using the definitions (4.43) and (4.62) of the integral operators $\mathbf{L}'$ and $\mathbf{Q}$, the first equation of the system (4.88) reads

$$b + \mathbf{M}'b = 2[\nu,[\nu,\operatorname{curl} F_+]] \quad \text{on} \quad \partial D \tag{5.46}$$

for $b := [\nu,[\nu,\operatorname{curl} E]]$ where we have set

$$F(x) := \operatorname{curl}\int_{\partial D}\Phi_k(x, y)c(y)\,ds(y) - \int_{\partial D}\Phi_k(x, y)\gamma(y)\nu(y)\,ds(y),$$

$$x \in \mathbf{R}^3\setminus\partial D. \tag{5.47}$$

Now, for the right-hand side of (5.46) by using Theorem 5.5, the first vector Green's theorem, and the jump relation of Corollary 2.25, we find that for any solution a to the homogeneous adjoint equation that

$$\langle[\nu,[\nu,\operatorname{curl} F], a\rangle = \int_{\partial D}(\nu, E, \operatorname{curl} F_+)\,ds$$

$$= \int_{\partial D}\{(\nu, E, \operatorname{curl} F_-) + (\nu, E)\operatorname{div} F_-\}\,ds$$

$$= \int_D(E, \Delta F)\,dx = -k^2\int_D(E, F)\,dx$$

$$= -k^2\int_D(E, F_0)\,dx + o(k^2)$$

where E denotes a Neumann field in D and F_0 is the limit of (5.47) as $k \to 0$. Note that c and γ also depend on k. Then by Theorem 1.32, we can conclude that

$$\|b - b_0\|_{\alpha, \partial D} \to 0, \qquad k \to 0, \tag{5.48}$$

for the solution of (5.46).

The second equation of the system (4.88) is

$$\mu + \mathbf{K}'\mu = 2(\nu, F_+) - (\nu, \mathbf{S}[\nu, b]) \tag{5.49}$$

for $\mu := (\nu, E)$. Using Stokes' and Gauss' theorems for any solution ψ to the homogeneous adjoint equation, we find for the right-hand side

$$f := 2(\nu, F_+) - (\nu, \mathbf{S}[\nu, b])$$

of (5.49) that

$$\langle f, \psi \rangle = -2\int_D v \operatorname{div} A \, dx$$

where by Theorem 5.2 v denotes a function that is constant on the components of D and

$$A(x) := \int_{\partial D} \Phi_k(x, y)\{\gamma(y)\nu(y) + [\nu(y), b(y)]\}\, ds(y), \qquad x \in \mathbb{R}^3 \setminus \partial D. \tag{5.50}$$

From the integral equation (5.46), the transformation (2.86), and Theorem 2.17 we see that

$$[\nu(x), \operatorname{curl} A_-(x)] = \left[\nu(x), k^2 \int_{\partial D} \Phi_k(x, y) c(y)\, ds(y) + \operatorname{grad} \int_{\partial D} \Phi_k(x, y) \operatorname{Div} c(y)\, ds(y)\right], \qquad x \in \partial D. \tag{5.51}$$

Now denote the components of D by $D_1, \ldots, D_m$ and noting that the function v is a linear combination of the characteristic functions for the D_j, determine a Dirichlet field Y in $\mathbb{R}^3 \setminus D$ such that

$$\int_{\partial D_l} (\nu, Y)\, ds = \int_{D_l} v \, dx, \quad l = 1, \ldots m. \tag{5.52}$$

By Theorems 5.2 and 5.7, the basis given by (5.26) has the property that the matrix

$$\int_{\partial D_l} (\nu, Y_j)\, ds, \qquad j,l=1,\dots,m$$

is regular. Hence Y is uniquely determined by (5.52). Consider the interior Neumann problem

$$\Delta w = 0 \quad \text{in} \quad D$$

$$\frac{\partial w}{\partial \nu} = (\nu, Y) - \frac{1}{6} v (\nu, \operatorname{grad}|x|^2) \quad \text{on} \quad \partial D$$

for Laplace's equation. Since $\Delta|x|^2 = 6$, by (5.52) the solvability condition for this boundary-value problem is satisfied. Since the boundary values belong to $C^{0,\alpha}(\partial D)$, we conclude by the single-layer approach to the Neumann problem that for any solution we have $w \in C^{1,\alpha}(\bar{D})$. Define $u(x) := w(x) + \frac{1}{6} v|x|^2$. Then u satisfies

$$\Delta u = v \quad \text{in} \quad D$$

and

$$\frac{\partial u}{\partial \nu} = (\nu, Y) \quad \text{on} \quad \partial D,$$

and hence by the second Green's theorem

$$\int_D v \operatorname{div} A\, dx = \int_D u \Delta \operatorname{div} A\, dx + \int_{\partial D} \operatorname{div} A_- (\nu, Y)\, ds - \int_{\partial D} u \frac{\partial}{\partial \nu} \operatorname{div} A_-\, ds.$$

Therefore

$$\int_D u \Delta \operatorname{div} A\, dx = -k^2 \int_D u \operatorname{div} A\, dx = -k^2 \int_D u \operatorname{div} A_0\, dx + o(k^2)$$

where A_0 denotes the limit of (5.50) as $k \to 0$. Using the jump relations of Corollary 2.25, the condition (5.40), and the first vector Green's theorem, we now see that

$$\begin{aligned}\int_{\partial D} \operatorname{div} A_- (\nu, Y)\, ds &= \int_{\partial D} (\gamma + \operatorname{div} A_+)(\nu, Y)\, ds \\ &= \int_{\partial D} \gamma(\nu, Y)\, ds - \int_{\mathbf{R}^3 \setminus D} (\Delta A, Y)\, dx \\ &= k^2 \left[e + \int_{\mathbf{R}^3 \setminus D} (A_0, Y)\, dx \right] + o(k^2)\end{aligned}$$

for some constant $e \in \mathbf{C}$ according to (5.40). Using Stokes' theorem and the boundary value (5.51), we have

$$\int_{\partial D} u(\nu, \operatorname{curl}\operatorname{curl} A_-)\, ds = \int_{\partial D} (\nu, \operatorname{curl} A_-, \operatorname{grad} u)\, ds$$

$$= k^2 \int_{\partial D} (\nu, \mathbf{S}_0 c_0, \operatorname{grad} u)\, ds + o(k^2),$$

and therefore

$$\int_{\partial D} u \frac{\partial \operatorname{div} A_-}{\partial \nu}\, ds = k^2 \left[\int_{\partial D} \{u(\nu, A_0) + (\nu, \mathbf{S}_0 c_0, \operatorname{grad} u)\}\, ds \right] + o(k^2).$$

Summarizing, we see that

$$\int_D v \operatorname{div} A\, dx = ck^2 + o(k^2)$$

for some constant $c \in \mathbf{C}$ and therefore the right-hand side of equation (5.49) satisfies condition (1.11) of Theorem 1.32. Hence

$$\|\mu - \mu_0\|_{\alpha, \partial D} \to 0, \qquad k \to 0, \tag{5.53}$$

for the solution of (5.49).

From the convergence (5.39), (5.48), and (5.53) of the boundary data we now obtain the convergence (5.41) of E, $\operatorname{div} E$, and $\operatorname{curl} E$ by the representation Theorem 4.13 and the regularity Theorems 2.12, 2.17, and 2.24. For $\operatorname{curl} E$ we also make use of the transformation (2.86). The differential equation and asymptotic properties of the limiting field E_0 follow from the limiting form of the representation Theorem 4.13. The boundary conditions for E_0 are obvious.

In a similar way, convergence for the exterior magnetic boundary-value problem can also be shown (Gülzow [1]).

Theorem 5.10. The existence of constants $d_1, \ldots, d_p$ and functions d_0, δ_0 such that the limits

$$\|d - d_0\|_{\alpha, \partial D} \to 0, \qquad k \to 0,$$

$$\|\delta - \delta_0\|_{\alpha, \partial D} \to 0, \qquad k \to 0, \tag{5.54}$$

and

$$\frac{1}{k^2} \int_{\partial D} (\nu, d, Z_j)\, ds \to d_j, \qquad k \to 0, \tag{5.55}$$

$j=1,\ldots,p$, exist are necessary and sufficient for the convergence

$$\begin{aligned} &\|H-H_0\|_{\alpha,\mathbb{R}^3\setminus D}\to 0, \qquad k\to 0,\\ &\|\operatorname{div} H-\operatorname{div} H_0\|_{\alpha,\mathbb{R}^3\setminus D}\to 0, \qquad k\to 0,\\ &\|\operatorname{curl} H-\operatorname{curl} H_0\|_{\alpha,\mathbb{R}^3\setminus D}\to 0, \qquad k\to 0, \end{aligned} \tag{5.56}$$

of the solution H of the exterior magnetic boundary-value problem (4.25). The limiting field solves the exterior magnetic boundary-value problem for the vector Laplace equation

$$\Delta H_0 = 0 \quad \text{in} \quad \mathbb{R}^3\setminus D \tag{5.57}$$

satisfying the boundary condition

$$[[\operatorname{curl} H_0, \nu], \nu] = d_0, \qquad (\nu, H_0) = \delta_0 \quad \text{on} \quad \partial D \tag{5.58}$$

and at infinity

$$H_0(x) = o(1), \qquad |x|\to\infty, \tag{5.59}$$

uniformly for all directions $x/|x|$. It is uniquely determined by the additional properties

$$\int_{\mathbb{R}^3\setminus D} (H_0, Z_j)\, dx = d_j, \tag{5.60}$$

$j=1,\ldots,p$.

Finally, we want to apply Theorem 5.9 to the exterior Maxwell problem.

Theorem 5.11. The existence of constants $d_1,\ldots,d_p$ and functions c_0, δ_0 such that the limits

$$\begin{aligned} &\|c-c_0\|_{\alpha,\partial D}\to 0, \qquad k\to 0,\\ &\left\|\frac{i}{k}\operatorname{Div} c-\delta_0\right\|_{\alpha,\partial D}\to 0, \qquad k\to 0, \end{aligned} \tag{5.61}$$

and

$$\frac{i}{k}\int_{\partial D} (c, Z_j)\, ds \to d_j, \qquad k\to 0, \tag{5.62}$$

exist are necessary and sufficient for the convergence

$$\|E-E_0\|_{\alpha,\mathbb{R}^3\setminus D}\to 0, \qquad \|H-H_0\|_{\alpha,\mathbb{R}^3\setminus D}\to 0, \qquad k\to 0, \tag{5.63}$$

of the solutions E and H of the exterior Maxwell problem (4.19). The limiting field E_0 is a harmonic field

$$\operatorname{div} E_0 = 0, \qquad \operatorname{curl} E_0 = 0 \quad \text{in} \quad \mathbb{R}^3 \setminus \bar{D} \tag{5.64}$$

satisfying the boundary condition

$$[\nu, E_0] = c_0 \quad \text{on} \quad \partial D \tag{5.65}$$

and at infinity

$$E_0(x) = o(1), \qquad |x| \to \infty. \tag{5.66}$$

It is uniquely determined by

$$\int_{\mathbb{R}^3 \setminus D} (E_0, Y_j)\, dx = 0, \tag{5.67}$$

$j = 1, \ldots, m$. The limiting field H_0 is a harmonic field

$$\operatorname{div} H_0 = 0, \qquad \operatorname{curl} H_0 = 0 \quad \text{in} \quad \mathbb{R}^3 \setminus \bar{D} \tag{5.68}$$

satisfying the boundary condition

$$(\nu, H_0) = \delta_0 \quad \text{on} \quad \partial D \tag{5.69}$$

and at infinity

$$H_0(x) = o(1), \qquad |x| \to \infty. \tag{5.70}$$

It is uniquely determined by

$$\int_{\mathbb{R}^3 \setminus D} (H_0, Z_j)\, dx = d_j, \tag{5.71}$$

$j = 1, \ldots, p$.

Proof. ***Necessity.*** The necessity of (5.61) follows from (5.63) and the boundary condition (4.19) with the help of the identity $\operatorname{Div}[\nu, E] = -(\nu, \operatorname{curl} E) = -ik(\nu, H)$. The necessity of (5.62) and the property (5.71) follow from the identity

$$\begin{aligned}
\frac{i}{k} \int_{\partial D} (c, Z_j)\, ds &= \frac{i}{k} \int_{\partial D} (\nu, E, Z_j)\, ds \\
&= \frac{1}{k^2} \int_{\partial D} \{(\nu, Z_j, \operatorname{curl} H) + (\nu, Z_j) \operatorname{div} H\}\, ds \\
&= -\frac{1}{k^2} \int_{\mathbb{R}^3 \setminus D} (\Delta H, Z_j)\, dx \\
&= \int_{\mathbb{R}^3 \setminus D} (H, Z_j)\, dx.
\end{aligned}$$

Sufficiency. The convergence of E is obtained from Theorem 5.9 for the special case $\gamma = 0$.

We now consider the magnetic field H. Using the same notations as in the proof of Theorem 5.9, we see from the representation Theorem 4.5 and the transformation (2.86) that

$$H(x) = \frac{i}{k}\operatorname{curl}\int_{\partial D}\Phi_k(x,y)[\nu(y),b(y)]\,ds(y) - ik\int_{\partial D}\Phi_k(x,y)c(y)\,ds(y)$$

$$-\frac{i}{k}\operatorname{grad}\int_{\partial D}\Phi_k(x,y)\operatorname{Div}c(y)\,ds(y), \qquad x\in\mathbb{R}^3\setminus\bar{D}.$$

Hence, noting (5.61), the convergence of H will follow if we can show the convergence of b/k as $k\to 0$. Dividing the integral equation (5.46) by k and using the analysis following (5.46), we see that in order to show convergence of b/k, we only have to prove that

$$\lim_{k\to 0}\frac{1}{k}\int_D(E,F)\,dx \tag{5.72}$$

exists for all Neumann vector fields E in D. Using (5.47) with $\gamma = 0$ we apply Gauss' theorem and interchange the order of integration to arrive at

$$\int_D(E,F)\,dx = \int_{\partial D}(c,A)\,ds$$

where

$$A(x) := \int_{\partial D}\Phi_k(x,y)[E(y),\nu(y)]\,ds(y), \qquad x\in\mathbb{R}^3\setminus\partial D. \tag{5.73}$$

By the representation formula (5.11) we see that the limit A_0 of (5.73) as $k\to 0$ satisfies

$$\operatorname{div}A_0 = 0, \qquad \operatorname{curl}A_0 = 0 \quad \text{in} \quad \mathbb{R}^3\setminus D.$$

Now let w be the solution of the exterior Neumann problem for the Laplace equation

$$\Delta w = 0 \quad \text{in} \quad \mathbb{R}^3\setminus D$$

satisfying

$$\frac{\partial w}{\partial \nu} = (\nu, A_0) \quad \text{on} \quad \partial D$$

and

$$w(x) = o(1), \qquad |x| \to \infty.$$

Then $Z := A_0 - \operatorname{grad} w$ is a Neumann field in $\mathbb{R}^3 \setminus D$ and from Gauss' theorem (2.73) we obtain

$$\int_{\partial D} (c, A_0)\, ds = \int_{\partial D} \{(c, Z) + w \operatorname{Div} c\}\, ds.$$

The existence of the limit (5.72) now follows from (5.61) and (5.62).

To conclude the proof of the theorem, we have to verify the stated properties of the limiting fields E_0 and H_0. From $\operatorname{div} E = 0$ for all $k \neq 0$ we obtain $\operatorname{div} E_0 = 0$. Then $\operatorname{div} \operatorname{curl} E_0 = 0$, $\operatorname{curl} \operatorname{curl} E_0 = -\Delta E_0 + \operatorname{grad} \operatorname{div} E_0 = 0$ in $\mathbb{R}^3 \setminus D$ and $(\nu, \operatorname{curl} E_0) = -\operatorname{Div} c_0 = 0$ on ∂D because of (5.61). Therefore $\operatorname{curl} E_0$ is a Neumann field in $\mathbb{R}^3 \setminus D$. Since $\operatorname{curl} E = ikH$ we see that

$$\int_D (\operatorname{curl} E_0, Z_j)\, ds = 0,$$

$j = 1, \ldots, p$, whence $\operatorname{curl} E_0 = 0$ in $\mathbb{R}^3 \setminus D$ follows. Then E_0 is harmonic and the remaining properties (5.65), (5.66), and (5.67) follow from Theorem 5.9.

We now consider H_0. Arguing as above, we see that the field $\operatorname{curl} H_0$ is harmonic in $\mathbb{R}^3 \setminus D$ and from $[\nu, \operatorname{curl} H] = -ikc$ on ∂D we can conclude that $\operatorname{curl} H_0$ is a Dirichlet field in $\mathbb{R}^3 \setminus D$. Then since $\operatorname{curl} H = -ikE$ we have

$$\int_D (\operatorname{curl} H_0, Y_j)\, ds = 0,$$

$j = 1, \ldots, m$, whence $\operatorname{curl} H_0 = 0$ in $\mathbb{R}^3 \setminus D$ follows. Therefore H_0 is harmonic. The boundary condition (5.69) follows from $(\nu, H) = (1/ik)(\nu, \operatorname{curl} E) = (i/k)(\operatorname{Div} c)$ on ∂D and (5.71) has already been shown.

The above results on the continuity of solutions to the exterior electromagnetic boundary-value problems as $k \to 0$ were first obtained for the special case of simply connected domains (i.e., $p = 0$) by Werner [1], [3]. Through the use of methods considerably more complicated than that of this chapter, Werner has extended his work to multiply connected domains (Werner [4]). The above analysis based on the use of singular perturbation properties for integral operators follows the work of Kress [4], [7].

We wish to mention that by applying the general theory of Section 1.4, it is also possible to obtain continuity results on the solutions to the interior boundary-value problems for the scalar and vector Helmholtz equation in the vicinity of the interior eigenvalues (Wilde [1]).

6

THE INVERSE SCATTERING PROBLEM: EXACT DATA

Up until now we have been considering the *direct problem* of scattering theory, that is, given the shape of the obstacle and boundary data to determine the field outside the obstacle. In this chapter we shall begin our study of the *inverse problem* of scattering theory, that is, given the far-field pattern to determine either the shape or the surface impedance of the obstacle. We note that these are not the only types of inverse problems that arise in scattering theory, although it is probably safe to say that they are the ones of primary practical importance, and for this and pedagogical reasons we have decided to restrict ourselves to these special cases. We have furthermore decided not to discuss high frequency methods such as physical or geometric optics. Our motivation for this decision was twofold: first, the use of integral equation methods in inverse scattering theory is basically restricted to low or intermediate values of the frequency, and second, the mathematical difficulties in rigorously establishing the validity of high frequency methods in inverse scattering theory is formidable and beyond the intended aim of this book. For a survey of some of the formal approaches to high frequency methods in the inverse scattering problem we refer the interested reader to Chapter 9 of Jones [2], and for an indication of the mathematical techniques needed to justify such a formal analysis the reader can consult Majda [1].

The inverse scattering problem is considerably more difficult to solve than the direct scattering problem. This is due to two main reasons: the problem is (1) nonlinear and (2) improperly posed. Furthermore, as will be seen in the sequel, in order to treat the inverse problem it is necessary to be able to solve the direct problem for arbitrary domains and frequencies. Since this problem has been satisfactorily resolved in the first five chapters, we can now turn our attention to the inverse problem itself. Of the two basic problems, nonlinearity and improperly posedness, it is the latter that presents the more basic difficulty. Indeed, for a given measured far-field pattern we shall see shortly that no solution exists in general to the inverse scattering problem, and if a solution does exist it does not depend continuously on the measured data. Hence before we can begin to construct a solution to the inverse scattering problem we must answer the question of what we mean by a "solution." This question will be answered in Chapters 7 and 8. However, for the time being it is worthwhile recalling the remark of Lanczos: "A lack of information cannot be remedied by any mathematical trickery." Hence in order to determine what we mean by a solution it will be necessary to introduce "nonstandard" information that reflects the physical situation we are trying to model. Having resolved the question of what is meant by a solution, we then have to actually construct this solution, and this is complicated not only by the fact that the problem is nonlinear, but also by the fact that the above-mentioned nonstandard information has been incorporated into the mathematical model.

Before we can begin to deal with the problem of constructing a solution to the inverse scattering problem, we must first examine the simple problem of why the problem is improperly posed to begin with. In particular, what kind of functions are far-field patterns of scattered waves, and given such a function what can be said of the location of the sources that generated such a pattern? This will be the subject matter of the first part of this chapter. We shall show that a necessary condition for a function to be a far-field pattern can be expressed in terms of entire functions of exponential type, and hence we begin our discussion with a brief introduction to those parts of the theory of entire functions that are relevant to the development of this condition, in particular the definitions of order and type, the connection between the growth of an entire function and its Taylor coefficients, the indicator function, and finally Polya's theorem. We shall then use these results to establish Müller's theorem (Müller [4], Hartman and Wilcox [1]) that establishes necessary conditions for a function to be a far-field pattern. By deriving a reflection principle for the Helmholtz equation it is possible to considerably sharpen Müller's results in the case of axial symmetry, and we shall do this following the presentation of Colton [1]. Since no boundary conditions are imposed, the above analysis applies to both acoustic and electromagnetic scattering.

We conclude this chapter by presenting various theorems on the uniqueness of the solution of the inverse scattering problem due to Schiffer (cf. Lax and Phillips [1]), Colton and Kirsch [1], and Colton and Kress [1]. The proofs of these theorems are all based on the unique continuation property of solutions to the Helmholtz equation. Note that the improperly posed nature of the inverse scattering problem does not affect the question of uniqueness, since in this case our aim is to show that if two far-field patterns are the same then the corresponding obstacles or surface impedances are also the same, and it is assumed a priori that the far-field patterns are known exactly and correspond to some scattering obstacle or surface impedance.

6.1 ENTIRE FUNCTIONS OF EXPONENTIAL TYPE

An entire function f of a complex variable z is an analytic function that is holomorphic in the entire complex z plane and hence can be represented in the form

$$f(z) = \sum_{n=0}^{\infty} a_n z^n \tag{6.1}$$

where the series is convergent for all values of z. In the theory of entire functions an important role is played by the growth of such functions as $|z|$ tends to infinity, and to this end we introduce the function

$$M_f(r) := \max_{|z|=r} |f(z)|.$$

Note that it follows from the maximum modulus principle that M_f is a monotonically increasing function of r. An entire function f is said to be of finite order if there exists a positive constant k such that

$$M_f(r) < e^{r^k}$$

for r sufficiently large. The greatest lower bound ρ of such numbers k is called the order of f, that is,

$$\rho = \overline{\lim_{r \to \infty}} \frac{\log\log M_f(r)}{\log r}.$$

Similarly, we define the type σ of an entire function f of order ρ to be the greatest lower bound of the positive numbers A such that

$$M_f(r) < e^{Ar^\rho}$$

for r sufficiently large, that is,

$$\sigma = \overline{\lim_{r \to \infty}} \frac{\log M_f(r)}{r^\rho}.$$

The following theorem characterizes the order and type of an entire function in terms of its Taylor series coefficients a_n defined by (6.1).

Theorem 6.1. Let f be an entire function of finite order ρ and finite type σ and have the Taylor series expansion (6.1). Then

$$\rho = \overline{\lim_{n\to\infty}} \frac{n\log n}{\log\frac{1}{|a_n|}}$$

$$(\sigma e\rho)^{1/\rho} = \overline{\lim_{n\to\infty}} \left(n^{1/\rho}|a_n|^{1/n}\right).$$

Proof. From Cauchy's inequality we have that

$$|a_n| \leqslant \frac{M_f(r)}{r^n}.$$

For r sufficiently large there exist positive constants A and k such that

$$M_f(r) < e^{Ar^k}$$

and hence

$$|a_n| < e^{Ar^k} r^{-n}. \tag{6.2}$$

The minimum value (for $r > 0$) of the right-hand side of (6.2) is achieved when $r = (n/Ak)^{1/k}$ and hence

$$|a_n| < \left(\frac{eAk}{n}\right)^{n/k} \tag{6.3}$$

for n sufficiently large.

We now assume that (6.3) holds for n sufficiently large and deduce an estimate for $M_f(r)$. For $n > n_0 = [2^k eAkr^k]$ and r sufficiently large we have from (6.3) that

$$|a_n z^n| < 2^{-n}$$

and hence

$$|f(z)| < \sum_{n=0}^{n_0} |a_n| r^n + 2^{-n_0}.$$

Therefore, if we define $\mu(r)$ by

$$\mu(r) := \max_n |a_n| r^n$$

then

$$M_f(r) < (1+2^k eAkr^k)\mu(r)+2^{-n_0}. \tag{6.4}$$

Without loss of generality assume that f is not a polynomial. Then the index of the largest term in the series (6.1) tends to infinity as r tends to infinity, and hence for r sufficiently large

$$\mu(r) < \max_n \left(\frac{eAk}{n}\right)^{n/k} r^n.$$

Since the maximum on the right hand side is attained for $n = Akr^k$ we have

$$\mu(r) < e^{Ar^k} \tag{6.5}$$

for r sufficiently large. Expressions (6.4) and (6.5) now imply that

$$M_f(r) < (2+2^k eAkr^k)e^{Ar^k}$$

for r sufficiently large.

The above analysis now allows us to conclude that the order ρ of f is the greatest lower bound of the numbers k such that (6.3) is valid, and the type σ is equal to the greatest lower bound of the numbers A for which (6.3) is valid for $k = \rho$. This statement now immediately implies the theorem.

The theory of entire functions of order one and finite type $\sigma > 0$ is of special interest to us and we shall call such functions entire functions of exponential type. The theory of entire functions of exponential type is particularly rich, due mainly to the exploitation of the Phragmén–Lindelöf indicator function h_f defined by

$$h_f(\theta) := \overline{\lim_{r\to\infty}} \frac{\log|f(re^{i\theta})|}{r}.$$

The importance of the indicator function in the study of entire functions of exponential type lies in the fact that it has a simple geometric interpretation in terms of convex sets. This property is based on the following simple, but fundamental, theorem.

Theorem 6.2. The indicator function of an entire function of exponential type satisfies the relation

$$h_f(\theta_1)\sin(\theta_2-\theta_3)+h_f(\theta_2)\sin(\theta_3-\theta_1)+h_f(\theta_3)\sin(\theta_1-\theta_2) \leqslant 0$$

for all $\theta_1 \leqslant \theta_2 \leqslant \theta_3$, $\theta_3-\theta_2<\pi$, $\theta_2-\theta_1<\pi$.

Proof. We first note that if a and b are real numbers and $F(z) = e^{(a-ib)z}$, then the indicator function of F is given by $H(\theta) = a\cos\theta + b\sin\theta$ and if

$H(\theta_1)=h_1$, $H(\theta_2)=h_2$, we can write H as

$$H(\theta)=\frac{h_1\sin(\theta_2-\theta)+h_2\sin(\theta-\theta_1)}{\sin(\theta_2-\theta_1)}. \tag{6.6}$$

Now let f be an entire function of exponential type with indicator function h_f satisfying $h_f(\theta_1)=h_1$, $h_f(\theta_2)=h_2$. Let $\delta>0$ and define H_δ by

$$H_\delta(\theta)=a_\delta\cos\theta+b_\delta\sin\theta$$

where α_δ, b_δ are chosen such that $H_\delta(\theta_1)=h_1+\delta$, $H_\delta(\theta_2)=h_2+\delta$. Then the function

$$\phi(z)=f(z)e^{-(a_\delta-ib_\delta)z}$$

has the indicator function

$$h_\phi(\theta)=h_f(\theta)-H_\delta(\theta). \tag{6.7}$$

Now note that along the rays $\arg z=\theta_1$, $\arg z=\theta_2$, we have from (6.7) that ϕ tends to zero as $|z|$ tends to infinity. Hence if $\theta_2-\theta_1<\pi$, it follows from the Phragmén–Lindelöf theorem (cf. Boas [1] or Levin [1]) that ϕ is bounded in the sector $\theta_1\leqslant\arg z\leqslant\theta_2$ which implies that $h_\phi(\theta)\leqslant 0$ in this region, that is,

$$h_f(\theta)\leqslant H_\delta(\theta),\qquad \theta_1\leqslant\theta\leqslant\theta_2. \tag{6.8}$$

The statement of the theorem for the special case when $\theta_3-\theta_1<\pi$ now follows from (6.8) if we let $\delta\to 0$, use (6.6), and relabel the indices on the angles. To prove the theorem in general let $H(\theta)=a\cos\theta+b\sin\theta$ where a and b are chosen such that $H(\theta_1)=h_f(\theta_1)$ and $H(\theta_2)=h_f(\theta_2)$ and note that the inequality of the theorem becomes an equality for this function. Choose θ_4 such that $\theta_2<\theta_4<\theta_1+\pi$. Then by an argument similar to that leading to (6.8) we can conclude that if $h_f(\theta_4)<H(\theta_4)$ then $h_f(\theta_2)<H(\theta_2)$, a contradiction. Hence $h_f(\theta_4)\geqslant H(\theta_4)$, and a similar argument applied to θ_2, θ_4 and θ_3 shows that $h_f(\theta_3)\geqslant H(\theta_3)$. Since

$$h_f(\theta_1)\sin(\theta_2-\theta_3)+h_f(\theta_2)\sin(\theta_3-\theta_1)+H(\theta_3)\sin(\theta_1-\theta_2)=0$$

and $\sin(\theta_1-\theta_2)<0$, the theorem now follows in the general case.

In order to see the connection between Theorem 6.2 and convex sets in the plane, it is first necessary to review some of the basic properties of plane convex sets. We recall that a convex set is a nonempty set that contains any line segment joining two points in the set. The intersection of all convex sets containing a given set is the smallest convex set containing the given set and is called the convex hull of the set. The supporting function of a bounded convex set G is the function k defined by

$$k(\theta):=\sup_{x+iy\in G}(x\cos\theta+y\sin\theta)$$

$$=\sup_{z\in G}\operatorname{Re}(ze^{-i\theta}).$$

The lines l_θ defined by

$$x\cos\theta+y\sin\theta-k(\theta)=0$$

are called the supporting lines of G and we note that all points of G lie on one side of l_θ. If G is closed, then it follows from the definition of k and the compactness of G that each line l_θ has a point in common with G. It is furthermore easily seen that the value of the supporting function at an angle θ is equal to the distance from the supporting line to the origin.

Theorem 6.3. A necessary and sufficient condition that a function k be the supporting function of a bounded convex set is that the following conditions hold:

(a) $k(\theta+2\pi)=k(\theta)$

(b) $k(\theta_1)\sin(\theta_2-\theta_3)+k(\theta_2)\sin(\theta_3-\theta_1)+k(\theta_3)\sin(\theta_1-\theta_2)\leqslant 0$

for all $\theta_1\leqslant\theta_2\leqslant\theta_3$, $\theta_3-\theta_2<\pi$, $\theta_2-\theta_1<\pi$.

Proof. Let k be the supporting function of a bounded convex set G and let $z=x+iy$ be a point on the boundary of G lying on the supporting line l_{θ_2}. Then

$$x\cos\theta_1+y\sin\theta_1-k(\theta_1)\leqslant 0$$

$$x\cos\theta_2+y\sin\theta_2-k(\theta_2)=0$$

$$x\cos\theta_3+y\sin\theta_3-k(\theta_3)\leqslant 0.$$

If we multiply the first of these equations by $\sin(\theta_3-\theta_2)>0$, the second by $\sin(\theta_1-\theta_3)$, and the third by $\sin(\theta_2-\theta_1)>0$, and add we obtain (b). Condition (a) is obviously a necessary condition on k.

Now suppose we have a function k satisfying conditions (a) and (b). Without loss of generality, let $\theta_2=\pi/2$ and construct the line $l_{\pi/2}$ defined by

$$y=k\left(\frac{\pi}{2}\right). \tag{6.9}$$

Now note that the union of the intersections of the half planes

$$x\cos\theta_1 + y\sin\theta_1 - k(\theta_1) > 0, \qquad -\frac{\pi}{2} < \theta_1 < \frac{\pi}{2} \tag{6.10}$$

with the line $l_{\pi/2}$ is an interval $(b, +\infty)$, and the union of the intersections of the half planes

$$x\cos\theta_3 + y\sin\theta_3 - k(\theta_3) > 0, \qquad \frac{\pi}{2} < \theta_3 < \frac{3\pi}{2} \tag{6.11}$$

with the line $l_{\pi/2}$ is an interval $(-\infty, a)$. Suppose a point of $l_{\pi/2}$ belonged to both the intervals $(-\infty, a)$ and $(b, +\infty)$. Then the coordinates of such a point would satisfy (6.9) to (6.11) for some angles θ_1, θ_3, and multiplying these equations by $\sin(\theta_1 - \theta_3)$, $\sin(\theta_3 - (\pi/2))$, and $\sin((\pi/2) - \theta_1)$, respectively, and adding would lead to a contradiction to condition (b). Hence $a \leqslant b$, and all points of the interval $a \leqslant x \leqslant b$ on $l_{\pi/2}$ belong to all the half planes

$$x\cos\theta + y\sin\theta - k(\theta) \leqslant 0, \qquad 0 \leqslant \theta \leqslant 2\pi.$$

Hence the intersection of these half planes is not empty and is therefore some bounded convex set G. By construction the supporting lines of this convex set are given by

$$x\cos\theta + y\sin\theta - k(\theta) = 0, \qquad 0 \leqslant \theta \leqslant 2\pi$$

and hence k is the supporting function of this set.

It now follows from Theorems 6.2 and 6.3 that the indicator function of an entire function of exponential type is the supporting function of some bounded convex set. This convex set is called the indicator diagram of the given function. We are now in a position to prove a remarkable connection between the indicator diagram of an entire function of exponential type and the location of the singularities of the Borel transform of this function. To define the Borel transform, let f be an entire function of exponential type having the Taylor series expansion

$$f(z) = \sum_{n=0}^{\infty} \frac{c_n}{n!} z^n.$$

Then the Borel transform of f is defined by

$$\phi(z) := \sum_{n=0}^{\infty} c_n z^{-n-1}.$$

We note that it follows from Theorem 6.1 that if f is of type σ, then

$$\sigma = \overline{\lim_{n\to\infty}} |c_n|^{1/n}$$

and hence ϕ is analytic for $|z| > \sigma$. The smallest convex set containing all the singularities of ϕ is called the conjugate diagram of f. This set clearly lies inside $|z| \leq \sigma$. The following theorem is due to Polya (cf. Levin [1]).

Theorem 6.4 (Polya's Theorem). The conjugate diagram of an entire function of exponential type is the reflection of the indicator diagram with respect to the real axis or, more concisely, the conjugate indicator diagram.

Proof. We first note that it follows from termwise integration that

$$f(z) = \frac{1}{2\pi i} \oint_C e^{\zeta z} \phi(\zeta)\, d\zeta \tag{6.12}$$

where C is any contour containing the conjugate diagram. We now want to show that (6.12) can be inverted by the formula

$$\phi(z) = \int_0^\infty e^{-z\zeta} f(\zeta)\, d\zeta \tag{6.13}$$

where the integration is along a ray $\zeta = te^{-i\theta}$. To this end we first note that it follows from the inequality (valid for t sufficiently large and $\varepsilon > 0$)

$$|f(te^{-i\theta})| < \exp\left(\left[h_f(-\theta) + \frac{\varepsilon}{2}\right] t\right)$$

that the integral in (6.13) is absolutely and uniformly convergent for

$$\operatorname{Re} ze^{-i\theta} > h_f(-\theta) + \varepsilon.$$

It suffices to show that (6.13) is valid in a part of this domain, for example, in

$$\operatorname{Re} ze^{-i\theta} > 3\sigma.$$

But for such values of z we have

$$|e^{-zte^{-i\theta}}| < e^{-3\sigma t} \tag{6.14}$$

and from the inequality

$$\frac{|c_n|}{n!} < \frac{M_f(r)}{r^n}$$

it follows that

$$\left| \sum_{k=n+1}^{\infty} \frac{c_k}{k!} t^k \right| < \frac{M_f(r)}{1 - \frac{t}{r}} \left(\frac{t}{r}\right)^{n+1}$$

or, setting $r = 2t$,

$$\left| \sum_{k=n+1}^{\infty} \frac{c_k}{k!} t^k \right| < \left(\frac{1}{2} \right)^n e^{2(\sigma+\varepsilon)t}. \tag{6.15}$$

It now follows from (6.14) and (6.15) that the series

$$\sum_{k=0}^{\infty} e^{-z\zeta} \frac{c_k \zeta^k}{k!}$$

converges uniformly on the ray $\zeta = te^{-i\theta}$, $t > 0$, and termwise integration now yields (6.13).

We shall now use (6.12) and (6.13) to prove the theorem. Let k_f be the supporting function of the conjugate diagram. In (6.12) let the distance between C and the conjugate diagram be less than $\varepsilon > 0$. We then have

$$\begin{aligned} |f(re^{i\theta})| &< M_\varepsilon \exp\left[r \max_{\zeta \in C} \operatorname{Re}(\zeta e^{i\theta}) \right] \\ &\leqslant M_\varepsilon \exp\left[(k_f(-\theta) + \varepsilon) r \right] \end{aligned}$$

where

$$M_\varepsilon = \frac{1}{2\pi} \max_{\zeta \in C} |\phi(\zeta)|.$$

This implies that $h_f(\theta) \leqslant k_f(-\theta)$. On the other hand, ϕ is analytic in the domain $\operatorname{Re} ze^{-i\theta} > h_f(-\theta) + \varepsilon$ and hence $k_f(\theta) \leqslant h_f(-\theta)$, that is, $k_f(-\theta) = h_f(\theta)$ and the theorem is proved.

For further information on the theory of entire functions of exponential type the reader is referred to Levin [1] or Boas [1], and for applications of Polya's theorem to the problem of polynomial expansion of analytic functions see Boas and Buck [1]. In the next section of this chapter we shall apply the above results on entire functions, and in particular Polya's theorem, to the problem of trying to determine the class of functions that are far-field patterns corresponding to scattering by bounded obstacles.

6.2 FAR-FIELD PATTERNS AND THEIR CLASSIFICATION

Let u be a solution of the three-dimensional Helmholtz equation

$$\Delta u + k^2 u = 0$$

in the exterior of the sphere of radius R_0 such that u satisfies the Sommerfeld radiation condition

$$\left(\operatorname{grad} u(x), \frac{x}{|x|}\right) - iku(x) = o\left(\frac{1}{|x|}\right), \qquad |x| \to \infty.$$

For convenience we shall assume in this and subsequent chapters that the wave number k is strictly positive. We recall from Corollary 3.7 that in this case u has the asymptotic behavior

$$u(x) = \frac{e^{ikr}}{r} F(\theta, \phi) + O\left(\frac{1}{r^2}\right)$$

where (r, θ, ϕ) are the spherical coordinates of x and F is known as the far-field pattern of u.

Our aim in this section is to use the theory of entire functions of a complex variable to classify those functions that can be far-field patterns of a radiating solution of the Helmholtz equation. The first result that we shall prove is a theorem due to Müller [4] (see also Hartman and Wilcox [1]) that gives necessary conditions for a function to be a far-field pattern corresponding to the scattering of an incoming wave by a bounded obstacle. We note that in Müller's original presentation the following theorem was presented as a necessary and sufficient condition for a function to be a far-field pattern. The reason we are calling it only a necessary condition is that for us, in contrast to Müller, a far-field pattern means the far-field pattern corresponding to an actual physical scattering problem, that is, u is required to satisfy certain boundary data on the boundary of the scattering obstacle. We wish to make this assumption implicit in our use of the term far-field pattern to emphasize the problem of actually characterizing such functions in a complete and satisfactory manner. Indeed, as we shall see in Chapter 8, it is this problem that causes many of the difficulties in solving the inverse scattering problem, that is, it is, in general, not possible to determine whether or not a given measured function corresponds to a far-field pattern for some bounded obstacle. In mathematical terms what is needed is a characterization of the range of the mapping taking bounded domains into their corresponding far-field patterns, and this problem is one of the many intriguing open problems in inverse scattering theory. As the following theorem indicates, an answer to this problem will probably require the use of deep results in the theory of entire functions. For partial results in this direction, we refer the reader to Colton and Kirsch [3].

Since the theorems that follow do not depend in an essential way on the wave number k as long as it is greater than zero, without loss of generality we set $k = 1$, that is, u is a solution of

$$\Delta u + u = 0 \tag{6.16}$$

in the exterior of the sphere of radius R_0 satisfying the Sommerfeld radiation condition

$$\left(\operatorname{grad} u(x), \frac{x}{|x|}\right) - iu(x) = o\left(\frac{1}{|x|}\right), \qquad |x| \to \infty. \tag{6.17}$$

Theorem 6.5 (Müller's Theorem). Let F be the far-field pattern corresponding to an obstacle situated inside the sphere $|x| = R_0$. Then there exists a harmonic function h defined in all of $\mathbb{R}^3$ such that

(a) $h(1, \theta, \phi) = F(\theta, \phi)$

(b) $\int_0^{2\pi} \int_0^{\pi} |h(r, \theta, \phi)|^2 \sin\theta \, d\theta \, d\phi$ is an entire function of r of exponential type less than or equal to R_0.

Proof. It follows from Theorem 3.6 that F can be expanded in a uniformly convergent Legendre series

$$F(\theta, \phi) = \sum_{n=0}^{\infty} \sum_{m=-n}^{n} a_{nm} P_n^{|m|}(\cos\theta) e^{im\phi}.$$

Following the proof of Lemma 3.14 and using the asymptotic behavior of the spherical Hankel functions, we see that for $r \geqslant R_0$, u has the expansion

$$u(x) = \sum_{n=0}^{\infty} \sum_{m=-n}^{n} a_{nm} i^{n+1} h_n^{(1)}(r) P_n^{|m|}(\cos\theta) e^{im\phi} \tag{6.18}$$

where the series is absolutely and uniformly convergent in this region, that is, F is the far-field pattern corresponding to the outgoing wave function u. In particular

$$\int_0^{2\pi} \int_0^{\pi} |u(r, \theta, \phi)|^2 \sin\theta \, d\theta \, d\phi = 4\pi \sum_{n=0}^{\infty} \sum_{m=-n}^{n} \frac{|a_{nm}|^2 |h_n^{(1)}(r)|^2 (n+|m|)!}{(2n+1)(n-|m|)!}$$

is bounded for $r \geqslant R_0$. Hence if we define

$$b_n = 4\pi \sum_{m=-n}^{n} \frac{|a_{nm}|^2 (n+|m|)!}{(2n+1)(n-|m|)!} \tag{6.19}$$

we can conclude from the asymptotic relation

$$h_n^{(1)}(r) = \frac{-i\Gamma(n+\frac{1}{2})2^n}{\sqrt{\pi}\, r^{n+1}} \left[1 + O\left(\frac{1}{n}\right)\right]$$

that $b_n[\Gamma(n+\frac{1}{2})2^nR_0^{-n}]^2$ is bounded, that is, using Stirling's formula,

$$\overline{\lim_{n\to\infty}}\, n^2|b_n|^{1/n} \leqslant \tfrac{1}{4}e^2R_0^2. \tag{6.20}$$

We now define the function h by

$$h(r,\theta,\phi) := \sum_{n=0}^{\infty}\sum_{m=-n}^{n} a_{nm}r^nP_n^{|m|}(\cos\theta)e^{im\phi} \tag{6.21}$$

and note that the absolute convergence of (6.18) implies that (6.21) converges and defines a harmonic function in all of $\mathbb{R}^3$ such that condition (a) of the theorem is satisfied. Furthermore,

$$\int_0^{2\pi}\int_0^{\pi}|h(r,\theta,\phi)|^2\sin\theta\,d\theta\,d\phi = \sum_{n=0}^{\infty} b_nr^{2n}$$

where the b_n are defined by (6.19). From (6.20) we can now conclude from Theorem 6.1 that (6.21) defines an entire function of order one and type at most R_0, that is, of exponential type less than or equal to R_0.

We shall now show that Müller's theorem can be considerably sharpened in the case when u is axially symmetric (Colton [1], [4], Sleeman [1]). This is basically due to the fact that the harmonic function in Müller's theorem is now independent of ϕ, which we indicate by writing h as $h=h(r,\theta)$, and note that h is uniquely determined by the function $h(r,0)$. Hence instead of considering the function

$$\int_0^{2\pi}\int_0^{\pi}|h(r,\theta,\phi)|^2\sin\theta\,d\theta\,d\phi$$

as in Müller's theorem we can consider $h(2iz,0)$ that is also of exponential type less than or equal to R_0. Since in this case we are not averaging h over spheres as in Müller's theorem, we might expect that sharper results are possible. By using the theory of the indicator diagram we shall show that this is indeed the case. However, in order to accomplish this goal we first must establish a reflection principle for solutions of the Helmholtz equation vanishing on a sphere (Colton [2], [4]) and Gilbert's envelope method as applied to solutions of the axially symmetric potential equation (Gilbert [1], [2]).

We begin by proving the above-mentioned reflection principle. Let u be a solution of the Helmholtz equation (6.16) defined in $D\setminus B$ where D is a bounded starlike domain containing the closed ball $B=\{x|\ |x|\leqslant a\}$. On the boundary of B we assume that u continuously assumes the boundary data

$$u=0 \quad \text{on} \quad \partial B. \tag{6.22}$$

We shall obtain a reflection principle for solutions of (6.16), (6.22) by using an integral operator that maps harmonic functions defined in $D\setminus B$ and vanishing on $r=a$ onto solutions of (6.16), (6.22). Our presentation follows that of Colton [2]. We begin by looking for a solution of (6.16) in the form

$$u(r,\theta,\phi)=h(r,\theta,\phi)+\int_a^r K(r,s)h(s,\theta,\phi)\,ds \tag{6.23}$$

where $h\in C^2(D\setminus B)\cap C(\overline{D\setminus B})$ is a harmonic function vanishing on $r=a$. Substituting (6.23) in to (6.16) and integrating by parts shows that (6.23) will be a solution of (6.16) provided K satisfies the initial value problem

$$r^2\left[K_{rr}+\frac{2}{r}K_r+K\right]=s^2\left[K_{ss}+\frac{2}{s}K_s\right] \tag{6.24a}$$

$$K(r,r)=-\frac{1}{4r}(r^2-a^2) \tag{6.24b}$$

$$K(r,a)=0. \tag{6.24c}$$

If we now set

$$\begin{aligned}\xi&=\log r\\ \eta&=\log s\end{aligned} \tag{6.25}$$

we can transform (6.24) into the form

$$M_{\xi\xi}-M_{\eta\eta}+e^{2\xi}M=0 \tag{6.26a}$$

$$M(\xi,\xi)=-\tfrac{1}{4}(e^{2\xi}-a^2) \tag{6.26b}$$

$$M(\xi,\log a)=0 \tag{6.26c}$$

where $M(\xi,\eta):=e^{(1/2)(\xi+\eta)}K(e^{\xi},e^{\eta})$. Equation (6.26) is a Goursat problem for a hyperbolic equation and by using the method of successive approximations (cf. Garabedian [2]), it can easily be shown that a unique, analytic solution to (6.26) exists in the cone $\xi\leqslant\eta$, $\eta\leqslant\log a$, or $\xi\geqslant\eta$, $\eta\geqslant\log a$. Hence we have established the existence of the operator (6.23). Since (6.23) is a Volterra integral equation for h if u is given, it is easy to show that if $u\in C^2(D\setminus B)\cap C(\overline{D\setminus B})$ is any solution of (6.16), (6.22) then u can be represented in the form (6.23) for some harmonic function defined in $D\setminus B$ and vanishing on $r=a$.

Before using (6.23) to establish a reflection principle for solutions of (6.16), we shall establish the existence of another integral operator similar to (6.23) that is needed to obtain our desired result on the far-field patterns of axially symmetric solutions of the Helmholtz equation. This operator is of the same form as (6.23), that is,

$$u(r,\theta,\phi)=h(r,\theta,\phi)+\int_a^r \tilde{K}(r,s)h(s,\theta,\phi)\,ds \tag{6.27}$$

but where now $h\in C^2(D\setminus B)\cap C^1(\overline{D\setminus B})$ is a harmonic function satisfying

$$h_r+\frac{1}{2a}h=0 \tag{6.28}$$

on $r=a$. In order for the right-hand side of (6.27) to be a solution of the Helmholtz equation (6.16) we now must require $\tilde{K}$ to be a solution of the initial value problem

$$r^2\left[\tilde{K}_{rr}+\frac{2}{r}\tilde{K}_r+\tilde{K}\right]=s^2\left[\tilde{K}_{ss}+\frac{2}{s}\tilde{K}_s\right] \tag{6.29a}$$

$$\tilde{K}(r,r)=-\frac{1}{4r}(r^2-a^2) \tag{6.29b}$$

$$\tilde{K}_s(r,a)+\frac{1}{2a}\tilde{K}(r,a)=0. \tag{6.29c}$$

By using the change of variables (6.25) and setting $\tilde{M}(\xi,\eta):=e^{(1/2)(\xi+\eta)}$ $\tilde{K}(e^{\xi},e^{\eta})$ we can now reduce (6.29) to the initial value problem

$$\tilde{M}_{\xi\xi}-\tilde{M}_{\eta\eta}+e^{2\xi}\tilde{M}=0 \tag{6.30a}$$

$$\tilde{M}(\xi,\xi)=-\tfrac{1}{4}(e^{2\xi}-a^2) \tag{6.30b}$$

$$\tilde{M}_\eta(\xi,\log a)=0. \tag{6.30c}$$

In order to construct a solution to (6.30) we introduce the function E defined as the unique solution of the characteristic initial value problem

$$E_{\xi\xi}-E_{\eta\eta}+e^{2\xi}E=0 \tag{6.31a}$$

$$E(\xi,\xi)=-\tfrac{1}{4}(e^{2\xi}-a^2) \tag{6.31b}$$

$$E(\xi,-\xi+2\log a)=-\tfrac{1}{4}(e^{2\xi}-a^2). \tag{6.31c}$$

It again follows by the method of successive approximations that a unique analytic solution of (6.31) exists in the cone $\xi\leqslant\eta$, $\eta+\xi\leqslant 2\log a$, or $\xi\geqslant\eta$, $\eta+\xi\geqslant 2\log a$. It is now easily verified that

$$\tilde{M}(\xi,\eta)=\tfrac{1}{2}[E(\xi,\eta)+E(\xi,-\eta+2\log a)]$$

defines a solution of (6.30). Hence we have established the existence of the integral operator (6.27). It is again easily verified that if $u \in C^2(D \setminus B) \cap C^1(\overline{D \setminus B})$ is any solution of (6.16) satisfying

$$u_r + \frac{1}{2a}u = 0$$

on $r = a$, then u can be represented in the form (6.27) for some harmonic function satisfying (6.28).

We are now in a position to prove the following reflection principle for solutions of the Helmholtz equation vanishing on a sphere in $\mathbb{R}^3$.

Theorem 6.6. Let $u \in C^2(D \setminus B) \cap C(\overline{D \setminus B})$ be a solution of (6.16), (6.22) and let D^* denote the set obtained by inverting $D \setminus B$ across the boundary of B, that is, $(r, \theta, \phi) \in D^*$ if and only if $(a^2/r, \theta, \phi,) \in D \setminus B$. Then u can be analytically continued as a solution of (6.16) into $(D \setminus B) \cup D^*$.

Proof. We first represent u in the form (6.23) where h is an harmonic function in $D \setminus B$ vanishing on $r = a$. Then $u \in C^2(D \setminus B) \cap C(\overline{D \setminus B})$ implies that $h \in C^2(D \setminus B) \cap C(\overline{D \setminus B})$ and hence by the Schwarz reflection principle for harmonic functions (Colton [4]), h is harmonic in $(D \setminus B) \cup D^*$. Hence by (6.23) u can also be continued into this region.

Before passing on to our next topic, Gilbert's envelope method, we pause to make two observations on Theorem 6.6. The first is that, in contrast to the Schwarz reflection principle for harmonic functions, the domain of dependence of u at a point outside B is a line segment inside B (instead of a point in the case of harmonic functions). Second, Theorem 6.6 remains valid if u only vanishes on a portion σ of the boundary of B, in which case D^* is replaced by the "truncated cone" $\{(r, \theta, \phi): (a^2/r, \theta, \phi) \in D \setminus B, (a, \theta, \phi) \in \sigma\}$.

In order to sharpen Müller's theorem in the case of axial symmetry it is necessary to determine the location of the singularities of axially symmetric harmonic functions in terms of the regularity properties of their axial values. To this end we shall need the following theorem.

Theorem 6.7 (The Envelope Method). Let F be defined by

$$F(z) := \oint_C K(z, \zeta)\, d\zeta$$

where K is an analytic function of its two independent complex variables except for possible singularities lying on the set $G_0 = \{(z, \zeta): S(z, \zeta) = 0\}$ where S is analytic and C is a simple closed contour. Then F is analytic for all points z such that $(z, \zeta) \notin G_0 \cap G_1$ for any ζ where $G_1 = \{(z, \zeta): \partial S(z, \zeta)/\partial \zeta = 0\}$.

Proof. Let F be analytic at $z = z_0$ and hence in a neighborhood $N(z_0)$ of z_0. Let γ be a path beginning at $z = z_0$. Then F can be analytically continued along γ as long as no point of γ corresponds to a singularity of the integrand on C. Suppose now that F has been continued along γ to a point $z=z_1$ corresponding to a singularity of the integrand K at $(z,\zeta)=(z_1,\alpha)$. However, if $\partial S(z,\alpha)/\partial\zeta \neq 0$ we can locally write

$$S(z_1,\zeta)\approx(\zeta-\alpha)(\partial S(z_1,\zeta)/\partial\zeta)$$

and deform C about the point $\zeta=\alpha$ by allowing it to follow a portion of the circle $|\zeta-\alpha|=\varepsilon$ for ε sufficiently small. This implies that F is analytic at $z=z_1$ and the theorem is proved.

We now apply the envelope method to solutions of the axially symmetric potential equation

$$\frac{\partial^2 h}{\partial z^2}+\frac{\partial^2 h}{\partial r^2}+\frac{1}{r}\frac{\partial h}{\partial r}=0. \tag{6.32}$$

A simple power series argument shows that if h is an analytic solution of (6.32) in some neighborhood of the origin, then h is an even function of r and is uniquely determined by its axial values $f(z):=h(z,0)$. Furthermore, we can explicitly represent h in terms of f by the integral representation $h=\mathbf{A}[f]$ where

$$\mathbf{A}[f]:=\frac{1}{2\pi i}\oint_C f(\sigma)\frac{d\zeta}{\zeta}$$

where C is a simple closed contour surrounding the origin and $\sigma=z+ir/2(\zeta+\zeta^{-1})$. The following theorem is due to Gilbert [1], [2]; see also Erdélyi [1] and Henrici [1].

Theorem 6.8. If the only singularities of f in the complex plane are at $z=\alpha$, then the only possible singularities of h on its first Riemann sheet are at $z+ir=\alpha$ and $z-ir=\alpha$.

Proof. We write $h=\mathbf{A}[f]$. Then by the envelope method the only possible singularities of h are in the set $G=G_0\cap G_1$ where

$$G_0=\left\{(z,r,\zeta)|\,(z-\alpha)\zeta+\frac{ir}{2}(\zeta^2+1)=0\right\}$$

$$G_1=\{(z,r,\zeta)|\,(z-\alpha)+ir\zeta=0\}.$$

Eliminating ζ gives

$$G=G_0\cap G_1=\{(z,r)|(z-\alpha)^2+r^2=0\}$$

which implies $z+ir=\alpha$ or $z-ir=\alpha$.

We note that singularities of h can exist on other sheets of the Riemann surface of h if h has branch points (Gilbert [1]). Furthermore, it is possible to show that $z+ir=\alpha$ and $z-ir=\alpha$ are indeed singular points of h if f is singular at $z=\alpha$ (Gilbert [1], [2]).

We now return to Müller's theorem and make the assumption that u is an axially symmetric solution of (6.16) for $r \geqslant R_0$, that is, u is independent of the angle ϕ. Then the far-field pattern has an expansion of the form

$$F(\theta)=\sum_{m=0}^{\infty} a_n P_n(\cos\theta) \tag{6.33}$$

and for $r \geqslant R_0$

$$u(r,\theta)=\sum_{n=0}^{\infty} a_n i^{n+1} h_n^{(1)}(r) P_n(\cos\theta) \tag{6.34}$$

where the series is absolutely and uniformly convergent. The harmonic function h in Müller's theorem is given by

$$h(r,\theta)=\sum_{n=0}^{\infty} a_n r^n P_n(\cos\theta)$$

and is uniquely determined by the function

$$h(z,0)=\sum_{n=0}^{\infty} a_n z^n \tag{6.35}$$

since $P_n(1)=1$. From (6.19), (6.20) and Theorem 6.1 we see that $h(z,0)$ is an entire function of order one and type at most $R_0/2$, which implies $h(2iz,0)$ is an entire function of exponential type less than or equal to R_0. Considering the complex z plane as superimposed over the Euclidean plane, we shall show that if G is the indicator diagram of $h(2iz,0)$ and G^* is its conjugate, then u, as defined by (6.34), can be continued as a solution of (6.16) into the exterior of $G \cup G^*$. Note how this is a considerable strengthening of Müller's theorem: In terms of $h(2iz,0)$ Müller's theorem says that if the scattering obstacle is inside $r=R_0$, then this function must be of exponential type less than or equal to R_0, whereas the result we are about to show now goes further and relates the indicator diagram of $h(2iz,0)$ to the domain of regularity of u.

Theorem 6.9. Let u be an axially symmetric solution of (6.16) regular for $r \geqslant R_0$ and satisfying the radiation condition (6.17). Let F be the far-field pattern of u and G the indicator diagram of $h(2iz,0)$ where h is related to the far-field pattern F by (6.33), (6.35). Then if G^* denotes the conjugate of G, u is a solution of (6.16) in the exterior of $G \cup G^*$.

Proof. We first want to show that the function defined by

$$g(z) := \sum_{n=0}^{\infty} a_n i^{n+1} h_n^{(1)}(R_0)\left(\frac{z}{R_0}\right)^{-n-1} \tag{6.36}$$

can be analytically continued into the exterior of G^*. We note that from (6.34) the series (6.36) converges and defines an analytic function for $|z| > R_0$. Hence to show g is analytic in the exterior of G^*, by Polya's theorem and the Hadamard multiplication of singularities theorem (cf. Colton [4]), it suffices to show that the singularities of

$$G(z) := \sum_{n=0}^{\infty} \frac{h_n^{(1)}(R_0)}{n!2^n}\left(\frac{z}{R_0}\right)^{-n} \tag{6.37}$$

lie on the closed interval $[0,1]$. But we can actually sum the series (6.37) (cf. Erdélyi et al. [1]) to get

$$G(z) = -\frac{i}{R_0}\left(1-\frac{1}{z}\right)^{-1/2}\exp\left[iR_0\left(1-\frac{1}{z}\right)^{1/2}\right],$$

that is, the only singularities of G are branch points at zero and one.

We now construct an axially symmetric harmonic function v such that on the axis of symmetry v is equal to $g(r)$:

$$v(r,\theta) := \sum_{n=0}^{\infty} a_n i^{n+1} h_n^{(1)}(R_0)\left(\frac{r}{R_0}\right)^{-n-1} P_n(\cos\theta).$$

Then $(1/r)v(R_0^2/r,\theta)$ is an axially symmetric harmonic function in a neighborhood of the origin and hence from Theorem 6.8 we can conclude that v is harmonic in the exterior of $G \cup G^*$. Note that by construction we have $v(R_0,\theta) = u(R_0,\theta)$.

Now define the axially symmetric harmonic function h by

$$h(r,\theta) = \frac{1}{2}\left[v(r,\theta) + \frac{R_0}{r}v\left(\frac{R_0^2}{r},\theta\right)\right].$$

Then h is harmonic in the domain exterior to $G \cup G^*$ and interior to the inversion of $G \cup G^*$ across the circle $r = R_0$, and satisfies

$$h_r + \frac{1}{2R_0}h = 0$$

for $r = R_0$. If we now use the operator (6.27) to define U by

$$U(r,\theta) := h(r,\theta) + \int_{R_0}^{r} \tilde{K}(r,s)h(s,\theta)\,ds$$

we see that U is a solution of (6.16) in the same region that h is harmonic and $U(R_0, \theta) = h(R_0, \theta)$. Hence $w = U - u$ is a solution of (6.16) in the region bounded by $r = R_0$ and the inversion of $\partial G \cup \partial G^*$ across this circle and satisfies $w(R_0, \theta) = 0$. Therefore by Theorem 6.6 we can continue w into the complement of $G \cup G^*$ with respect to the disk $r \leqslant R_0$. Since U is already known to be regular in this region, we can conclude that u must also be regular there, and since we already know that u is regular for $r \geqslant R_0$, the theorem is proved.

6.3 UNIQUENESS OF SOLUTIONS TO THE INVERSE SCATTERING PROBLEM

We now want to establish the uniqueness of the solution to the inverse scattering problems that we shall be considering in the next two chapters. We shall first formulate these problems for acoustic waves and then for electromagnetic waves. Although we shall formulate our problems in $\mathbb{R}^3$, they may also be considered in $\mathbb{R}^2$ with obvious modifications. In what follows D shall always denote a bounded domain with C^2 boundary ∂D with λ and ψ being continuous functions defined on ∂D. $u^i(x) = e^{ik(x, \alpha)}$ will denote a fixed plane wave solution of the Helmholtz equation

$$\Delta u + k^2 u = 0 \tag{6.38}$$

and (E^i, H^i) a plane wave solution of the time-harmonic Maxwell equations

$$\operatorname{curl} H + ikE = 0, \qquad \operatorname{curl} E - ikH = 0 \tag{6.39}$$

where $k > 0$ and the superscript i denotes "incident wave." As before, ν will denote the unit outward normal to ∂D.

Problem A1

Let $u = u^i + u^s$ be a solution of (6.38) in the exterior of D such that u^s satisfies the radiation condition

$$\left(\operatorname{grad} u^s(x), \frac{x}{|x|}\right) - iku^s(x) = o\left(\frac{1}{|x|}\right), \qquad |x| \to \infty \tag{6.40}$$

and $u = 0$ on ∂D. From a knowledge of the far-field pattern of u^s, determine D.

We are purposely being vague about what is meant by a "knowledge" of the far-field pattern since we desire to be flexible in this regard. In general it will mean knowing the far-field pattern at least for an interval of angles and a fixed set of frequencies, the precise requirements to be stated explicitily in each theorem.

Problem A2

Let $u = u^i + u^s$ be a solution of (6.38) in the exterior of a given domain D such that u^s satisfies the radiation condition (6.40) and $(\partial u/\partial \nu) + \lambda u = 0$ on ∂D where $\operatorname{Im} \lambda \geq 0$. From a knowledge of the far-field pattern of u^s, determine $\lambda = \lambda(x)$, $x \in \partial D$.

Problem E1

Let $E = E^i + E^s$, $H = H^i + H^s$ be a solution of (6.39) in the exterior of D such that (E^s, H^s) satisfies the radiation condition

$$\left[H^s(x), \frac{x}{|x|} \right] - E^s(x) = o\left(\frac{1}{|x|} \right); \qquad |x| \to \infty \tag{6.41}$$

and $[\nu, E] = 0$ on ∂D. From a knowledge of the far-field pattern for E^s, determine D.

Problem E2

Let $E = E^i + E^s$, $H = H^i + H^s$ be a solution of (6.39) in the exterior of a given domain D such that (E^s, H^s) satisfies the radiation condition (6.41) and

$$[\nu, [\nu, H]] - \psi[\nu, E] = 0 \quad \text{on} \quad \partial D \tag{6.42}$$

where $\operatorname{Re} \psi > 0$. From a knowledge of the far-field pattern for E^s, determine $\psi = \psi(x)$, $x \in \partial D$.

Although we shall not discuss it in this book, the Neumann boundary-value problem corresponding to Problem A1 can be treated in an almost identical manner to the Dirichlet problem considered here. Furthermore, as is well known (cf. Baker and Copson [1]), the two-dimensional version of Problem E1 can be reduced to the two-dimensional problem corresponding to Problem A1. Perhaps it is worthwhile to point out at this time that our aim in this book is not to consider all possible problems that can be treated by our methods, but rather to discuss them for the "canonical" problems A1, A2, E1, E2, and then leave it to the reader to adjust the techniques given here to those problems that are minor variations of these canonical problems.

The following uniqueness theorem for Problem A1 is due to Schiffer (Lax and Phillips [1]) and is based on the analyticity of solutions to the Helmholtz equation (Theorem 3.5).

Theorem 6.10. Let F be the far-field pattern of u^s in Problem A1. Then D is uniquely determined by a knowledge of F on some surface element of the unit sphere and k on any interval of the positive real axis.

Proof. Since we have shown in Section 6.2 that F is an analytic function, knowing F on a surface element of the unit sphere implies that F is known on the entire unit sphere by analytic continuation. Now suppose there existed two obstacles D_1 and D_2 having the same far-field pattern F. Consider first the case when $\bar{D}_1$ and $\bar{D}_2$ are disjoint. Then since F uniquely determines u^s outside a ball containing D_1 and D_2 in its interior, we can conclude by analytic continuation that u^s is an entire solution of (6.38) satisfying the radiation condition (6.40). But this implies $u^s \equiv 0$ (cf. Theorem 3.40), that is, $u^i = 0$ on ∂D_1. But this is a contradiction since u^i is a plane wave. Now consider the case when $D = D_1 \cap D_2$ has a nonempty interior. Then by the above arguments u is a solution of (6.38) in $D_1 \setminus \bar{D}$ or $D_2 \setminus \bar{D}$ satisfying $u = 0$ on the boundary, that is, k^2 is an eigenvalue of the Laplacian in this region. But since the set of eigenvalues for this problem is discrete, we again arrive at a contradiction. The case when D_1 and D_2 are tangent can be easily handled by simply considering D_1 instead of $D_1 \setminus \bar{D}$ or $D_2 \setminus \bar{D}$ and this completes the proof of the theorem.

The above method of proof immediately carries over to the case of Problem E1, and for completeness we simply state the theorem without proof.

Theorem 6.11. Let F be the far-field pattern of E^s in Problem E1. Then D is uniquely determined by a knowledge of F on some surface element of the unit sphere and k on any interval of the positive real axis.

We now turn to the uniqueness theorems for Problems A2 and E2. In contrast to the case of Problems A1 and E1 it turns out in this case that different proofs must be derived for each of these problems, although both ultimately depend upon an application of Holmgren's uniqueness theorem. A further interesting contrast to Problems A1 and E1 is that in this case it is only necessary to know the far-field patterns for a single fixed value of the wave number k instead of an interval of k values. As we have just stated, both proofs depend upon Holmgren's uniqueness theorem, and hence for convenience we state this theorem here as it applies to the situation in which we are interested. For a proof of Holmgren's theorem we refer the reader to Colton [4] or Garabedian [2].

Theorem 6.12 (Holmgren's Uniqueness Theorem). Let $u \in C^2(R^2 \setminus \bar{D}) \cap C^1(R^3 \setminus D)$ be a solution of (6.38) in the exterior of D such that u has zero Cauchy data on a surface element of ∂D. Then u is identically zero.

We first consider Problem A2 (Colton and Kirsch [1]).

Theorem 6.13. Let F be the far-field pattern of u^s in Problem A2. Then λ is uniquely determined by a knowledge of F on some surface element of the unit sphere and fixed $k > 0$.

Proof. As in Theorem 6.11 we can assume F is known on the entire unit sphere. Now suppose there were two solutions λ_1 and λ_2 of Problem A2. We want to show that $\lambda_1(x) = \lambda_2(x)$ for all $x \in \partial D$. Let u_1 and u_2 be the solutions of (6.38) corresponding to λ_1 and λ_2. Then we can conclude from Corollary 3.9 that $u_1 = u_2$ outside of D and hence $\partial u_1 / \partial \nu = \partial u_2 / \partial \nu$ on ∂D. Then from the boundary condition satisfied by u_1 and u_2 we have

$$(\lambda_1 - \lambda_2)u_1 = 0 \quad \text{on} \quad \partial D. \tag{6.43}$$

We now note that if $u_1 \equiv 0$ on a surface element S of ∂D, then from the boundary condition satisfied by u_1 we would have $\partial u_1/\partial \nu \equiv 0$ on S and hence by Holmgren's uniqueness theorem $u_1 \equiv 0$ in $\mathbb{R}^3 \setminus D$. But this is a contradiction since $u_1 = u^i + u_1^s$ and u_1^s satisfies the radiation condition but u^i does not. Hence u_1 cannot vanish on any surface element of ∂D and therefore if $x \in \partial D$ there exists a sequence of points $x_n \to x$ such that $u_1(x_n) \neq 0$. Then from (6.43) we have $\lambda_1(x_n) = \lambda_2(x_n)$ and since λ_1 and λ_2 are continuous we have $\lambda_1(x) = \lambda_2(x)$. Since x was an arbitrary point on ∂D this completes the proof of the theorem.

Note that in the proof of Theorem 6.13 no direct use was made of the fact that $\operatorname{Im} \lambda \geqslant 0$. This restriction was incorporated in the formulation of Problem A2 only for the purpose of consistency since this condition is required for the proof of the existence and uniqueness of a solution to the corresponding direct scattering problem (Theorems 3.37 and 3.38).

We now conclude this section and the chapter by a proof of the uniqueness of the solution to Problem E2 (Colton and Kress [2]). As with Problem A2 no use will be made of the fact that $\operatorname{Re} \psi > 0$, although this condition is required for the proof of the existence and uniqueness of the direct scattering problem (Theorems 4.45 and 4.46). The necessity to modify the proof of Theorem 6.13 in order to treat Problem E2 arises from the fact that the boundary condition satisfied by (E, H) does not explicitly involve the Cauchy data of E and H.

Theorem 6.14. Let F be the far-field pattern of E^s in Problem E2. Then ψ is uniquely determined by a knowledge of F on some surface element of the unit sphere and fixed $k > 0$.

Proof. We can again assume by analyticity that F is known on the entire unit sphere. Suppose there were two solutions ψ_1 and ψ_2 of Problem E2. We again want to show that $\psi_1(x) = \psi_2(x)$ for all $x \in \partial D$. Let (E_1, H_1) and (E_2, H_2) be the total fields corresponding to ψ_1 and ψ_2. Then by Corollary 3.9 we can conclude that $E_1 = E_2$ and $H_1 = H_2$ in $\mathbb{R}^3 \setminus D$. Hence from the boundary condition (6.42), we have that

$$(\psi_1 - \psi_2)[\nu, E_1] = 0 \quad \text{on} \quad \partial D. \tag{6.44}$$

Suppose $[\nu, E_1] = 0$ on a surface element $S \subset \partial D$. Then since from (6.42)

$$[\nu, [\nu, H_1]] = 0 \quad \text{on} \quad S$$

we have that H_1 is normal to S and hence using (2.75)

$$ik(\nu, E_1) = -(\nu, \operatorname{curl} H_1) = \operatorname{Div}[\nu, H_1] = 0 \quad \text{on} \quad S.$$

Therefore $E_1 = 0$ on S. We note that from the regularity result contained in Theorem 3.27 we can now conclude that $E_1 \in C^1(\mathbb{R}^3 \setminus \bar{D} \cup S)$. Now let $x_0 \in S$ and without loss of generality assume x_0 is the origin and that the outward normal to ∂D at x_0 points along the positive x_3-axis where $x = (x_1, x_2, x_3)$. Recall that

$$\operatorname{curl} E_1 = \left(\left(\frac{\partial E^{(3)}}{\partial x_2} - \frac{\partial E^{(2)}}{\partial x_3} \right), \left(\frac{\partial E^{(1)}}{\partial x_3} - \frac{\partial E^{(3)}}{\partial x_1} \right), \left(\frac{\partial E^{(2)}}{\partial x_1} - \frac{\partial E^{(1)}}{\partial x_2} \right) \right)$$

$$\operatorname{div} E_1 = \frac{\partial E^{(1)}}{\partial x_1} + \frac{\partial E^{(2)}}{\partial x_2} + \frac{\partial E^{(3)}}{\partial x_3}$$

where $E_1 = (E^{(1)}, E^{(2)}, E^{(3)})$ and note that from Maxwell's equations and the regularity of E_1 in $\mathbb{R}^3 \setminus \bar{D} \cup S$ we have $\operatorname{div} E_1 = 0$ at x_0. Hence since $\operatorname{curl} E_1$ is normal to ∂D at x_0, we have, using the facts that ∂D is in class C^2 and $E_1 = 0$ on S, that

$$\frac{\partial E^{(1)}}{\partial x_3} = \frac{\partial E^{(2)}}{\partial x_3} = \frac{\partial E^{(3)}}{\partial x_3} = 0 \quad \text{on} \quad x_0 .$$

Since x_0 was an arbitrary point of S, it follows that each component of E_1 vanishes along with its normal derivative on S. Since E_1 is a solution of the vector Helmholtz equation it now follows from Holmgren's uniqueness theorem that $E_1 \equiv 0$ in $\mathbb{R}^3 \setminus \bar{D}$. But this is a contradiction since E^i does not satisfy the radiation condition. Hence $[\nu, E_1] \neq 0$ on any surface element $S \subset \partial D$. We can now conclude from (6.44) exactly as in the proof of Theorem 6.13 that $\psi_1(x) = \psi_2(x)$ for all $x \in \partial D$, and this completes the proof of the theorem.

7

IMPROPERLY POSED PROBLEMS AND COMPACT FAMILIES

Our aim in this chapter is to introduce methods for dealing with linear improperly posed problems of the type that arise in studying inverse scattering problems. The approaches that we shall focus on are Tikhonov's selection method, Ivanov's idea of a quasi-solution, and the Backus–Gilbert method. These are perhaps the simplest approaches for dealing with a wide variety of improperly posed problems and are based on the idea of making use of a priori assumptions in order to restrict the solution sought after to lie in a compact set. The selection method and the concept of a quasi-solution have the added advantage of being easily adaptable to studying nonlinear improperly posed problems of the type we shall discuss in Chapter 8. There are, of course, more general methods for studying improperly posed problems than the ones we discuss here, for example, Tikhonov's regularization method (cf. Tikhonov and Arsenin [1]). However, our aim in this chapter is to outline only a few simple procedures that are all based on the concept of compactness and are well suited to treating improperly posed problems in scattering theory rather than attempting to give any kind of a survey of the numerous methods that can be used to deal with such problems. For an idea of alternate approaches the reader is referred to the previously mentioned book of Tikhonov and Arsenin as well as the survey papers by Bertero, De Mol, and Viano [1], Angell and Nashed [1], and Nashed [1].

After our discussion of the above methods for treating improperly posed problems we shall illustrate their applicability by considering two inverse problems arising in scattering theory. The first of these is to determine the surface impedance of an obstacle from a knowledge of the far-field pattern of the scattered acoustic wave (Problem A2 of Chapter 6) and the second is to determine the shape of the scattering obstacle from a knowledge of the scattering cross section, given the fact that the obstacle is acoustically soft (Problem A1 of Chapter 6). In this second problem we shall assume that an initial approximation to the shape is known, thus linearizing the problem. The full nonlinear problem will be dealt with in Chapter 8.

The final section of this chapter is concerned with compact families of univalent functions and is designed to connect the subject matter of this chapter with that of Chapter 8. The theory of conformal mappings and univalent functions exhibits a rich interplay between geometry and analysis and hence it is not surprising that this area of mathematics is particularly useful in the investigation of two-dimensional inverse scattering problems. In this chapter we shall present the basic theorems on normal families of univalent functions and in particular introduce an appropriate compact family that will be exploited in Chapter 8 to study the problem of determining the shape of a bounded two-dimensional obstacle from a knowledge of the far-field pattern.

7.1 A PRIORI ASSUMPTIONS AND THE SOLUTION OF IMPROPERLY POSED PROBLEMS

Let $\mathbf{A}: X \to Y$ be a bounded (and hence continuous) linear operator mapping the Banach space X into the Banach space Y and let $Z \subset Y$ denote the range of $\mathbf{A}$. Assume further that $\mathbf{A}^{-1}$ exists on Z, although no assumption is made on the continuity of $\mathbf{A}^{-1}$. (In what follows the assumption of the existence of $\mathbf{A}^{-1}$ can be removed by considering $\mathbf{A}$ as a set valued map; however, for the sake of simplicity, we shall not pursue this generalization here.) Our aim is to "solve" the operator equation

$$\mathbf{A}x = y, \qquad y \in Y \tag{7.1}$$

such that the solution x is stable under small changes of the right-hand side y. Note that unless $y \in Z$, no solution to (7.1) exists, and even if $y \in Z$, small perturbations of y in Z can cause large perturbations of the solution x, since $\mathbf{A}^{-1}$ is not in general continuous. The approach that we shall now present to "solve" (7.1) is based on restricting the class of admissible solutions to lie a priori in a compact set $X_0 \subset X$ and to define the "solution" of (7.1) to be that element $x \in X$ such that $\|\mathbf{A}x - y\|$ is a minimum. Restricting the class of admissible solutions to lie in a compact set is known as Tikhonov's selection method.

We begin our analysis by proving the following theorem that demonstrates how continuity of the inverse operator can be restored by restricting the domain of $\mathbf{A}$ to be a compact set.

Theorem 7.1. Let X_0 be a compact set of a Banach space X and $\mathbf{A}$ a continuous operator (not necessarily linear) defined on X_0 such that $\mathbf{A}^{-1}$ exists on $\mathbf{A}(X_0)$. Then $\mathbf{A}^{-1}$ is continuous on $\mathbf{A}(X_0)$.

Proof. Since the continuous image of a compact set is compact, $\mathbf{A}(X_1)$ is a compact set for any closed subset $X_1 \subset X_0$, that is, the inverse image under the mapping $\mathbf{A}^{-1}$ of an arbitrary closed set $X_1 \subset X_0$ is closed. But this statement implies that $\mathbf{A}^{-1}$ is continuous on $\mathbf{A}(X_0)$.

By itself Theorem 7.1 is not of much use in solving (7.1) since in most cases of practical importance it is not possible to decide if $y \in \mathbf{A}(X_0)$ or not. In particular, if y is arrived at through measurements subject to a certain amount of experimental error we cannot in general claim that this "noisy" data lies in $\mathbf{A}(X_0)$. In this case we define a quasi-solution of (7.1) to be any element $x_0 \in X_0$ such that

$$\|\mathbf{A}x_0 - y\| = \inf_{x \in X_0} \|\mathbf{A}x - y\|. \tag{7.2}$$

Note that since X_0 is compact the infimum on the right-hand side of (7.2) exists, and that there may be more than one quasi-solution to equation (7.1). The following theorem gives sufficient conditions for a quasi-solution to be unique and depend continuously on $y \in Y$.

Theorem 7.2. Let $\mathbf{A}$: $X \to Y$ be a continuous linear operator mapping the Banach space X into the Banach space Y. Assume further that the compact set $X_0 \subset X$ is convex and that the Banach space Y is strictly convex. Then the quasi-solution of (7.1) is unique and depends continuously on the element y.

Proof. Let $x_0 \in X_0$ be a quasi-solution and set $y_0 = \mathbf{A}x_0$. Since X_0 is convex, so is $\mathbf{A}(X_0)$, and y_0 is the projection of y on $\mathbf{A}(X_0)$. Since Y is strictly convex, this projection is unique. Let $\mathbf{P}$: $Y \to \mathbf{A}(X_0)$ denote this projection operator. Then $\mathbf{P}$ is well defined, continuous, and we can write $x_0 = \mathbf{A}^{-1}y_0 = \mathbf{A}^{-1}\mathbf{P}y$. By Theorem 7.1 $\mathbf{A}^{-1}$ is continuous on $\mathbf{A}(X_0)$, and hence we can conclude that $\mathbf{A}^{-1}\mathbf{P}$ is continuous on Y. The theorem is now proved.

From the point of view of applications it is, of course, important to provide a constructive procedure for approximating a quasi-solution to (7.1). We shall present one such procedure now, where the operator $\mathbf{A}$ and spaces X_0 and Y are as in Theorem 7.2. Let $X_1 \subset X_2 \subset X_3 \ldots$ be a sequence of nested compact sets in X_0 such that $\cup_{n=1}^{\infty} X_n = X_0$ and let Q_n denote the set of all quasi-solutions of $\mathbf{A}x = y$ with x restricted to the set X_n.

Theorem 7.3. Let $x_0 \in X_0$ be the unique quasi-solution of $\mathbf{A}x = y$, $y \in Y$, where $\mathbf{A}$, X_0, and Y are as in Theorem 7.2. Then if $x_n \in Q_n$, $\lim_{n\to\infty} x_n = x_0$.

Proof. Let $Z_0 = \mathbf{A}(X_0)$ and $Z_n = \mathbf{A}(X_n)$. Denote by $d(y, Z_n)$ the distance between y and Z_n, that is,

$$d(y, Z_n) := \inf_{z \in Z_n} \|y - z\|.$$

Then $d(y, Z_1) \geqslant \cdots d(y, Z_n) \geqslant \cdots d(y, Z_0) = \|y - \mathbf{A}x_0\|$ where x_0 is the unique quasi-solution of $\mathbf{A}x = y$. Then since $\cup_{n=1}^{\infty} X_n = X_0$ we have

$$\lim_{n \to \infty} d(y, Z_n) = \|y - \mathbf{A}x_0\|.$$

But $d(y, Z_n) = d(y, \mathbf{A}(Q_n))$ and hence

$$\lim_{n \to \infty} d(y, \mathbf{A}(Q_n)) = \|y - \mathbf{A}x_0\|.$$

Since $\mathbf{A}(Q_n)$ is a closed subset of the compact set Z_n it is compact, and hence there exists a sequence (z_n), $z_n \in \mathbf{A}(Q_n)$, such that

$$\|z_n - y\| = \inf_{z \in \mathbf{A}(Q_n)} \|z - y\|.$$

Since Z_0 is compact, $Z_1 \subset Z_2 \subset \cdots \subset Z_n \subset \cdots \subset Z_0$, the sequence (z_n) has a limit point in Z_0, denoted by z_0. Let $(z_{n(k)})$ be a subsequence of (z_n) such that

$$\lim_{k \to \infty} \|z_{n(k)} - z_0\| = 0.$$

Then from the above we have that

$$\begin{aligned} \|y - z_0\| &= \lim_{k \to \infty} \|y - z_{n(k)}\| \\ &= \lim_{k \to \infty} d\big(y, \mathbf{A}(Q_{n(k)})\big) \\ &= \|y - \mathbf{A}x_0\|, \end{aligned}$$

and from the uniqueness of the quasi-solution x_0 we can conclude that

$$z_0 = \mathbf{A}x_0.$$

Since z_0 was an arbitrary limit point of (z_n), we have

$$\lim_{n \to \infty} \|z_n - \mathbf{A}x_0\| = 0$$

and hence

$$\begin{aligned} \lim_{n \to \infty} \|x_n - x_0\| &= \lim_{n \to \infty} \|\mathbf{A}^{-1}\mathbf{A}x_n - \mathbf{A}^{-1}\mathbf{A}x_0\| \\ &= \lim_{n \to \infty} \|\mathbf{A}^{-1}(z_n - \mathbf{A}x_0)\| \\ &= 0 \end{aligned}$$

due to the fact that from Theorem 7.1, $\mathbf{A}^{-1}$ is continuous on Z_0 and $\mathbf{A}^{-1}0 = 0$.

We note that if X_n is taken as an n-dimensional subset, the problem of approximating the quasi-solution x_0 of $\mathbf{A}x = y$ reduces to the problem of minimizing the functional $\|\mathbf{A}x - y\|$ for $x \in X_n$, that is, to finding the minimum of a function of n variables.

Although the approach described above for solving the improperly posed operator equation is quite general, it suffers from several defects. In particular the assumption on the existence of $\mathbf{A}^{-1}$ is often not valid for practical inverse problems. In what follows we shall present a method for solving a rather specialized class of improperly posed equations of the form $\mathbf{A}x = y$ where it is no longer assumed that $\mathbf{A}^{-1}$ exists. Indeed, the nonuniqueness of the solution will be seen to play a central role in deciding what is meant by a solution and how to approximate it. The method we have in mind is due to Backus and Gilbert [1] and is designed to treat linear moment problems of the form

$$\mu_n = \int_0^1 \phi(x) g_n(x)\, dx \tag{7.3}$$

where the g_n are known linearly independent continuous real-valued functions defined on $[0,1]$, the μ_n are given constants, and ϕ is a real-valued continuous function defined on $[0,1]$ that is to be determined from the relations (7.3) where it is only assumed that a finite number of the μ_n are known, that is, $\mu_0, \mu_1, \ldots, \mu_N$ for some integer N. Note that the solution of (7.3) is nonunique since if $\tilde{\phi}$ is any function orthogonal to $g_0, g_1, \ldots, g_N$ then

$$\int_0^1 \tilde{\phi}(x) g_n(x)\, dx = 0.$$

Note also that ϕ does not depend continuously on the μ_n, $n = 0, 1, \ldots, N$. Indeed, by the Riemann–Lebesgue lemma

$$\lim_{m \to \infty} \int_0^1 \sin mx\, g_n(x)\, dx = 0$$

for each fixed n, and hence the numbers μ_n, $n = 0, 1, \ldots, N$, tending to zero does not imply that the "solution" of (7.3) tends to zero unless further restrictions are placed on the class of admissible solutions to (7.3). As we shall see in the next section, moment problems of the form (7.3) arise in the study of inverse scattering problems where the μ_n correspond to $N+1$ distinct measurements of the far-field pattern or to the scattering amplitudes corresponding to $N+1$ distinct incoming waves. In these cases the region of integration in (7.3) is no longer $[0,1]$ but rather the surface of a domain in $\mathbb{R}^2$ or $\mathbb{R}^3$; however, this difference has no effect on the validity of the Backus–Gilbert method and the one-dimensional interval of integration in (7.3) is taken purely for the sake of notational simplicity.

We first consider the case when the μ_n, $n = 0, 1, \ldots, N$, are known exactly. The Backus–Gilbert method for solving (7.3) in this case is to first choose functions a_n such that

$$\delta_N(x, y) := \sum_{n=0}^{N} g_n(x) a_n(y) \tag{7.4}$$

is a delta sequence or, more specifically, such that for each y

$$\int_0^1 \delta_N(x, y)\, dx = 1 \tag{7.5}$$

and

$$\int_0^1 (x-y)^2 [\delta_N(x, y)]^2\, dx \tag{7.6}$$

is minimized. We denote this minimum value by $\varepsilon = \varepsilon(N, y)$. Note that (after possibly defining a new unknown function $\tilde{\phi} = \phi e^{\alpha x}$ for α a constant) we can assume without loss of generality that

$$c := \int_0^1 g_0(x)\, dx \neq 0,$$

that is, there exist numbers a_n such that (7.5) is valid. Furthermore, if we assume the unknown function ϕ in (7.3) lies in a class U of functions satisfying (7.3) and that in addition are Lipschitz continuous with uniformly bounded Lipschitz constant, that is,

$$|\phi(x) - \phi(y)| \leqslant M|x-y|$$

where M is a positive constant independent of ϕ, then from the inequality

$$\begin{aligned}
\left|\frac{\mu_0}{c} - \phi(y)\right| &= \left|\frac{1}{c}\int_0^1 \phi(x) g_0(x)\, dx - \phi(y)\right| \\
&= \left|\frac{1}{c}\int_0^1 [\phi(x) - \phi(y)] g_0(x)\, dx\right| \\
&\leqslant \frac{M}{c}\int_0^1 |x-y|\, |g_0(x)|\, dx \\
&=: M_0
\end{aligned}$$

we can conclude that

$$|\phi(y)| \leqslant M_0 + \left|\frac{\mu_0}{c}\right|$$

where the positive constant M_0 is independent of ϕ. Hence from the Arzéla–Ascoli theorem (Theorem 1.12) we can conclude that the set U is compact in $C[0,1]$. Assuming a priori that $\phi \in U$, an approximation to the solution of (7.3) is now given by

$$\phi_N(x) = \sum_{n=0}^{N} \mu_n a_n(x). \tag{7.7}$$

Indeed, we have the estimates

$$\begin{aligned} |\phi_N(y) - \phi(y)| &= \left| \int_0^1 \phi(x)\delta_N(x,y)\,dx - \phi(y) \right| \\ &= \left| \int_0^1 [\phi(x) - \phi(y)]\delta_N(x,y)\,dx \right| \\ &\leqslant \int_0^1 M|x-y|\,|\delta_N(x,y)|\,dx \\ &\leqslant M\left[\int_0^1 |x-y|^2 |\delta_N(x,y)|^2\,dx \right]^{1/2} \\ &\leqslant M\varepsilon^{1/2}. \end{aligned}$$

Thus an approximation to ϕ can be found provided we know the constant M and can construct the functions a_n. One approach for doing this is as follows. For each y we can write (7.6) as

$$\int_0^1 (x-y)^2 [\delta_N(x,y)]^2\,dx = \sum_{n,m=0}^{N} S_{nm}(y) a_n a_m \tag{7.8}$$

where

$$S_{nm}(y) := \int_0^1 (x-y)^2 g_n(x) g_m(x)\,dx$$

$$a_n = a_n(y)$$

and rewrite (7.5) as

$$\sum_{n=0}^{N} a_n \int_0^1 g_n(x)\,dx = 1. \tag{7.9}$$

Hence our problem is to minimize the quadratic function (7.8) subject to the linear constraint (7.9) or, in matrix notation,

$$\begin{aligned} &\text{minimize} \quad a^T\mathbf{S}a \\ &\text{subject to} \quad g^Ta = 1 \end{aligned} \tag{7.10}$$

where the superscript denotes transpose, $\mathbf{S} = (S_{nm})$, $a = (a_0, a_1, \dots, a_N)^T$, and

$$g = \left(\int_0^1 g_0(x)\,dx, \int_0^1 g_1(x)\,dx, \dots, \int_0^1 g_N(x)\,dx\right)^T.$$

Geometrically $a^T\mathbf{S}a = \gamma$ represents a family of ellipsoids and $g^Ta = 1$ a hyperplane. Hence problem (7.10) is to find the smallest ellipsoid of the above family having a nonempty intersection with the hyperplane $g^Ta = 1$. Analytically the solution to (7.10) can be found by introducing the Lagrange multiplier λ and solving the system

$$\mathbf{S}a = \lambda g$$

$$g^Ta = 1. \tag{7.11}$$

The solution of (7.11) is given by

$$\lambda = \frac{1}{g^T\mathbf{S}^{-1}g}$$

$$a = \frac{1}{g^T\mathbf{S}^{-1}g}\mathbf{S}^{-1}g. \tag{7.12}$$

The invertibility of $\mathbf{S}$ follows from the fact that $\mathbf{S}$ is positive definite.

We now turn our attention to the case when the μ_n are not known exactly, but only to within a certain experimental error. This is often the case in practice, where in a typical situation we have many estimates for each μ_n, each one the result of combining many independent observations. In particular we shall assume we know the means $\bar{\mu}_n$ and the variance matrix $\mathbf{E} = (\mathbf{E}_{nm})$ where the bar denotes mean value,

$$\begin{aligned} &\mathbf{E}_{nm} = \overline{\Delta_n\Delta_m}; \quad && n, m = 0, 1, \dots, N, \\ &\mu_n = \bar{\mu}_n + \Delta_n; \quad && n = 0, 1, \dots, N, \\ &\bar{\Delta}_n = 0; \quad && n = 0, 1, \dots, N. \end{aligned} \tag{7.13}$$

From (7.7) we have that for fixed x

$$\phi_N(x) = a^T\mu \tag{7.14}$$

where $\mu = (\mu_0, \mu_1, \ldots, \mu_N)^T$ and hence from (7.12)

$$\overline{\phi_N(x)} = a^T\bar{\mu}$$

$$= \frac{g^T\mathbf{S}^{-1}\bar{\mu}}{g^T\mathbf{S}^{-1}g}.$$

If we now write

$$\phi_N(x) = \overline{\phi_N(x)} + \Delta\phi_N$$

and define e by

$$e = \overline{(\Delta\phi_N)^2}$$

$$= a^T\mathbf{E}a \tag{7.15}$$

we can interpret e as a measure of the error made in approximating ϕ_N from the means $\bar{\mu}_n$. The important point to notice here is that in computing $\overline{\phi_N(x)} = a^T\bar{\mu}$ it is possible that additions take place in such a manner that e is large even though the relative errors Δ_n are small. In this case it may be advantageous to allow $a^T\mathbf{S}a$ to be slightly larger than its minimum value if in so doing the accuracy of the approximation to ϕ_N can be significantly improved. More precisely, we can pose the following optimization problem: For fixed γ, minimize e subject to the constraints that $g^Ta = 1$ and $a^T\mathbf{S}a \leqslant \gamma$. Let $\gamma_{\min}$ denote the minimum value of $a^T\mathbf{S}a$ subject to the constraint $g^Ta = 1$. Then the following situations can arise: If $\gamma < \gamma_{\min}$, there is no solution to the above-defined optimization problem, whereas if $\gamma \geqslant \gamma_{\min}$ either e is minimized for a value of a such that $a^T\mathbf{S}a = \gamma$ or at a value of a such that $a^T\mathbf{S}a < \gamma$. Since the parameter γ is at our disposal, the case of primary interest is when e is minimized for a value of a such that $a^T\mathbf{S}a = \gamma$ where $\gamma > \gamma_{\min}$. In this case our optimization problem can be solved by introducing Lagrange multipliers λ_1 and λ_2 and solving the system

$$\mathbf{E}a + \lambda_1\mathbf{S}a = \lambda_2 g$$

$$g^Ta = 1$$

$$a^T\mathbf{S}a = \gamma. \tag{7.16}$$

Having computed the solution of (7.16) we now evaluate $e = a^T\mathbf{E}a$ and compare its value to that corresponding to $\gamma = \gamma_{\min}$. Note that for each value of γ there may be several solutions of (7.16) of which we hope only one will minimize e.

For more details on the above procedure, we refer the reader to Backus and Gilbert [1] and Burridge [1].

7.2 LINEARIZED IMPROPERLY POSED PROBLEMS IN SCATTERING THEORY

We shall now turn our attention to the application of the Backus–Gilbert method to two inverse problems arising in the scattering of acoustic waves. The first problem we shall consider is that of determining the surface impedance $\lambda = \lambda(x)$ from a knowledge of the far-field pattern for a fixed wave number k (Problem A2 of Chapter 6).

From Corollary 3.7 we have

$$u^s(x) = \frac{e^{ik|x|}}{|x|} F(\hat{x}) + O\left(\frac{1}{|x|^2}\right) \tag{7.17}$$

where F is the far-field pattern, and the problem we want to consider is that of determining λ from a knowledge of F for all angles $\hat{x} = x/|x|$ and fixed wave number k. From Theorem 6.13 we know that λ is uniquely determined by F. However, in practice F is determined from measurements that are subject to a certain amount of experimental error, and hence the measured F may not even be in the class of far-field patterns corresponding to Problem A2, that is, in general no solution exists to our problem. We note also that the problem under consideration is nonlinear, that is, F does not depend linearly on the function λ. In what follows we shall show that the first difficulty can be handled by using our previously derived results on improperly posed problems, whereas the second difficulty can be circumvented by using Green's function to reduce the nonlinear problem to that of solving two linear moment problems where the kernel of the second depends upon the solution of the first (cf. Colton and Kirsch [1]).

In order to carry out the above program we need to restrict the class of admissible impedances λ to lie in a compact set. More specifically, let $C^+(\partial D)$ denote the cone in $C(\partial D)$ consisting of all functions λ such that $\operatorname{Im} \lambda \geqslant 0$ and let the set U be defined by

$$U = \{\lambda \in C^+(\partial D) \mid |\lambda(x)| \leqslant M_1, \ |\lambda(x) - \lambda(y)| \leqslant M_2 |x - y|\}$$

where M_1 and M_2 are fixed constants. Then since U is bounded, closed, and equicontinuous, by the Arzéla–Ascoli theorem U is compact in $C(\partial D)$. Now let $\lambda \in U$ and let G denote the radiating Green's function for the Helmholtz equation defined in the exterior of D, that is, G is a fundamental solution of the Helmholtz equation satisfying the Sommerfeld radiation condition and the boundary condition

$$G(x, y) = 0, \qquad x \in \partial D, \qquad y \in \mathbb{R}^3 \setminus \bar{D}. \tag{7.18}$$

Then using Green's formula we can represent u^s in the form

$$u^s(x) = \frac{1}{4\pi} \int_{\partial D} u^s(y) \frac{\partial G}{\partial \nu(y)}(x, y)\, ds(y). \tag{7.19}$$

We now note that for $|x|$ large G has the asymptotic behavior

$$G(x,y)=\frac{e^{ik|x|}}{|x|}\left[e^{-ik(\hat{x},y)}+g(\hat{x},y)\right]+O\left(\frac{1}{|x|^2}\right)$$

where, as a function of y, g is a solution of the Helmholtz equation in the exterior of D, satisfies the Sommerfeld radiation condition, and on ∂D assumes the boundary data

$$g(\hat{x},y)=-e^{-ik(\hat{x},y)},\qquad y\in\partial D.$$

Letting x tend to infinity in (7.19) now leads to the relation

$$F(\hat{x})=\frac{1}{4\pi}\int_{\partial D}u^s(y)\frac{\partial}{\partial\nu(y)}\left[e^{-ik(\hat{x},y)}+g(\hat{x},y)\right]ds(y).$$

Suppose now that F is known for the observation angles $\hat{x}_0,\hat{x}_1,\ldots,\hat{x}_N$, and define

$$\mu_n=F(\hat{x}_n)$$

for $n=0,1,\ldots,N$. Then we have the improperly posed generalized moment problem

$$\mu_n=\frac{1}{4\pi}\int_{\partial D}u^s(y)\frac{\partial}{\partial\nu(y)}\left[e^{-ik(\hat{x}_n,y)}+g(\hat{x}_n,y)\right]ds(y)\tag{7.20}$$

for u^s on ∂D. However, by using regularity results analogous to Theorem 3.27 we have that $u^s\in C^1(\mathbf{R}^3\setminus D)$ and hence if we have an a priori bound on the velocity grad u^s we have a bound on the Lipschitz constant for u^s. Such a bound in fact follows from potential theoretic arguments using the fact that $\lambda\in U$ and ∂D is in class C^2. With this information at our disposal we can now apply the Backus–Gilbert method to (7.20) and arrive at an approximation to u^s on ∂D.

Having determined an approximation to u^s on ∂D we have an approximation to u on ∂D, and an application of Green's formula now gives

$$u^s(x)=\frac{1}{4\pi}\int_{\partial D}\left[u(y)\frac{\partial}{\partial\nu(y)}\frac{e^{ik|x-y|}}{|x-y|}-\frac{e^{ik|x-y|}}{|x-y|}\frac{\partial u(y)}{\partial\nu}\right]ds(y).$$

Letting x tend to infinity and making use of the impedance boundary condition satisfied by u now gives us the moment problem

$$\gamma_n=\frac{1}{4\pi}\int_{\partial D}\lambda(y)\left[u(y)e^{-ik(\hat{x}_n,y)}\right]ds(y)\tag{7.21}$$

where

$$\gamma_n = \frac{1}{4\pi}\int_{\partial D} u^s(y)\frac{\partial}{\partial\nu(y)} g(\hat{x}_n, y)\, ds(y)$$

$$-\frac{1}{4\pi}\int_{\partial D} u^i(y)\frac{\partial}{\partial\nu(y)} e^{-ik(\hat{x}_n, y)}\, ds(y). \tag{7.22}$$

Note that errors in computing u on ∂D are equivalent to errors of known magnitude in the computation of γ_n since a bound on λ is assumed known. Hence under the assumption that $\lambda \in U$ and a sufficient number of measurements of the far-field pattern have been made we can use the Backus–Gilbert method to determine an approximation to the unknown impedance λ.

We note that from a practical point of view the above procedure is complicated by the fact that one must know both the Green's function for D as well as an a priori bound on grad u^s. This last restriction can be removed if one has low frequency data available (Colton [6]).

We now turn our attention to the problem of determining the shape of the scattering obstacle from a knowledge of the far-field pattern, given the fact that the scattering obstacle is acoustically "soft," that is, Problem A1 of Chapter 6. We note that in order to guarantee uniqueness it is in general necessary to know $F = F(\hat{x}; k)$ for an interval of k values, $k \in [k_0, k_1] \subset \mathbb{R}$ (Theorem 6.11). We also observe that this problem is nonlinear, for example, if D is starlike and described by $r = f(\theta, \phi)$ then f is not a linear function of the far-field pattern F. In order to arrive at a linear moment problem amenable to the Backus–Gilbert method, we shall assume that an initial guess D_0 to D is known and then derive a linear moment problem for the first variation of D_0. Under this assumption we shall show that an improved approximation to D can be determined from a knowledge of the scattering cross section σ defined by

$$\sigma := \lim_{r\to\infty}\int_{|x|=r} |u^s(x)|^2\, ds$$

$$= \int_{|\hat{x}|=1} |F(\hat{x})|^2\, ds, \tag{7.23}$$

that is, it is only necessary to know the amplitude, and not the phase, of the far-field pattern F in order to arrive at an improved approximation to the initial guess D_0. However, we shall need to know the scattering cross section, corresponding to waves incident upon D from $N+1$ different directions, that is, $\sigma_n = \sigma(k, \alpha_n)$, $n = 0, 1, \ldots, N$ where α_n is the direction of propagation of the incoming plane wave $u^i = u^i(x; k, \alpha_n)$. Our presentation will be based on a variational principle due to Garabedian [1] and was first presented in Colton and Kirsch [2] and Angell, Colton, and Kirsch [1]. For related results, including numerical examples, we refer the reader to Roger [1], Sleeman [3], and Kirsch [5].

We emphasize the fact that our analysis is based on having an initial approximation D_0 to the domain D. The situation where this is not the case (i.e., the full nonlinear problem) will be discussed in the next chapter.

We begin by deriving the above-mentioned variational formula of Garabedian. Suppose ∂D is obtainable from ∂D_0 by shifting ∂D_0 an infinitesimal positive amount $\delta\nu$ along the inner normal to ∂D_0. Let $u_+ = u_+(x; k, \alpha_n)$ and $u_- = u(x; k, \alpha_n)$ be the total fields due to the scattering by D_0 of the incoming plane waves $e^{ik(x,\alpha_n)}$ and $e^{-ik(x,\alpha_n)}$, respectively, and let u_+^s and u_-^s denote the corresponding scattered fields. Denote by w_+, w_-, and so on, the corresponding quantities associated with ∂D and let $\sigma = \sigma(k, \alpha_n)$ denote the scattering cross section corresponding to D and $\sigma^0 = \sigma^0(k, \alpha_n)$ the scattering cross section corresponding to D_0. Then from Green's theorem, the radiation condition, and the boundary conditions $u_+ = u_- = 0$ on ∂D_0, $w_+ = w_- = 0$ on ∂D, we have (where ν denotes the unit outward normal to ∂D or ∂D_0)

$$\begin{aligned}
\sigma - \sigma^0 &= \lim_{r\to\infty} \int_{|x|=r} \left[|w_+^s|^2 - |u_+^s|^2 \right] ds \\
&= \operatorname{Im} \frac{1}{k} \int_{\partial D} \overline{w}_+^s \frac{\partial w_+^s}{\partial \nu} ds - \operatorname{Im} \frac{1}{k} \int_{\partial D_0} \overline{u}_+^s \frac{\partial u_+^s}{\partial \nu} ds \\
&= \operatorname{Im} \frac{1}{k} \int_{\partial D} w_-^s \frac{\partial w_+}{\partial \nu} ds - \operatorname{Im} \frac{1}{k} \int_{\partial D_0} u_+^s \frac{\partial u_-}{\partial \nu} ds \\
&= \operatorname{Im} \frac{1}{k} \int_{\partial D} \left(w_-^s \frac{\partial w_+}{\partial \nu} - w_+ \frac{\partial w_-^s}{\partial \nu} \right) ds \\
&\quad - \operatorname{Im} \frac{1}{k} \int_{\partial D_0} \left(u_+^s \frac{\partial u_-}{\partial \nu} - u_- \frac{\partial u_+^s}{\partial \nu} \right) ds \\
&= \operatorname{Im} \frac{1}{k} \int_{\partial D} \left(w_-^s \frac{\partial e^{ik(x,\alpha_n)}}{\partial \nu} - e^{ik(x,\alpha_n)} \frac{\partial w_-^s}{\partial \nu} \right) ds \\
&\quad - \operatorname{Im} \frac{1}{k} \int_{\partial D_0} \left(u_+^s \frac{\partial e^{-ik(x,\alpha_n)}}{\partial \nu} - e^{-ik(x,\alpha_n)} \frac{\partial u_+^s}{\partial \nu} \right) ds \\
&= \operatorname{Im} \frac{1}{k} \int_{\partial D_0} \left(w_- \frac{\partial u_+}{\partial \nu} - u_+ \frac{\partial w_-}{\partial \nu} \right) ds \\
&= \operatorname{Im} \frac{1}{k} \left\{ \int_{\partial D_0} w_- \frac{\partial u_+}{\partial \nu} ds - \int_{\partial D} w_- \frac{\partial u_+}{\partial \nu} ds \right\} \\
&= -\operatorname{Im} \frac{1}{k} \iint_{D_0 \setminus D} \left(\operatorname{grad} w_- \operatorname{grad} u_+ - k^2 w_- u_+ \right) dx
\end{aligned}$$

where we have assumed for the time being that ∂D_0 is analytic and $\delta\nu$ is small enough such that u_+ can be continued across ∂D_0 into the exterior of D. Introducing the notation $\delta\sigma = \sigma - \sigma^0$ and splitting the above volume integral into an integration over ∂D_0 and an integration normal to it, we now see that up to first-order terms in $\delta\nu$ we have (again using the fact that $u_+ = u_- = 0$ on ∂D_0) the Hadamard variational formula

$$\delta\sigma = -\operatorname{Im}\frac{1}{k}\int_{\partial D_0}\frac{\partial u_-}{\partial\nu}\frac{\partial u_+}{\partial\nu}\,\delta\nu\, ds. \tag{7.24}$$

Although (7.24) was derived under the assumption that $\delta\nu$ was positive and ∂D_0 was analytic, it can be shown that the result is valid for small shifts $\delta\nu$ of either sign as well as domains D_0 with twice continuously differentiable boundary.

We now assume that k is fixed and σ is measured for $N+1$ different directions $\alpha_0, \alpha_1, \ldots, \alpha_N$. Since D_0 is given, we can compute u_-, u_+, and σ^0, and hence $\delta\sigma$ is known (to within a certain experimental error). Hence (7.24) defines an improperly posed moment problem for $\delta\nu$ and is amenable to solution by the Backus–Gilbert method, provided suitable a priori restrictions are made on $\delta\nu$. In particular it is reasonable to assume that not only is $\delta\nu$ small, but is also a slowly varying function of arclength on ∂D_0, that is, as a function of arclength s, positive constants ε and M are known such that

1. $\max_{\partial D_0}|\delta\nu(s)| \leqslant \varepsilon$.
2. $|\delta\nu(s_1) - \delta\nu(s_2)| \leqslant M|s_1 - s_2|$.

Under these assumptions we can apply the Backus–Gilbert method to (7.24), thus yielding an approximation to $\delta\nu$ and hence a refined approximation to ∂D. It is now possible to repeat this procedure using this new approximation in place of ∂D_0, although the amount of labor involved in such an iterative process rapidly becomes prohibitive.

The major drawback in the above procedure for finding the shape of the scattering obstacle is that it is necessary to have a reasonably accurate initial guess D_0 to the shape. If this strong a priori information is not available, then the only alternative remaining if one wants to use this method is to compare the measured scattering cross section to those arising from given canonical figures, for example, spheres or ellipsoids, and to choose that figure whose scattering cross section is closest to the one measured. The drawback of this approach is that the shape of the obstacle does not depend continuously on the far-field pattern unless a priori constraints are imposed, that is, small perturbations of the scattering cross section can lead to large perturbations of the scattering obstacle. This problem will be dealt with in the next chapter where we shall show how to stabilize this problem and thereby construct accurate initial approximations to the unknown scattering obstacle.

7.3 NORMAL FAMILIES OF UNIVALENT FUNCTIONS

The problem of determining the shape of a scattering obstacle from a knowledge of the far-field pattern is basically one of deducing geometric information on the shape of the obstacle from the analytic knowledge of the far-field pattern. Furthermore, if the analytic data are imprecise, we have reasons to suspect that the problem can be stabilized by restricting the class of scattering obstacles to lie in a compact set. (This will indeed be verified in the next chapter.) In view of these facts it can be expected that for the case of infinite cylinders (i.e., scattering problems in $\mathbb{R}^2$), geometric function theory, in particular the theory of compact families of univalent functions, will provide a useful tool in the analysis of such problems. This will indeed turn out to be the case, and hence in this section we shall provide a brief introduction to the theory of compact families of univalent functions, delaying its application to the next chapter. For more details concerning the material of this section, we refer the reader to the monographs by Nehari [1] and Pommerenke [1], both of which have influenced our own presentation.

As in the case of compact families of continuous functions in $\mathbb{R}^n$ (Theorem 1.12) the concepts of equicontinuity and uniform boundedness play a central role in our theory. In what follows f will always denote an analytic function of a complex variable z defined in some domain D of the complex plane. It is also always assumed that f is single valued in D.

Definition 7.4. Let $f \in G$ where G is a class of analytic functions defined in a domain D. Then the functions in G are said to be equicontinuous in D if for every $\varepsilon > 0$ and closed subdomain D_0 of D there exists a positive number $\delta = \delta(\varepsilon, D_0)$, independent of f, such that for any $z_1, z_2 \in D_0$ we have

$$|f(z_1) - f(z_2)| < \varepsilon$$

if $|z_1 - z_2| < \delta$.

Definition 7.5. Let $f \in G$ where G is a class of analytic functions defined in a domain D. Then the functions in G are said to be locally uniformly bounded if for every $z_0 \in D$ there exists a positive constant $M = M(z_0)$ and a neighborhood $N = N(z_0)$ of z_0, where M and N are independent of f, such that

$$|f(z)| \leqslant M(z_0)$$

for $z \in N(z_0)$.

Note that it is easily verified that if a class of functions is locally uniformly bounded in D it is also uniformly bounded in any closed subdomain of D.

Theorem 7.6. Let G be a class of analytic, locally uniformly bounded functions defined in a domain D. Then the functions in G are equicontinuous.

Proof. Let $\overline{\Omega}_r$ be a closed disk in D of radius r. Then if $z_1, z_2 \in \Omega_r$, we have from Cauchy's integral theorem that

$$f(z_1)-f(z_2)=\frac{1}{2\pi i}\int_{\partial\Omega_r}\left(\frac{1}{\xi-z_1}-\frac{1}{\xi-z_2}\right)f(\xi)\,d\xi$$

$$=\frac{z_1-z_2}{2\pi i}\int_{\partial\Omega_r}\frac{f(\xi)\,d\xi}{(\xi-z_1)(\xi-z_2)}.$$

Hence if $f \in G$, $|f| \leqslant M$ on $\partial\Omega_r$, and we restrict $z_1, z_2 \in \Omega_{r/2}$, we have

$$|f(z_1)-f(z_2)| \leqslant \frac{4M|z_1-z_2|}{r}. \tag{7.25}$$

Now let D_0 be a closed subdomain of D. Let $(\Omega^{(k)})$ be a finite subcovering of D_0 of disks $\Omega^{(k)} \subset D$ with centers at ξ_k and radii $r_k/4$. Let $|f| \leqslant M_k$ on $\Omega^{(k)}$, $r = \min r_k$, $M = \max M_k$, and for given $\varepsilon > 0$ let $\delta = \min\{r/4, \varepsilon r/4M\}$. Then if $|z_1 - z_2| < \delta$ and $|z_2 - \xi_k| < r_k/4$ we have $|z_1 - \xi_k| < \delta + r_k/4 \leqslant r_k/2$. Hence from (7.25) we have

$$|f(z_1)-f(z_2)| \leqslant \frac{4M_k\delta}{r_k} \leqslant \frac{4M\delta}{r} \leqslant \varepsilon,$$

that is, the functions in G are equicontinuous.

Theorem 7.7. Let G be a class of analytic, equicontinuous functions defined in a domain D. Then if the functions in G are uniformly bounded at one point in D, they are locally uniformly bounded.

Proof. Let $z_0 \in D$ such that $|f(z_0)| \leqslant M$ for all $f \in G$ where M is a positive constant. Let $z \in D$, D_0 a closed subdomain containing z and z_0, and C a contour joining z_0 to z and contained in D_0. Let L be the length of C. By equicontinuity we have that for every $\varepsilon > 0$ there exists a δ such that if $|z_1 - z_2| < \delta$ then $|f(z_1) - f(z_2)| < \varepsilon$ for $z_1, z_2 \in D_0$. Hence if $m > L/\delta$ we can find points $z_1, z_2, \ldots, z_{m-1}$ on C such that

$$|f(z_1)-f(z_0)| < \varepsilon$$
$$|f(z_2)-f(z_1)| < \varepsilon$$
$$\vdots$$
$$|f(z)-f(z_{m-1})| < \varepsilon,$$

that is,

$$|f(z)-f(z_0)| < m\varepsilon,$$

or $|f(z)| < |f(z_0)| + m\varepsilon \leqslant M + m\varepsilon$. Hence in a neighborhood of z we have

$$|f(z)| < M + (m+1)\varepsilon,$$

and this establishes the theorem.

We now turn our attention to compact families of analytic functions.

Definition 7.8. A family G of analytic functions defined in a domain D is called normal if from any sequence (f_n), $f_n \in G$, it is possible to find a subsequence that is uniformly convergent in any closed subdomain of D.

It is customary in the above definition to include the case in which the sequence tends to infinity, that is, for any positive constant M, $|f_n(z)| > M$ for all z in the closed subdomain of D provided n is large enough.

Definition 7.9. A normal family G is said to be compact if the limits of all convergent sequences of functions in G are also functions in G.

The main result on normal families of analytic functions is the following theorem due to Montel.

Theorem 7.10 (Montel's Theorem). If the functions of a family G are analytic and locally uniformly bounded in a domain D, then G is a normal family in D.

Proof. Since the functions in G are locally uniformly bounded they are uniformly bounded on any closed subdomain of D, and by Theorem 7.6 the functions in G are equicontinuous. Hence by the Arzéla–Ascoli theorem G is normal.

Corollary 7.11. The class of analytic functions that are analytic and uniformly bounded in a domain D is compact.

Proof. By Montel's theorem this class is clearly normal. Moreover, if (f_n) is a uniformly convergent sequence of functions from this class such that $|f_n(z)| \leqslant M$, then the limit function f also satisfies this inequality and since the uniform limit of analytic functions is analytic, f is analytic. Hence the class of functions under consideration is compact.

Theorem 7.12 (Hurwitz's Theorem). Let (f_n) be a sequence of analytic functions defined in a domain D such that f_n converges uniformly (on compact subsets) to a nonconstant (analytic) function defined in D. Then if $z_0 \in D$ such that $f(z_0) = 0$, for every $\varepsilon > 0$ there must be a zero of f_n in the disk $|z - z_0| < \varepsilon$, provided n is sufficiently large.

Proof. Let ε be sufficiently small such that $|z - z_0| \leqslant \varepsilon$ is contained in D and f does not vanish in this disk (except of course at $z = z_0$). Since f is continuous there exists a positive constant m such that $|f(z)| > m$ on $|z - z_0| = \varepsilon$. Furthermore, by the uniform convergence of the sequence (f_n), $|f(z)-$

$f_n(z)| < m$ on $|z - z_0)| = \varepsilon$ for n sufficiently large, that is,

$$|f(z) - f_n(z)| < m < |f(z)|, \qquad |z - z_0| = \varepsilon.$$

Hence by Rouche's theorem

$$f_n(z) = f(z) + [f_n(z) - f(z)]$$

has the same number of zeros in $|z - z_0| < \varepsilon$ as f does, that is, exactly one zero. This proves the theorem.

Definition 7.13. An analytic function f defined in a domain D is said to be univalent in D if for $z_1, z_2 \in D$, $z_1 \neq z_2$, we have $f(z_1) \neq f(z_2)$.

Corollary 7.14. Let (f_n) be a sequence of univalent functions defined in a domain D such that f_n converges in D to a nonconstant analytic function f. Then f is also univalent in D.

Proof. Suppose on the contrary that $f(z_1) = f(z_2)$ for $z_1, z_2 \in D$, $z_1 \neq z_2$, and consider the sequence (g_n) where

$$g_n(z) := f_n(z) - f_n(z_1).$$

Then since f_n is univalent, $g_n(z) \neq 0$ except at $z = z_1$. But the limit function $g(z) = f(z) - f(z_1)$ vanishes at $z = z_2$ which implies by Hurwitz's theorem that g_n must have a zero within an arbitrarily small neighborhood of $z = z_2$, provided n is large enough. But this contradicts the fact that $g_n(z) = 0$ only for $z = z_1$. Hence f must be univalent.

We now want to consider a particular class of univalent functions that will be basic to our discussion of the inverse scattering problem in Chapter 8. However, before doing this we need to slightly extend our definition of univalency as given in Definition 7.13 by allowing f to have a simple pole in D, that is, f can be meromorphic, and allowing D to be a domain of the extended z plane, that is, D can contain the point $z = \infty$. With this extension of Definition 7.13, we now want to consider the class of functions univalent in $\Delta = \{z \mid |z| > 1\}$ and analytic in this region except for a simple pole at infinity. (Δ is considered as a subset of the extended z plane.) In particular, functions in this class have Laurent expansions of the form

$$f(z) = az + a_0 + \frac{a_1}{z} + \cdots, \qquad |z| > 1. \tag{7.26}$$

We shall make the further restriction that there exist positive constants α, β, and γ such that

$$0 < \alpha \leqslant a \leqslant \beta < \infty$$

$$|a_0| \leqslant \gamma. \tag{7.27}$$

The motivation for considering this class of functions is that if D is a bounded simply connected domain in the complex plane, then by the Riemann mapping theorem there exists a unique function of the form (7.26) with $a > 0$ that maps Δ univalently onto $\mathbb{R}^2 \setminus \bar{D}$. In this case a is known as the mapping radius or transfinite diameter of D. We shall see in the next chapter that conditions (7.27) are met if ∂D is required to lie in a given annulus centered at the origin, and we shall show presently that conditions (7.27) are sufficient to guarantee the compactness of this class of functions. Hence this class of functions will allow us to characterize a compact family of scattering obstacles by simple a priori geometric information.

Definition 7.15. The class of meromorphic functions that are univalent in Δ and have the Laurent expansion (7.26) where a and a_0 are subject to (7.27) is denoted by $\Sigma(\alpha, \beta, \gamma)$.

We note that if a univalent function has a pole at $z = z_0$ then z_0 must be a simple pole in order that the function be univalent.

Our first result on functions lying the class $\Sigma(\alpha,\beta,\gamma)$ relates the area of the compliment of the image of Δ under the mapping (7.26) to the Laurent coefficients of f. In particular let $f \in \Sigma(\alpha, \beta, \gamma)$ such that f: $\Delta \to \mathbb{R}^2 \setminus \bar{D}$ and let E_r be the image under f of the set $\{z \mid |z| \geqslant r > 1\}$. Define

$$H(r) = \mathbf{C} \setminus E_r$$

where $\mathbf{C}$ denotes the complex z plane. Then $H(r)$ is bounded by an analytic Jordan curve $C(r)$ and the area of D is given by the formula

$$\text{area } D = \lim_{r \to 1} \text{area } H(r).$$

Theorem 7.16 (Area Theorem). Let $f \in \Sigma(\alpha, \beta, \gamma)$. Then

$$\text{area } D = \pi\left(a^2 - \sum_{n=1}^{\infty} n|a_n|^2\right).$$

Proof. Let $w = u + iv = f(z)$. We first apply Green's theorem to $H(r)$ to find that

$$\begin{aligned}
\text{area } H(r) &= \iint_{H(r)} du\, dv \\
&= \frac{1}{4}\iint_{H(r)} \Delta(u^2 + v^2)\, du\, dv \\
&= \frac{1}{4}\int_{C(r)} \frac{\partial}{\partial \nu}(u^2 + v^2)\, ds
\end{aligned}$$

where ν denotes the unit outward normal to $C(r)$ and s denotes arclength.

Since $\nu = ((dv/ds), (-du/ds))$ we have

$$\text{area } H(r) = \frac{1}{4}\int_{C(r)} \frac{\partial}{\partial \nu}(u^2 + v^2)\, ds$$

$$= \frac{1}{2}\int_{C(r)} u\, dv - v\, du$$

$$= -\frac{i}{2}\int_{C(r)} (u - iv)(du + i\, dv),$$

where we have used the fact that

$$\int_{C(r)} u\, du = \int_{C(r)} v\, dv = 0.$$

Recalling that $C(r)$ is the image of $|z| = r$ under the mapping f we now have that

$$\text{area } H(r) = -\frac{i}{2}\int_0^{2\pi} \overline{f(re^{i\theta})} \frac{\partial f(re^{i\theta})}{\partial \theta}\, d\theta$$

$$= \pi a^2 r^2 - \pi \sum_{n=1}^{\infty} n|a_n|^2 r^{-2n} \tag{7.28}$$

where we have used the uniform convergence of the series (7.26) in $|z| \geqslant r > 1$ to integrate termwise. Since the area of $H(r)$ is nonnegative we can conclude from (7.28) that for any integer N

$$\sum_{n=1}^{N} n|a_n|^2 r^{-2n} \leqslant a^2 r^2. \tag{7.29}$$

Hence letting $r \to 1$ in (7.29) and then letting $N \to \infty$ we have that

$$\sum_{n=1}^{\infty} n|a_n|^2$$

converges. We can therefore take the limit $r \to 1$ in (7.28) and thus conclude the validity of the theorem.

Corollary 7.17. Let $f \in \Sigma(\alpha, \beta, \gamma)$. Then $|a_1| \leqslant \beta$.

Proof. Since area $D \geqslant 0$ we have from Theorem 7.16 that

$$a^2 - |a_1|^2 \geqslant a^2 - \sum_{n=1}^{\infty} n|a_n|^2 \geqslant 0$$

and the corollary follows from (7.27).

We now want to obtain an upper bound on the radius of the smallest disk centered at the origin and containing D in its interior where $f: \Delta \to \mathbb{R}^2 \setminus \overline{D}$, $f \in \Sigma(\alpha, \beta, \gamma)$.

Theorem 7.18. Let $f \in \Sigma(\alpha, \beta, \gamma)$ and $w \in D$. Then

$$|w| \leqslant 2\beta + \gamma.$$

Proof. The even function $z^{-2}(f(z^2) - w)$ is analytic and nonvanishing in Δ. Hence the odd function

$$\begin{aligned} f_1(z) &= a^{1/2} z \left[z^{-2}(f(z^2) - w)\right]^{1/2} \\ &= a^{1/2} z \left[a + \frac{(a_0 - w)}{z^2} + \cdots\right]^{1/2} \\ &= az + \frac{(a_0 - w)}{2z} + \cdots \end{aligned}$$

is analytic in Δ. Furthermore, if $f_1(z_1) = f_1(z_2)$ then $f(z_1^2) = f(z_2^2)$ and since f is univalent $z_2 = \pm z_1$. The minus sign is impossible since f_1 is nonvanishing in Δ and $f_1(-z_1) = -f_1(z_1) \neq f_1(z_1)$. Hence f_1 is univalent in Δ and in particular $f_1 \in \Sigma(\alpha, \beta, 0)$. It now follows from Corollary 7.17 that

$$\left|\frac{a_0 - w}{2}\right| \leqslant \beta$$

that is, by the triangle inequality

$$|w| \leqslant 2\beta + \gamma.$$

Corollary 7.19. Let $f \in \Sigma(\alpha, \beta, \gamma)$. Then

$$|f(z)| \leqslant (2\beta + \gamma)|z|$$

for $|z| > 1$.

Proof. Since $f(z)/z$ is analytic at infinity, by the maximum principle we have

$$\left|\frac{f(z)}{z}\right| \leqslant \overline{\lim_{|\xi| \to 1}} \left|\frac{f(\xi)}{\xi}\right| = \max_{\overline{D}} |w| \leqslant 2\beta + \gamma$$

for $|z| > 1$, and the result follows.

We are now in a position to prove our main result concerning the class $\Sigma(\alpha, \beta, \gamma)$, that is, that it is compact. A sequence (f_n) in $\Sigma(\alpha, \beta, \gamma)$ is said to be convergent if $(z^{-1} f_n)$ is uniformly convergent on compact subsets of Δ.

Theorem 7.20. The class $\Sigma(\alpha, \beta, \gamma)$ is compact.

Proof. Let (f_n) be any sequence in $\Sigma(\alpha, \beta, \gamma)$ and define $g_n(z) = z^{-1} f_n(z)$. Then the functions g_n are analytic in the exterior of the unit disk and by Corollary 7.19 are uniformly bounded. Hence, by Corollary 7.11, from the sequence (g_n) it is possible to find a subsequence that converges uniformly for $|z| \geqslant r > 1$ to a function g that is analytic for $|z| > 1$. Since for each n, $|g_n(\infty)| \geqslant \alpha > 0$, the limit function g also satisfies this inequality. Then $f(z) = zg(z)$ has a simple pole at infinity and a Laurent expansion of the form (7.26). Since each $f_n \in \Sigma(\alpha, \beta, \gamma)$, it is easily verified that the Laurent coefficients of g satisfy the inequalities (7.27). It remains to be shown that f is univalent in Δ. But from the above analysis $f(z) \neq \infty$ for $1 < |z| < \infty$, where $f(\infty) = \infty$. By Corollary 7.14, $f(z_1) \neq f(z_2)$ for z_1, z_2 in the (finite) z plane, $z_1 \neq z_2$. Hence f is univalent in Δ and the theorem is proved.

8

THE DETERMINATION OF THE SHAPE OF AN OBSTACLE FROM INEXACT FAR-FIELD DATA

In this chapter we shall consider the problem of determining the shape of an acoustically "soft" obstacle (i.e., Dirichlet boundary data) from a knowledge of the phase and amplitude of the far-field pattern or, alternatively, the scattering cross section. We note that from the optical theorem (cf. Stakgold [1], p. 304) the scattering cross section can be determined from a knowledge of the phase and amplitude of the far-field pattern at a single point, that is, the direction of propagation of the incident plane wave. We wish to emphasize that our choice of considering the inverse problem of determining the shape of an acoustically soft obstacle is somewhat arbitrary in the sense that our methods apply equally well to the problem of determining the shape of a "hard" obstacle (i.e., Neumann boundary data) as well as to the inverse scattering problem for electromagnetic waves. In addition we can easily treat the problem of determining the impedance of an obstacle of known geometry (cf. Colton and Kirsch [1], Colton and Kress [2]). However, because we plan to treat the closely related problem of optimal control of the surface impedance of an antenna in the next chapter, we have decided not to discuss this class of inverse problems at present. For a survey of inverse problems in acoustic scattering theory we refer the reader to Colton [7].

The problem we are going to consider in this chapter is different from the inverse problems in scattering theory considered in Chapter 7 in that it is not only improperly posed but also basically nonlinear, that is, it cannot be reduced to one or more linear problems. Hence we are faced with two problems: one the question of sufficient conditions to stabilize the improperly posed problem, and the other to derive appropriate methods for approximating the solution of the resulting stabilized nonlinear problem. We shall concentrate here on the first of these problems and to this end use the method of integral equations to reformulate our inverse scattering problem as an optimization problem. The problem can then be stabilized by restricting the class of admissible domains to lie a priori in a compact set.

In order to approximate the solution of the resulting constrained optimization problem a variety of approaches are available, the simplest being to obtain an initial approximation by minimizing the nonlinear functional over a finite dimensional subset and then to use a Newton-type procedure to iteratively improve this initial estimate (cf. Kirsch [5], Roger [1], Sleeman [3]). If the given data are assumed to be the scattering cross section, we can make use of the variational principle developed in Chapter 7 in order to accomplish the second of these two steps. Hence in principle the construction of the solution to our inverse scattering problem is relatively straightforward, although since our mapping is defined by means of an integral equation it means that this integral equation must be solved at each step of the iterative process.

We begin our discussion with a simple two-dimensional model problem, which, at the risk of a slight abuse of the English language, we shall describe as the case of partially inexact data. By this we shall mean that the first N Fourier coefficients of the far-field pattern are known exactly for all values of the wave number k for $0 < k < k_0$ where k_0 is an arbitrarily small positive constant and nothing is known of the remaining Fourier coefficients. From this knowledge we shall use the theory of univalent functions to construct an approximation to the (normalized) univalent function mapping the exterior of the unit circle onto the exterior of the unknown scattering obstacle and provide explicit estimates of the mean square error made in arriving at this approximation. From a practical point of view there are, of course, serious criticisms to be made of this approach, in particular the restriction of the method to two-dimensional problems as well as the need to know N Fourier coefficients exactly in the low frequency limit. However, this model problem provides one of the few examples of an essentially complete solution to a nonlinear inverse scattering problem and lends considerable insight to the more general and physically relevant modes to be discussed subsequently. Although we shall discuss only the case of an acoustically soft scattering obstacle, basing our results on Colton and Kleinman [1], the same approach is also valid for hard obstacles (Colton [3]) as well as the electromagnetic case in which elementary polarization effects are seen to play an important role (Colton [5]). Further extensions may be found in Sleeman [2], Hariharan [1], and Smith [1].

Following our discussion of the model problem, we shall use the theory of univalent functions as developed in Chapter 7 combined with integral equation methods in scattering theory due to Garabedian [1] to reformulate the two-dimensional inverse scattering problem as an optimization problem. Here we shall assume that the wave number k lies in the "resonant region" (i.e., outside the range where either low frequency or high frequency methods are applicable). The analysis in this section is based on the work of Colton and Kirsch [2].

The final section of this chapter is devoted to extending the two-dimensional results of Section 8.2 to the case of three dimensions (Angell, Colton, and Kirsch [1]). In order to do this conveniently we shall make the extra assumption that the (unknown) scattering obstacle is starlike (with respect to the origin).

We note that in what follows we always assume that the scattering obstacle is connected. The extension to the case of more than one scattering obstacle should present no basic difficulty, at least a far as Sections 8.2 and 8.3 are concerned.

8.1 A MODEL PROBLEM

We now consider the inverse scattering problem for a soft cylinder under the assumption that the first N Fourier coefficients of the far-field pattern are known exactly for small values of the wave number. Our aim is to obtain a nonlinear moment problem for the conformal mapping taking the exterior of the unit disk onto the exterior of the unknown scattering obstacle D and then to solve this moment problem. In order to arrive at this moment problem it is first necessary to use the method of integral equations to solve the exterior Dirichlet problem for the Helmholtz equation by iteration, and then to identify the low frequency approximation of the solution to the integral equation as the velocity potential of an incompressible fluid flow past the (unknown) obstacle D. In this connection our results here are related to those of Chapter 5, except that in the present case the low frequency limit is identically zero and we are interested in the first nonzero term in the low frequency expansion of the solution to the integral equation. The solution of the integral equation is complicated by the fact that it has a nontrivial nullspace and hence in order to arrive at our desired iterative process we must modify the kernel of the integral operator. We begin our discussion with this problem.

Let D be a bounded simply connected domain in $\mathbb{R}^2$ with C^2 boundary ∂D such that D contains the origin and let ν denote the outward unit normal to ∂D. We first consider two boundary-value problems for Laplace's equation that we denote by Problem I and Problem II. Problem I is to find $u \in C^2(\mathbb{R}^2 \setminus \bar{D}) \cap C(\mathbb{R}^2 \setminus D)$ such that

$$u(x) = \log\frac{1}{|x|} + u^s(x), \qquad x \in \mathbb{R}^2 \setminus D$$

$$\Delta u = 0 \quad \text{in} \quad \mathbb{R}^2 \setminus \bar{D} \tag{I}$$

$$u = 0 \quad \text{on} \quad \partial D$$

u^s is bounded as x tends to infinity.

Note that by Theorem 3.27 we can conclude that $u \in C^2(\mathbb{R}^2 \setminus \bar{D}) \cap C^1(\mathbb{R}^2 \setminus D)$.

Since u^s is bounded as x tends to infinity we can conclude that $\operatorname{grad} u^s(x) = O(1/|x|^2)$ and

$$\lim_{|x|\to\infty} u^s(x) = \alpha$$

exists where α is a constant. Then from Green's formula we have

$$\frac{1}{2\pi}\int_{\partial D}\left[\log\frac{1}{|x-y|}\frac{\partial u^s}{\partial \nu}(y) - u^s(y)\frac{\partial}{\partial \nu(y)}\log\frac{1}{|x-y|}\right] ds(y)$$

$$= \begin{cases} -u^s(x)+\alpha\ ; & x \in \mathbb{R}^2\setminus\bar{D} \\ -\dfrac{1}{2}u^s(x)+\alpha; & x\in\partial D \end{cases}$$

$$\frac{1}{2\pi}\int_{\partial D}\left[\log\frac{1}{|x-y|}\frac{\partial}{\partial \nu(y)}\log\frac{1}{|y|} - \log\frac{1}{|y|}\frac{\partial}{\partial \nu(y)}\log\frac{1}{|x-y|}\right] ds(y)$$

$$= \begin{cases} -\log\dfrac{1}{|x|}\ ; & x \in \mathbb{R}^2\setminus\bar{D} \\ -\dfrac{1}{2}\log\dfrac{1}{|x|}; & x\in\partial D \end{cases}$$

and hence for $x\in\mathbb{R}^2\setminus D$ we have

$$\frac{1}{2\pi}\int_{\partial D}\log\frac{1}{|x-y|}\frac{\partial u}{\partial \nu}(y)\, ds(y) = -u(x)+\alpha. \tag{8.1}$$

Since

$$\frac{1}{2\pi}\int_{\partial D}\frac{\partial u}{\partial \nu}(y)\, ds(y) = -1$$

we have that (8.1) is equivalent to

$$\frac{1}{2\pi}\int_{\partial D}\left[\log\frac{1}{|x-y|} - \frac{1}{2}\log\frac{1}{|x|}\right]\frac{\partial u}{\partial \nu}(y)\, ds(y) = -u(x)+\alpha+\frac{1}{2}\log\frac{1}{|x|} \tag{8.2}$$

and hence from the discontinuity properties of single-layer potentials (Theorem 2.19) we have

$$\frac{\partial u}{\partial \nu}(x)+\frac{1}{\pi}\int_{\partial D}\frac{\partial}{\partial \nu(x)}\left[\log\frac{1}{|x-y|} - \frac{1}{2}\log\frac{1}{|x|}\right]\frac{\partial u}{\partial \nu}(y)\, ds(y) = \frac{\partial}{\partial \nu}\log\frac{1}{|x|} \tag{8.3}$$

for $x \in \partial D$. If we can solve (8.3) for $\partial u/\partial \nu$ then u and α can be determined from (8.2) and the fact that $u=0$ on ∂D.

We now consider our second boundary-value problem for Laplace's equation, which is to find $u \in C^2(\mathbb{R}^2 \setminus \bar{D}) \cap C(\mathbb{R}^2 \setminus D)$ such that

$$\begin{aligned} u(x) = u^i(x) + u^s(x), \qquad x \in \mathbb{R}^2 \setminus D \\ \Delta u = 0 \quad \text{in} \quad \mathbb{R}^2 \setminus \bar{D} \\ u = 0 \quad \text{on} \quad \partial D \\ u^s \text{ is bounded as } x \text{ tends to infinity} \end{aligned} \tag{II}$$

where u^i is a known solution of Laplace's equation in all of $\mathbb{R}^2$. Again we can conclude that $u \in C^2(\mathbb{R}^2 \setminus \bar{D}) \cap C^1(\mathbb{R}^2 \setminus D)$. Applying Green's formulas as above, with $\log 1/|x|$ replaced by $u^i(x)$, we arrive at

$$-u^i(x) + \frac{1}{2\pi}\int_{\partial D} \log\frac{1}{|x-y|}\frac{\partial u}{\partial \nu}(y)\, ds(y) = -u(x) + \alpha \tag{8.4}$$

for $x \in \mathbb{R}^2 \setminus D$. If we now use the fact that in this case

$$\frac{1}{2\pi}\int_{\partial D} \frac{\partial u}{\partial \nu}(y)\, ds(y) = 0$$

we have

$$\frac{\partial u}{\partial \nu}(x) + \frac{1}{\pi}\int_{\partial D} \frac{\partial}{\partial \nu(x)}\left[\log\frac{1}{|x-y|} - \frac{1}{2}\log\frac{1}{|x|}\right]\frac{\partial u}{\partial \nu}(y)\, ds(y) = 2\frac{\partial u^i}{\partial \nu}(x) \tag{8.5}$$

for $x \in \partial D$. If this equation can be solved for $\partial u/\partial \nu$ then u and α can be determined from (8.4) and the fact that $u=0$ on ∂D.

In order to complete our discussion of Problems I and II we shall now show that (8.3) and (8.5) can be solved by iteration. It was for this purpose that the logarithmic term was added to (8.1) and (8.4) since the operator $\mathbf{K}_0'\colon C(\partial D) \to C(\partial D)$ defined by (cf. Definition 2.78)

$$(\mathbf{K}_0'\phi)(x) := \frac{1}{\pi}\int_{\partial D} \phi(y)\frac{\partial}{\partial \nu(x)}\log\frac{1}{|x-y|}\, ds(y) \tag{8.6}$$

has an eigenvalue at $\lambda = -1$ (Theorem 5.1) and hence without the extra logarithmic term, (8.3) and (8.5) would not be uniquely solvable. Note that in terms of the operator $\mathbf{K}_0'$ defined in (8.6) we can write (8.3) and (8.5) in the form

$$g(x) = \phi(x) + (\mathbf{K}_0'\phi)(x) - \frac{1}{2\pi}\frac{\partial}{\partial \nu}\log\frac{1}{|x|}\langle \phi, 1\rangle$$

where

$$\langle \psi, \phi \rangle = \int_{\partial D} \psi \phi \, ds$$

$$\phi(x) = \frac{\partial u(x)}{\partial \nu}$$

and $g(x) = (\partial/\partial\nu)\log 1/|x|$ and $2(\partial u^i(x)/\partial\nu)$, respectively.

Theorem 8.1. Let $\mathbf{L}_0$: $C(\partial D) \to C(\partial D)$ be defined by $(\mathbf{L}_0\phi)(x) := (\mathbf{K}_0'\phi)(x) - (1/2\pi)(\partial/\partial\nu)(\log 1/|x|)\langle \phi, 1\rangle$. Then $\sigma(L_0) \subset (-1,1)$.

Proof. Let $\mathbf{K}_0$ be the adjoint of $\mathbf{K}_0'$ with respect to the dual system $\langle C(\partial D), C(\partial D)\rangle$ and $N(\mathbf{I}+\mathbf{K}_0')$ the nullspace of $\mathbf{I}+\mathbf{K}_0'$. Then from Theorems 5.1 and 5.7 we have that $N(\mathbf{I}+\mathbf{K}_0') = \operatorname{span}(\phi_0)$ for some continuous function ϕ_0 such that $\langle \phi_0, 1\rangle \neq 0$. Note that for any ϕ in $C(\partial D)$ we have

$$\langle \mathbf{K}_0'\phi, 1\rangle = \langle \phi, \mathbf{K}_0 1\rangle = -\langle \phi, 1\rangle$$

and since

$$\int_{\partial D} \frac{\partial}{\partial \nu} \log \frac{1}{|x|} \, ds = -2\pi$$

we have

$$\langle \mathbf{L}_0\phi, 1\rangle = \langle \mathbf{K}_0'\phi, 1\rangle + \langle \phi, 1\rangle = 0. \tag{8.7}$$

Since $\mathbf{L}_0$ is compact, its spectrum consists only of a discrete set of eigenvalues together with the point $\lambda = 0$ (Theorem 1.34). We shall show that if λ is an eigenvalue of $\mathbf{L}_0$ then $\lambda \in (-1,1)$. Suppose $\mathbf{L}_0\phi = \lambda\phi$ where $\lambda \notin [-1,1)$. Then from (8.7)

$$\lambda\langle \phi, 1\rangle = \langle \mathbf{L}_0\phi, 1\rangle = 0$$

and hence $\langle \phi, 1\rangle = 0$ and $\mathbf{L}_0\phi = \mathbf{K}_0'\phi$. But $\sigma(\mathbf{K}_0') \subset [-1,1)$ (Theorem 5.1) and this contradicts the assumption that $\lambda \notin [-1,1)$. We now complete the proof of the theorem by showing that $\lambda = -1$ is not an eigenvalue. Suppose $\mathbf{L}_0\phi + \phi = 0$. Then from (8.7) we again have $\langle \phi, 1\rangle = 0$ and $\mathbf{L}_0\phi = \mathbf{K}_0'\phi$. Hence $\phi = \alpha\phi_0$ for some constant α. But then

$$0 = \langle \phi, 1\rangle = \langle \alpha\phi_0, 1\rangle = \alpha\langle \phi_0, 1\rangle$$

and since $\langle \phi_0, 1\rangle \neq 0$ we have $\alpha = 0$ and hence $\phi = 0$, a contradiction. Therefore $\lambda = -1$ is not an eigenvalue and $\sigma(\mathbf{L}_0) \subset (-1,1)$.

Theorem 8.1 implies that the integral equations (8.3) and (8.5) can be solved by successive approximations (Theorem 1.36).

We are now in a position to derive an iterative method for solving the exterior Dirichlet problem for the Helmholtz equation for low values of the wave number k. Our problem is to determine (by iteration) a function $u \in C^2(\mathbb{R}^2 \setminus \bar{D}) \cap C(\mathbb{R}^2 \setminus D)$ such that

$$u(x) = u^i(x) + u^s(x), \qquad x \in \mathbb{R}^2 \setminus D$$

$$\Delta u + k^2 u = 0 \quad \text{in} \quad \mathbb{R}^2 \setminus \bar{D}$$

$$u = 0 \quad \text{on} \quad \partial D$$

$$\left(\operatorname{grad} u^s(x), \frac{x}{|x|}\right) - iku^s(x) = o\left(\frac{1}{|x|^{1/2}}\right)$$

where u^i is a solution of the Helmholtz equation in all of $\mathbb{R}^2$. Noting that $u^s = O(1/|x|^{1/2})$ and proceeding exactly as in the case of Laplace's equation, we can establish the relationship

$$-u^i(x) + \frac{i}{4}\int_{\partial D} H_0^{(1)}(k|x-y|)\frac{\partial u}{\partial \nu}(y)\,ds(y) = -u(x) \tag{8.8}$$

for $x \in \mathbb{R}^2 \setminus D$ where $H_0^{(1)}$ denotes a Hankel function of the first kind. Choose R such that D is contained in a disk of radius R. Then since $u = 0$ on ∂D we have from Green's formula and the radiation condition satisfied by u^s at infinity that

$$\frac{i}{4}\int_{\partial D} H_0^{(1)}(k|y|)\frac{\partial u}{\partial \nu(y)}\,ds(y)$$

$$= \frac{i}{4}\int_{|y|=R}\left[H_0^{(1)}(k|y|)\frac{\partial u}{\partial \nu}(y) - u(y)\frac{\partial}{\partial \nu(y)}H_0^{(1)}(k|y|)\right]ds(y)$$

$$= \frac{i}{4}\int_{|y|=R}\left[H_0^{(1)}(k|y|)\frac{\partial u^i}{\partial \nu}(y) - u^i(y)\frac{\partial}{\partial \nu(y)}H_0^{(1)}(k|y|)\right]ds(y)$$

$$= u^i(0).$$

Hence (8.8) is equivalent to

$$u(x) + \frac{i}{4}\int_{\partial D}\left[H_0^{(1)}(k|x-y|) + \frac{\pi i}{4\log k}H_0^{(1)}(k|x|)H_0^{(1)}(k|y|)\right]\frac{\partial u}{\partial \nu}(y)\,ds(y)$$

$$= u^i(x) + \frac{\pi i u^i(0)}{4\log k}H_0^{(1)}(k|x|)$$

for $x \in \mathbb{R}^2 \setminus D$, and from the discontinuity properties of single-layer potentials (Theorem 2.19) we now have that

$$\frac{\partial u}{\partial \nu}(x) + \frac{i}{2}\int_{\partial D} \frac{\partial}{\partial \nu(x)}\left[H_0^{(1)}(k|x-y|) + \frac{\pi i}{4\log k} H_0^{(1)}(k|x|) H_0^{(1)}(k|y|)\right] \frac{\partial u}{\partial \nu}(y)\, ds(y)$$

$$= 2\frac{\partial u^i}{\partial \nu}(x) + \frac{\pi i u^i(0)}{2\log k} \frac{\partial}{\partial \nu(x)} H_0^{(1)}(k|x|) \tag{8.9}$$

for x on ∂D.

We now want to show that for k sufficiently small we can solve the integral equation (8.9) by iteration. To this end we define $\mathbf{L}_0$ as in Theorem 8.1 and let $\mathbf{L}_k$ be given by

$$(\mathbf{L}_k \psi)(x) := \frac{i}{2}\int_{\partial D} \frac{\partial}{\partial \nu(x)}\left[H_0^{(1)}(k|x-y|) + \frac{\pi i}{4\log k} H_0^{(1)}(k|x|) H_0^{(1)}(k|y|)\right] \psi(y)\, ds(y)$$

Then we have $\sigma(\mathbf{L}_0) \subset (-1,1)$ and from the low frequency behavior of Hankel's function that

$$\|\mathbf{L}_k - \mathbf{L}_0\| = O\left(\frac{1}{|\log k|}\right) \tag{8.10}$$

where $\|\cdot\|$ denotes the maximum operator norm. From (8.10) and Theorem 1.37 we can now conclude that (8.9) can be solved by iteration for low values of the wave number k.

We now turn our attention to the inverse problem with "partially inexact" data. In particular let $u = u^i + u^s$ be the solution of the exterior Dirichlet problem for the Helmholtz equation in the exterior of D (where ∂D has the smoothness requirements previously stated) such that $u = 0$ on ∂D, u^i (the "incoming wave") is an entire solution of the Helmholtz equation, and u^s (the "scattered wave") satisfies the radiation condition. We assume that D is unknown, but u^i is given along with the behavior of u^s at infinity (cf. Section 3.10)

$$u^s = \frac{1}{4} e^{i(kr+\pi/4)}\sqrt{\frac{2}{\pi k r}}\, F(\theta; k) + O\left(\frac{1}{r^{3/2}}\right)$$

where F is the far-field pattern and (r, θ) are polar coordinates. The factor multiplying $r^{-1/2}F(\theta; k)$ has been chosen purely for the sake of notational convenience, and the precise information assumed known about F will be made clear shortly. From (8.8) we have that

$$u^s(x) = -\frac{i}{4}\int_{\partial D} \frac{\partial u}{\partial \nu}(y) H_0^{(1)}(k|x-y|)\, ds(y)$$

and hence from the asymptotic behavior of Hankel's function we see that F is given by

$$F(\theta;k) = -\int_{\partial D} \frac{\partial u}{\partial \nu}(y)\exp[-ik\rho\cos(\theta-\phi)]\,ds(y) \tag{8.11}$$

where $x = re^{i\theta}$, $y = \rho e^{i\phi}$. Expanding F in a Fourier series we have

$$F(\theta;k) = \sum_{n=-\infty}^{\infty} b_n(k)e^{in\theta}$$

where

$$b_n(k) = -\frac{1}{2\pi}\int_{-\pi}^{\pi}\int_{\partial D}\frac{\partial u}{\partial \nu}(y)\exp[-in\theta - ik\rho\cos(\theta-\phi)]\,ds(y)\,d\theta$$

$$= -i^{-n}\int_{\partial D}\frac{\partial u}{\partial \nu}J_n(k\rho)e^{-in\phi}\,ds \tag{8.12}$$

and J_n denotes a Bessel function of order n.

Assumption. The coefficients b_n are known exactly for $n = 1,\ldots,N$ and all k, $|k| \leqslant k_0$ where N is a positive integer and k_0 is an arbitrarily small, but fixed, positive constant.

Given the N coefficients b_n, our aim is to compute an approximation to ∂D and to obtain error estimates for this approximation. (It will turn out that in order to do this we will need one additional piece of information.)

In order to accomplish our objectives we first need to obtain a low frequency approximation to $\partial u/\partial \nu$ evaluated on the (unknown) boundary ∂D (for a related calculation see MacCamy [1]). To this end we first assume $u^i(x) = e^{ikx_1}$ where $x = (x_1, x_2)$. Then from our previous analysis we have that for $|k| < k_0$, where k_0 is sufficiently small, and x on ∂D we can write the solution of (8.9) in the form

$$\frac{\partial u}{\partial \nu}(x) = \sum_{n=0}^{\infty}(-1)^n\mathbf{L}_k^n\left(\frac{1}{\log k}\frac{\partial}{\partial \nu}\log\frac{1}{|x|}\right) + O(k)$$

$$= \sum_{n=0}^{\infty}(-1)^n\mathbf{L}_0^n\left(\frac{1}{\log k}\frac{\partial}{\partial \nu}\log\frac{1}{|x|}\right) + O\left(\frac{1}{(\log k)^2}\right) \tag{8.13}$$

$$= \frac{1}{\log k}\frac{\partial u_0}{\partial \nu}(x) + O\left(\frac{1}{(\log k)^2}\right)$$

where for $x \in \mathbb{R}^2 \backslash D$ we can identify u_0 as the solution of Problem I for

Laplace's equation. If $z = f^{-1}(w)$, $w = x_1 + ix_2$, is the (unique) analytic function that conformally maps the exterior of the (unknown) obstacle D onto the exterior of the unit disk Ω such that at infinity f^{-1} has the Laurent expansion (where a is the mapping radius)

$$f^{-1}(w) = a^{-1}w + b + \frac{c}{w} + \frac{d}{w^2} + \cdots; \qquad a > 0,$$

then we can write

$$u_0(x) = -\log|f^{-1}(w)|. \tag{8.14}$$

Hence from (8.12) to (8.14) and the Taylor series expansion of Bessel's function we have that for $n = 1, 2, \ldots$

$$b_n(k) = \frac{i^{-n}k^n}{2^n n! \log k}\int_{\partial D}\frac{\partial}{\partial\nu}\log|f^{-1}(\rho e^{i\phi})|\rho^n e^{-in\phi}\,ds + O\left(\frac{k^n}{(\log k)^2}\right).$$

Hence if we define (for $n \geqslant 1$)

$$\mu_n := i^n 2^n n! \lim_{k\to 0}\frac{b_n(k)\log k}{k^n}$$

we can reformulate our inverse scattering problem as follows: Given μ_n, $n = 1, \ldots, N$, as defined above, to determine an approximation to ∂D (together with error estimates) from the relation

$$\mu_n = \int_{\partial D}\frac{\partial}{\partial\nu}\log|f^{-1}(\rho e^{i\phi})|\rho^n e^{-in\phi}\,ds,$$

or, taking the complex conjugate of both sides,

$$\bar{\mu}_n = \int_{\partial D}\frac{\partial}{\partial\nu}\log|f^{-1}(\rho e^{i\phi})|\rho^n e^{in\phi}\,ds. \tag{8.15}$$

We shall now proceed to the determination of (an approximation to) ∂D from (8.15). More specifically we shall show that from (8.15) we can compute the Laurent coefficients of the mapping $w = f(z)$, module the mapping radius a. In order to determine a we shall need to consider the far-field pattern arising from an incoming wave from the negative x_1 direction. But first we consider (8.15) and note that from the Cauchy–Riemann equations

$$\frac{\partial}{\partial\nu}\log|f^{-1}(\rho e^{i\phi})| = -\frac{\partial}{\partial s}\arg f^{-1}(\rho e^{i\phi}).$$

Hence from (8.15) we have

$$\bar{\mu}_n = -\int_{\partial D} \frac{\partial}{\partial s} \arg f^{-1}(\rho e^{i\phi})\rho^n e^{in\phi}\, ds$$

$$= -\int_{|z|=1} \frac{\partial \arg z}{\partial z}[f(z)]^n\, dz \tag{8.16}$$

$$= i\int_{|z|=1} \frac{1}{z}[f(z)]^n\, dz.$$

Note that the basic nonlinear nature of the inverse scattering problem is clearly seen from the relation (8.16). We now note that f has a Laurent expansion of the form

$$f(z) = az + a_0 + \frac{a_1}{z} + \cdots \tag{8.17}$$

and since ∂D is assumed to be smooth the series (8.17) is uniformly convergent on $|z|=1$. Hence from (8.16) and (8.17) we have

$$\bar{\mu}_1 = -2\pi a_0$$

$$\bar{\mu}_2 = -2\pi\left(a_0^2 + 2aa_1\right) \tag{8.18}$$

$$\vdots$$

and in general

$$\bar{\mu}_n = -2\pi n a^{n-1} a_{n-1} + \text{lower order coefficients.}$$

Hence, módule the mapping radius a, we can determine the Laurent coefficients a_n recursively in terms of the far-field data μ_n.

In order to determine the mapping radius we propose to use the information gained by measuring the far-field pattern arising from the scattering of the incident wave $u^i(x) = e^{-ikx_1}$, or combined with $u^i(x) = e^{ikx_1}$, the standing wave $u^i(x) = \sin kx_1$. If u^* is the solution of our scattering problem corresponding to the incident field $u^i(x) = \sin kx_1$, and $|k| < k_0$, then for x on ∂D we have

$$\frac{\partial u^*(x)}{\partial \nu} = \sum_{n=0}^{\infty} (-1)^n \mathbf{L}_k^n\left(2k\frac{\partial x_1}{\partial \nu}\right) + O(k^2)$$

$$= \sum_{n=0}^{\infty} (-1)^n \mathbf{L}_0^n\left(2k\frac{\partial x_1}{\partial \nu}\right) + O\left(\frac{k}{\log k}\right)$$

$$= k\frac{\partial u_0^*}{\partial \nu}(x) + O\left(\frac{k}{\log k}\right)$$

where for $x \in \mathbb{R}^2 \setminus D$ we can identify u_0^* as the solution of Problem II for Laplace's equation with $u^i(x) = x_1$. Then in terms of the conformal mapping f^{-1} we can write

$$u_0^*(x) = a\,\mathrm{Re}\left[f^{-1}(w) - \frac{1}{f^{-1}(w)}\right]$$

and if $b_n^*(k)$ are the Fourier coefficients of the far-field pattern we have that

$$\begin{aligned} b_n^*(k) &= -i^{-n}\int_{\partial D} \frac{\partial u^*}{\partial \nu} J_n(k\rho)e^{-in\phi}\,ds \\ &= -\frac{i^{-n}k^{n+1}}{2^n n!}\int_{\partial D} \frac{\partial u_0^*}{\partial \nu}\rho^n e^{-in\phi}\,ds + O\left(\frac{k^{n+1}}{\log k}\right). \end{aligned}$$

Hence if we define

$$\mu_n^* := i^n 2^n n! \lim \frac{b_n^*(k)}{k^{n+1}}$$

we have

$$\begin{aligned} \overline{\mu}_n^* &= -\int_{\partial D} \frac{\partial u_0^*}{\partial \nu}\rho^n e^{in\phi}\,ds \\ &= \int_{\partial D} \frac{\partial v_0^*}{\partial s}\rho^n e^{in\phi}\,ds \end{aligned}$$

where v_0^* is the harmonic conjugate of u_0^* defined by

$$v_0^* = a\,\mathrm{Im}\left[f^{-1}(w) - \frac{1}{f^{-1}(w)}\right].$$

Hence we now have

$$\begin{aligned} \overline{\mu}_n^* &= a\int_{\partial D} \frac{\partial}{\partial s}\left[f^{-1}(\rho e^{i\phi}) - \frac{1}{f^{-1}(\rho e^{i\phi})}\right]\rho^n e^{in\phi}\,ds \\ &= a\int_{|z|=1} \frac{\partial}{\partial z}\left[z - \frac{1}{z}\right][f(z)]^n\,dz \qquad (8.19) \\ &= a\int_{|z|=1}\left[1 + \frac{1}{z^2}\right][f(z)]^n\,dz. \end{aligned}$$

From (8.17) and (8.19) we now have that

$$\bar{\mu}_1^* = 2\pi a i(a_1 + a) \tag{8.20}$$

and therefore we can compute a from (8.18) and (8.20). Thus assuming we know $\mu_1,\ldots,\mu_N$ and $\bar{\mu}_1^*$ exactly we can define an approximation to the mapping $w = f(z)$ by

$$f_N(z) := az + a_0 + \frac{a_1}{z} + \cdots + \frac{a_{N-1}}{z^{N-1}} \tag{8.21}$$

and hence an approximation to ∂D by evaluating f_N on the unit circle. In order to compute the mean square error in this approximation we can appeal to the area theorem (Theorem 7.16) to arrive at

$$\begin{aligned}\frac{1}{2\pi}\int_0^{2\pi}|f(e^{i\theta}) - f_N(e^{i\theta})|^2\,d\theta &= \sum_{n=N}^{\infty}|a_n|^2\\ &\leqslant \frac{1}{N}\sum_{n=N}^{\infty} n|a_n|^2\\ &\leqslant \frac{a^2}{N}.\end{aligned}$$

Note that although for large N the mean square error is small, the pointwise error could nevertheless be large. An improvement on this error estimate can be obtained if it is known a priori that D is convex, since in this case we have (Pommerenke [1], p. 50) that

$$|a_n| \leqslant \frac{2a}{n(n+1)} \tag{8.22}$$

and hence

$$\begin{aligned}\frac{1}{2\pi}\int_0^{2\pi}|f(e^{i\theta}) - f_N(e^{i\theta})|^2\,d\theta &= \sum_{n=N}^{\infty}|a_n|^2\\ &\leqslant 4a^2\sum_{n=N}^{\infty}\frac{1}{n^2(n+1)^2}\\ &\leqslant 4a^2\left(\frac{N+1}{N}\right)^2\sum_{n=N}^{\infty}\frac{1}{(n+1)^4}\\ &\leqslant 4a^2\left(\frac{N+1}{N}\right)^2\int_N^{\infty}\frac{dx}{x^4}\\ &= \frac{4a^2(N+1)^2}{3N^5}.\end{aligned}$$

This provides a simple example of the fact that the more a priori information one has in trying to solve an inverse scattering problem, the better results one can expect if this extra information is built into the mathematical analysis, for example, in the present case by the estimate (8.22).

8.2 THE DETERMINATION OF THE SHAPE OF AN OBSTACLE IN $\mathbb{R}^2$

We now want to consider the inverse scattering problem for an infinite cylinder where no assumptions are made on knowing a finite number of Fourier coefficients of the far-field pattern in the low frequency limit. In particular we no longer assume that the wave number k is small and make the assumption that the far-field pattern is only known to within a certain error in the least squares sense. Our aim is to show that the continuous dependence of the boundary on the far-field data can be obtained by restricting the class of admissible boundary curves to lie a priori in a compact class of continuously differentiable simple closed curves, and to indicate a constructive method for obtaining an approximation to the boundary from the given far-field data. Our work is based on Colton and Kirsch [2].

We first formulate our inverse scattering problem in a more precise fashion. Let D be an acoustically soft simply connected domain bounded by a continuously differentiable simple closed curve ∂D and let $F(\theta; k, \alpha)$ be the far-field pattern corresponding to the scattering of a plane time-harmonic incident wave with wave number $k > 0$ moving in a direction making an angle α with the positive real axis. Our aim is to determine the shape of D from a knowledge of F, the precise information needed about F being made clear in the sequel. Mathematically we can formulate this problem as in Section 8.1 by letting u and u^s denote the velocity potential of the total and scattered fields, respectively:

$$u(x) = \exp[ik(x_1 \cos\alpha + x_2 \sin\alpha)] + u^s(x) \tag{8.23a}$$

$$\Delta u + k^2 u = 0 \quad \text{in} \quad \mathbb{R}^2 \setminus \bar{D} \tag{8.23b}$$

$$u = 0 \quad \text{on} \quad \partial D \tag{8.23c}$$

$$\left(\operatorname{grad} u^s(x), \frac{x}{|x|}\right) - iku^s(x) = o\left(\frac{1}{|x|^{1/2}}\right) \tag{8.23d}$$

$$\begin{aligned} F(\theta; k, \alpha) = \int_{\rho=\rho_0} \Big\{ & u^s \frac{\partial}{\partial \rho} \exp[-ik\rho\cos(\theta - \phi)] \\ & - \frac{\partial u^s}{\partial \rho} \exp[-ik\rho\cos(\theta - \phi)] \Big\}\, ds \end{aligned} \tag{8.23e}$$

where $x=(x_1,x_2)$, (r,θ) are the polar coordinates of the point x and (ρ,θ) are the polar coordinates of a point y on the circle $\rho=\rho_0$ containing D in its interior. The expression (8.23e) for the far-field pattern is chosen for our later convenience and is obtained by using Green's formula to represent u^s in terms of the radiating fundamental solution and then letting x tend to infinity. In particular (8.23e) is consistent with (8.11).

In order to establish the continuous dependence of ∂D on F (for suitably restricted boundary curves!) we first need to give a more precise definition of the mapping from ∂D to F. Following Garabedian [1], [2] we shall do this by reformulating the direct scattering problem (8.23a) to (8.23d) as an integral equation involving the conformal mapping taking the exterior $|z|>R$ of a disk of radius R onto the exterior of D. In particular we assume that this conformal mapping is normalized such that

$$\begin{aligned} w=x_1+ix_2&=z+\sum_{n=0}^{\infty}\frac{a_n}{z^n}\\ &=z+q(z), \end{aligned} \tag{8.24}$$

and note that in this case R is the mapping radius of D. Under this transformation the radiation condition (8.23d) remains the same, whereas the Helmholtz equation becomes

$$\Delta U+k^2|1+q'(z)|^2U=0 \tag{8.25}$$

in the $z=\xi+i\eta$ plane. Now let $G(z,\zeta)$ be the radiating Green's function for the Helmholtz equation in the exterior of the disk of radius R in the z plane and apply Green's formula to G and the solution $U(z)=u^s(w)$ in the ring $R<|z|<R_1$, where R_1 is an arbitrary large positive number. The result of such a calculation is that U satisfies

$$\begin{aligned} U(\zeta)-\iint_{R\leqslant|z|\leqslant R_1}&(\Delta G+k^2|1+q'(z)|^2G)U\,d\xi\,d\eta\\ &=\int_{|z|=R}U\frac{\partial G}{\partial \nu}ds-\int_{|z|=R_1}\left(U\frac{\partial G}{\partial \nu}-G\frac{\partial U}{\partial \nu}\right)ds \end{aligned}$$

where ν is the unit outward normal to the circles $|z|=R$ and $|z|=R_1$. If we now let R_1 tend to infinity and make use of the fact that G is a radiating solution of the Helmholtz equation, we have that U is a solution of the integral equation

$$U(\zeta)-k^2\int_{|z|>R}p(z)G(z,\zeta)U(z)\,d\xi\,d\eta=\int_{|z|=R}U(z)\frac{\partial G(z,\zeta)}{\partial \nu}ds \tag{8.26}$$

where U is known on $|z|=R$ and

$$p(z):=2\operatorname{Re} q'(z)+|q'(z)|^2.$$

Multiplying (8.26) by $\sqrt{p}$ and setting

$$V(z):=\sqrt{p(z)}\,U(z)$$

enables us to rewrite (8.26) in the form

$$g(\zeta)=V(\zeta)-k^2\int_{|z|>R} V(z)K(z,\zeta)\,d\xi\,d\eta \tag{8.27}$$

where

$$\begin{aligned} g(\zeta)&:=\sqrt{p(\zeta)}\int_{|z|=R} U(z)\frac{\partial G}{\partial \nu}(z,\zeta)\,ds \\ K(z,\zeta)&:=\sqrt{p(\zeta)}\,G(z,\zeta)\sqrt{p(z)}\,. \end{aligned} \tag{8.28}$$

It is easily verified that (8.27) is equivalent to the scattering problem (8.23a) to (8.23d) and since $|z|^{3/2}|\zeta|^{3/2}K(z,\zeta)$ is bounded at infinity and has a logarithmic singularity at $z=\zeta$, the symmetric kernel $K(z,\zeta)$ is square integrable in $|z|>R$, $|\zeta|>R$. Hence we can conclude from the uniqueness of a solution to (8.23a) to (8.23d) and the Fredholm alternative that (8.27) has a unique solution for any positive value of the wave number k. Solving (8.27) for V, transforming back to the w plane, and substituting into (8.23e) now provide us with a direct method (although computationally nontrivial!) for determining the far-field pattern F corresponding to a given domain D.

We now want to restrict the class of admissible boundary curves ∂D to lie in a compact set. Consider the class of all simple, closed, continuously differentiable curves containing the disk $r \leqslant a$ in their interior and contained in the disk $r \leqslant b$, where a and b are fixed positive constants. For a given curve ∂D in this family let $w=f(z)$ map the exterior of the unit disk Ω in the z plane conformally onto the exterior of ∂D in the w plane such that at infinity $f'(z)>0$. Then in terms of the mapping (8.24) we can write

$$f(z)=Rz+q(Rz)$$

and the curve ∂D permits the parametrization

$$z=f(e^{i\phi}),\qquad \phi\in[0,2\pi].$$

We further restrict our class of curves by requiring the mappings f to satisfy the following restrictions:

1. There exists a positive constant M_1 independent of f such that

$$\max_{\phi \in [0,2\pi]} \left| \frac{d}{d\phi} f(e^{i\phi}) \right| \leqslant M_1 .$$

2. There exists a positive constant M_2 independent of f such that

$$\left| \frac{d}{d\phi} f(e^{i\phi_1}) - \frac{d}{d\phi} f(e^{i\phi_2}) \right| \leqslant M_2 |e^{i\phi_1} - e^{i\phi_2}|$$

for $\phi_1, \phi_2 \in [0, 2\pi]$.

We denote the class of functions described above by $\Sigma(a, b)$ and consider it as a subset of the class of analytic functions defined in the exterior of Ω having a simple pole at infinity and continuously differentiable in $\mathbb{R}^2 \setminus \Omega$ with norm given by

$$\|f\|_1 := \max_{\phi \in [0,2\pi]} |f(e^{i\phi})| + \max_{\phi \in [0,2\pi]} \left| \frac{d}{d\phi} f(e^{i\phi}) \right| .$$

Since $|f(e^{i\phi})| \leqslant b$, it follows from condition (1), the mean value theorem, and the Arzéla-Ascoli theorem, that for any sequence (f_n) there exists a subsequence that is uniformly convergent on $|z| = 1$. From conditions (1), (2), and the Arzéla-Ascoli theorem again there exists a subsequence $(f_{n(j)})$ that is convergent to a function $f \in C^1[0, 2\pi]$ with respect to the norm $\|\cdot\|_1$.

We now want to show that f is the restriction to the unit circle of a univalent function defined in $|z| > 1$ and continuously differentiable for $|z| \geqslant 1$. We first note that since D contains a disk of radius $a > 0$ it follows from Theorem 7.16 that if $f \in \Sigma(a, b)$ then the mapping radius is uniformly bounded away from zero, and an elementary application of Cauchy's integral formula shows that the mapping radius and conformal center (i.e., the constant term in the Laurent expansion of f) are uniformly bounded for all $f \in \Sigma(a, b)$ (since b is finite and fixed). Hence from Theorem 7.20 we can conclude that $\Sigma(a, b)$ is a compact set of univalent functions with respect to the topology of uniform convergence on compact subsets of $|z| > 1$. This fact, and an application of the maximum principle for analytic functions (consider for example the sequence $(z^{-1} f_{n(j)}(z))$), now shows that the limit function f of the sequence $(f_{n(j)})$ with respect to $\|\cdot\|_1$ is the restriction to the unit circle of a univalent function defined in $|z| > 1$ and continuously differentiable for $|z| \geqslant 1$. In particular the image of $|z| = 1$ under the mapping f is a simple closed continuously differentiable curve.

We can now conclude that $\Sigma(a, b)$ is a compact family of univalent functions with respect to the topology induced by the norm $\|\cdot\|_1$. At the risk of making a slight abuse of notation we shall henceforth identify a given admissible curve ∂D by its associated conformal mapping in the set $\Sigma(a,b)$ and denote this by either $\partial D \in \Sigma(a,b)$ or $f \in \Sigma(a,b)$.

Theorem 8.2. Let F be the far-field pattern associated with a given domain D where $\partial D \in \Sigma(a,b)$. Then the mapping $\partial D \to F$ is a continuous mapping from $\Sigma(a,b)$ into $C[0,2\pi]$,

Proof. We shall only give an outline of the proof, and refer the reader to Smith [1] for details of the tedious derivation of the explicit estimates in a closely related case. Suppose (∂D_n) is a sequence of simple closed curves such that $\partial D_n \to \partial D$ in $\Sigma(a,b)$ and consider the integral equation (8.27). By an appropriate change of variables we can make the region of integration in (8.27) and (8.28) (and the domains of definition of the corresponding known and unknown functions) the same for both the limiting and the perturbed problems. From (8.28) we can conclude that with respect to the L^2 norm over $|\zeta|>R$ and $|\zeta|>R$, $|z|>R$, respectively, $\|g_n-g\|\to 0$ and $\|K_n-K\|\to 0$ as $n\to\infty$, where the subscripts indicate the functions corresponding to D_n. The integral equation (8.27) corresponding to the domain D_n can be written in the form

$$g+(g_n-g)=(\mathbf{I}-\mathbf{T}-(\mathbf{T}_n-\mathbf{T}))V_n \tag{8.29}$$

where the subscripts again identify the functions and operators corresponding to D_n. Since $(\mathbf{I}-\mathbf{T})^{-1}$ exists as a bounded operator we have

$$(\mathbf{I}-\mathbf{T})^{-1}g+(\mathbf{I}-\mathbf{T})^{-1}(g_n-g)=V_n-(\mathbf{I}-\mathbf{T})^{-1}(\mathbf{T}_n-\mathbf{T})V_n. \tag{8.30}$$

Then since $V=(\mathbf{I}-\mathbf{T})^{-1}g$ and $\|\mathbf{T}_n-\mathbf{T}\|\to 0$ as $n\to\infty$ we can conclude from (8.30) that

$$\|V-V_n\|\to 0 \quad \text{as} \quad n\to\infty. \tag{8.31}$$

Since V and V_n are known to be continuous for $|\zeta|\geqslant R$, it can easily be verified by using (8.31) and the integral equation (8.27) that the pointwise limits

$$\lim_{n\to\infty} V_n(\zeta)=V(\zeta)$$

$$\lim_{n\to\infty} \operatorname{grad} V_n(\zeta)=\operatorname{grad} V(\zeta)$$

are uniformly valid for ζ on compact subsets of $|\zeta|>R$. Now let $\rho=\rho_0$ be a circle containing D and (D_n) in its interior. Transforming back to the $w=x_1+ix_2$ plane we have that for $\rho=\rho_0$ the scattered fields u_n^s and u^s (corresponding to V_n and V, respectively) satisfy the uniform pointwise limit relations

$$\lim_{n\to\infty} u_n^s = u^s$$

$$\lim_{n\to\infty} \operatorname{grad} u_n^s = \operatorname{grad} u^s,$$

and hence from (8.23e) the theorem follows.

Note that F depends continuously on ∂D as a mapping from $\Sigma(a,b)$ into $C[0,2\pi]$, that is, loosely speaking, F depends continuously on smooth deformations of ∂D.

In order to establish the continuity of the inverse mapping $F \to \partial D$ it is now tempting to use Theorem 7.1 that states that a one-to-one continuous mapping of a compact set is a homeomorphism. However, this is not applicable in our situation for two reasons. The first is that it is not clear that the mapping $\partial D \to F$ is one-to-one for fixed k (cf. Theorem 6.11). More seriously (as discussed in Chapter 6) a characterization of the range of this mapping is unknown. In particular there is no way of telling if a given measurement of F lies in the range or not. For these reasons we shall reformulate our inverse scattering problem as an optimization problem. This will allow us to give a precise statement on what is meant by saying that ∂D depends continuously on F for $\partial D \in \Sigma(a,b)$. To fix our ideas let $g \in L^2[0,2\pi]$ be a measured far-field pattern and let F be a far-field pattern associated with a domain D, $\partial D \in \Sigma(a,b)$. Then we define the optimization problem denoted by P_g as follows: Minimize

$$C_g(f) = \int_0^{2\pi} |F(\theta;k,\alpha) - g(\theta;k,\alpha)|^2\, d\theta$$

for fixed k and α, subject to the constraint that $f \in \Sigma(a,b)$. Note that if $\partial D^* \in \Sigma(a,b)$ is a solution of the inverse scattering problem corresponding to the far-field pattern g then the mapping f^* taking the unit circle onto ∂D^* is a solution of P_g since $C_g(f^*)=0$. Conversely, if f^* is a solution of P_g and $C_g(f^*)=0$ then ∂D^* is a solution of the inverse scattering problem. If $C_g(f^*)>0$ then the inverse scattering problem is not solvable, but ∂D^* is a best approximation in $\Sigma(a,b)$ in the sense that $C_g(f)$ is minimal.

Theorem 8.3. There exists a solution f^* of P_g.

Proof. Let (f_n) be a minimizing sequence, that is, $\lim_{n\to\infty} C_g(f_n) = \inf_{f\in\Sigma} C_g(f)$. Since $\Sigma(a,b)$ is compact there exists a convergent subsequence $(f_{n(j)})$ such that $f_{n(j)} \to f^*$ where $f^* \in \Sigma(a,b)$. By Theorem 8.2 and the continuity of C_g it now follows that $C_g(f^*) = \inf_{f\in\Sigma} C_g(f)$, that is, f^* is a solution of P_g.

In general we cannot expect the solution of the optimization problem P_g to be unique. Let $\Phi^*(g)$ be the set of all solutions f^* of P_g and denote the numerical value of $C_g(f^*)$ by C_g^*. Then we have the following result on the continuous dependence of $\partial D \in \Sigma(a, b)$ on the far-field data.

Theorem 8.4. The set $\Phi^*(g)$ is graph compact, that is, if $g_n \to g$ in $L^2[0, 2\pi]$, $f_n^* \in \Phi^*(g_n)$, then there exists a convergent subsequence of (f_n^*) and every limit point lies in $\Phi^*(g)$.

Proof. The functions f_n^* lie in a compact set and therefore there exists a convergent subsequence of (f_n^*). Let f^* be any limit point of (f_n^*) that is, there exists a subsequence $(f_{n(j)}^*)$ of (f_n^*) such that $f_{n(j)}^* \to f^*$. We want to show that f^* is optimal for P_g. Let $f^{**} \in \Phi^*(g)$. Then, using Theorem 8.2 and the continuity of the function C_g, we have $C_g^* \leqslant C_g(f^*) = \lim_{j \to \infty} C_{g_{n(j)}}(f_{n(j)}^*) = \lim_{j \to \infty} C_{g_{n(j)}}^* \leqslant \lim_{j \to \infty} C_{g_{n(j)}}(f^{**}) = C_g^*$, that is, $C_g^* = C_g(f^*)$, and this implies the theorem.

As previously mentioned, even if g is a far-field pattern for some $f \in \Sigma(a, b)$, we cannot conclude that the solution of P_g is unique.However, it follows from Theorem 6.11 that if in this case we look at the set $\Phi^*(g)$ for an interval of k values, that is, $\Phi_k^*(g)$, $k_0 \leqslant k \leqslant k_1$, then there exists a unique $f \in \bigcap_{k_0 \leqslant k \leqslant k_1} \Phi_k^*(g)$. Therefore in practice one could compute the solutions of P_g for sufficiently many values of k such that there is only one f lying in the intersection of all the solution sets. This f will map the unit circle onto ∂D^* where ∂D^* is the solution of the inverse scattering problem.

A modification of the arguments used in Theorem 8.4 shows that if ∂D is known a priori to lie in $\Sigma(a, b)$, then ∂D depends continuously on the scattering cross section σ defined by

$$\sigma := \lim_{r \to \infty} \int_{|x| = r} |u^s(x)|^2 \, ds,$$

where continuity is defined in the sense of Theorem 8.4. An initial approximation to the unknown boundary ∂D can now be obtained by choosing the ellipse in $\Sigma(a, b)$ such that the scattering cross section best fits the measured cross section. (An even simpler choice would be to consider circles in $\Sigma(a, b)$ instead of ellipses.) Improvements to this initial estimate can now be found by using the variational procedure discussed in Chapter 7, which is applicable to both two- and three-dimensional problems. We note at this point that the procedure in Chapter 7 is based on solving the variational formula (7.24) and the "solution" of this improperly posed problem can be obtained by methods other than the Backus–Gilbert method, for example, a least squares minimization procedure for suitably restricted variations $\delta\nu$ (cf. the following section). Finally, we note that the iterative use of (7.24) to find $\partial D \in \Sigma(a, b)$ is essentially the Newton–Kantorovich method.

8.3 THE DETERMINATION OF THE SHAPE OF AN OBSTACLE IN $\mathbb{R}^3$

We now conclude this chapter by describing how the results of the previous section can be extended to obstacles in $\mathbb{R}^3$ (Angell, Colton, and Kirsch [1]). Since conformal mapping methods are no longer applicable, we shall impose the simplifying assumption that the unknown scattering obstacle is strictly starlike with respect to the origin, that is, can be described in the form

$$x = f(\hat{x})\hat{x} \tag{8.32}$$

where $x \in \mathbb{R}^3$, $\hat{x} = x/|x|$, and f is required to be in a compact set of smooth functions defined on the unit sphere. As in the previous section we shall again need to use an integral equation formulation of the direct scattering problem that is applicable for all wave numbers and (sufficiently smooth) scattering obstacles (cf. Chapter 3). This need follows from the fact that the shape of the obstacle is unknown and hence the eigenvalues of the interior problem are unknown. Thus in order to guarantee that the integral equation formulation of the direct scattering problem has a unique solution we must use the ideas of Chapter 3 to formulate an integral equation that is uniquely solvable for a given wave number and any sufficiently smooth bounded domain (or equivalently a given domain and any value of the wave number).

We now proceed to make the above ideas more precise and show how they can be used as in the previous section to reformulate the inverse scattering problem as an optimization problem over a compact set. We begin by defining more specifically the compact set in which the function f defined in (8.32) is supposed to lie. Let

$$\Gamma = \{x \in \mathbb{R}^3 \mid |x| = 1\}$$

denote the unit sphere in $\mathbb{R}^3$ and let $C^{1,\alpha}(\Gamma)$ denote the space of Hölder continuously differentiable functions equipped with the Hölder norm $\|\cdot\|_{1,\alpha}$ (cf. Section 2.2). Then we require the function f in (8.32) to be in a compact subset U of the set $\mathfrak{F}$ defined by

$$\mathfrak{F} = \{f \in C^{1,\alpha}(\Gamma) \mid \|f\|_{1,\alpha} \leqslant b, f(\hat{x}) \geqslant a\}$$

where a and b are positive constants. In particular, a possible choice for U is for U to be a bounded subset of $C^{1,\beta}(\Gamma)$ for any $\beta, \alpha < \beta < 1$, such that U is closed in $C^{1,\alpha}(\Gamma)$ since the imbedding $C^{1,\beta}(\Gamma) \to C^{1,\alpha}(\Gamma)$ is compact (cf. Section 2.2).

Now let D be a bounded simply connected domain in $\mathbb{R}^3$ such that ∂D is described by (8.32) for $f \in U$, and let u^i be an entire solution of the Helmholtz equation. Then we can describe the direct scattering problem as the problem of finding $u = u^i + u^s$, $u \in C^2(\mathbb{R}^3 \setminus \bar{D}) \cap C(\mathbb{R}^3 \setminus D)$, such that

$$\Delta u + k^2 u = 0 \quad \text{in} \quad \mathbb{R}^3 \setminus \bar{D} \tag{8.33a}$$

$$u = 0 \quad \text{on} \quad \partial D \tag{8.33b}$$

$$\left(\operatorname{grad} u^s(x), \frac{x}{|x|} \right) - iku^s(x) = o\left(\frac{1}{|x|} \right) \tag{8.33c}$$

where k is assumed positive and the radiation condition (8.33c) holds uniformly in all directions. In order to reformulate (8.33) as an integral equation that is uniquely solvable for all values of the wave number k, we now have a variety of possibilities at our disposal (cf. Section 3.6). We choose to follow the method of Ursell [1] and represent the scattered wave u^s in the form

$$u^s(x) = \int_{\partial D} \psi(y) \frac{\partial}{\partial \nu(y)} G(x, y)\, ds(y), \qquad x \in \mathbb{R}^3 \setminus \bar{D} \tag{8.34}$$

where G is the dissipative Green's function for a ball contained in the interior of D, ν is the unit outward normal to ∂D, and ψ is a continuous density to be determined. For details of this approach the reader is referred to Ursell's original paper or Section 3.6 of this book. Letting x tend to ∂D in (8.34) now gives us the integral equation

$$\psi(x) + \int_{\partial D} \psi(y) \frac{\partial}{\partial \nu(y)} G(x, y)\, ds(y) = -u^i(x), \qquad x \in \partial D \tag{8.35}$$

for the determination of the unknown density ψ. Due to the choice of G, it can easily be verified that (8.35) has a unique solution ψ for any wave number k (Ursell [1] and Section 3.6). Hence we now have the desired reformulation of the direct scattering problem (8.33) as an integral equation (8.35). In particular, from (8.34) we see that u^s has the asymptotic behavior

$$u^s(x) = \frac{e^{ik|x|}}{|x|} F(\hat{x}) + O\left(\frac{1}{|x|^2} \right) \tag{8.36}$$

where $F(\hat{x}) = (F\psi)(\hat{x})$ is defined by

$$F(\hat{x}) := \int_{\partial D} \psi(y) \frac{\partial}{\partial \nu(y)} G_0(\hat{x}, y)\, ds(y) \tag{8.37}$$

with G_0 being an analytic function of its independent variables. (The explicit form of G_0 can easily be computed, but for our purposes the above-stated regularity result is sufficient.) As in Section 8.2 we can now formulate our inverse scattering problem as follows: If g is the measured far-field pattern, minimize the functional

$$C_g(f) = \|F - g\|_{L^2(\Gamma)} \tag{8.38}$$

subject to the constraint that ∂D is described by (8.32) where $f \in U$. Having discussed the existence and continuous dependence of the solution to this problem, we can easily treat the problem where the functional to be minimized is

$$I(f) = \sum_{j=1}^{N} |\sigma_j(f) - \tilde{\sigma}_j|^2$$

where $\tilde{\sigma}_j$ is the measured scattering cross section corresponding to an incident wave in the α_j direction and $\sigma_j(f)$ is the scattering cross section corresponding to a domain D described by (8.32) with $f \in U$ (cf. Section 8.2).

In order to establish the existence of a function $f \in U$ that minimizes the functional C_g over the set U we first need to use the integral equation (8.35) to study the map $f \to F$ as a mapping from U into $C(\Gamma)$. In particular we want to show that this mapping is continuous and hence C_g assumes its minimum value on the compact set U. To this end we use (8.32) to rewrite the integral equation (8.35) in the form

$$\phi(\hat{x}) + \int_\Gamma a_f(\hat{x}, \hat{y})\phi(\hat{y})\, ds(\hat{y}) = v_f(\hat{x}), \qquad \hat{x} \in \Gamma \tag{8.39}$$

where

$$\phi(\hat{x}) := \psi(f(\hat{x})\hat{x})$$

$$v_f(\hat{x}) := -u^i(f(\hat{x})\hat{x})$$

and

$$a_f(\hat{x}, \hat{y}) := \frac{\partial G}{\partial \nu(y)}(f(\hat{x})\hat{x}, f(\hat{y})\hat{y}) J_f(\hat{y})$$

where J_f denotes the Jacobian of the mapping (8.32). A long but straightforward calculation now establishes the following lemma, the proof of which we refer to Angell, Colton, and Kirsch [1].

Lemma 8.5. For any $\delta \leqslant \frac{1}{2}(1 - \alpha)$ there exists a constant γ such that

(a) $\qquad |a_f(\hat{x}, \hat{y})| \leqslant \gamma |\hat{x} - \hat{y}|^{\alpha - 2}$ for all $\hat{x}, \hat{y} \in \Gamma$ and $f \in U$,

(b) $$|a_f(\hat{x},\hat{y})-a_g(\hat{x},\hat{y})|\leqslant \gamma|\hat{x}-\hat{y}|^{\alpha-2}\|f-g\|_{1,\alpha}^{\delta}$$

$$\text{for all } \hat{x},\hat{y}\in\Gamma \quad\text{and}\quad f,g\in U.$$

We can now use Lemma 8.5 to establish the continuity of the mapping $f\to F$. From the definition of F in (8.37) and the fact that f is Hölder continuously differentiable, it is easily seen that it suffices to prove that the mapping $f\to\phi$ of U into $C(\Gamma)$ is continuous. Defining the operator $\mathbf{A}_f$ by

$$(\mathbf{A}_f\phi)(\hat{x}):=\int_\Omega a_f(\hat{x},\hat{y})\phi(\hat{y})\,ds(\hat{y}),\qquad \hat{x}\in\Gamma$$

we have the following result:

Theorem 8.6. Let $B(C(\Gamma),C(\Gamma))$ denote the space of bounded linear operators on $C(\Gamma)$ into itself equipped with the usual operator norm. Then the mapping $f\to\mathbf{A}_f$ from $C^{1,\alpha}(\Gamma)$ into $B(C(\Gamma),C(\Gamma))$ is Hölder continuous. Moreover the mapping $f\to\phi_f$ of U into $C(\Gamma)$ where ϕ_f is the unique solution of the integral equation (8.39) is Hölder continuous.

Proof. Let $f,g\in U$ and $\phi\in C(\Gamma)$. Then for any $x\in\Gamma$ we have, using Lemma 8.5,

$$|\mathbf{A}_f\phi(\hat{x})-\mathbf{A}_g\phi(\hat{x})|\leqslant\int_\Omega|a_f(\hat{x},\hat{y})-a_g(\hat{x},\hat{y})|\,|\phi(\hat{y})|\,ds(\hat{y})$$

$$\leqslant\gamma\|\phi\|_\infty\|f-g\|_{1,\alpha}^{\delta}\int_\Gamma|\hat{x}-\hat{y}|^{\alpha-2}\,ds(\hat{y})$$

where $\delta\leqslant\frac{1}{2}(1-\alpha)$ and $\|\cdot\|_\infty$ is the usual norm on $C(\Gamma)$. This establishes the first part of the theorem. To prove the second part of the theorem we note that it suffices to show that the operators $(\mathbf{I}+\mathbf{A}_f)^{-1}$, $f\in U$, are equibounded. Indeed, if this is the case, the results follows from the identity

$$\phi_f-\phi_g=(\mathbf{I}+\mathbf{A}_f)^{-1}\{(v_f-v_g)+(\mathbf{A}_g-\mathbf{A}_f)(\mathbf{I}+\mathbf{A}_g)^{-1}v_g\}$$

and the Hölder continuity of the mapping $f\to v_f$. To see that the family of operators $((\mathbf{I}+\mathbf{A}_f)^{-1})$ is equibounded, assume the contrary. Then there exist sequences $(v_n),(f_n)$, with $\|v_n\|_\infty\leqslant 1$ such that $\|(\mathbf{I}+\mathbf{A}_{f_n})^{-1}v_n\|\to\infty$. Since U is compact, we can assume without loss of generality that there is an $f\in U$ such that $f_n\to f\in U$. Then, setting

$$\phi_n=\frac{(\mathbf{I}+\mathbf{A}_{f_n})^{-1}v_n}{\|(\mathbf{I}+\mathbf{A}_{f_n})^{-1}v_n\|_\infty},$$

we have

$$\phi_n = (\mathbf{I}+\mathbf{A}_{f_n})\phi_n + (\mathbf{A}_f - \mathbf{A}_{f_n})\phi_n - \mathbf{A}_f\phi_n.$$

But $(\mathbf{I}+\mathbf{A}_{f_n})\phi_n \to 0$ and $\mathbf{A}_{f_n} \to \mathbf{A}_f$ in $B(C(\Gamma), C(\Gamma))$. Furthermore, the compactness of the operator $\mathbf{A}_f$ now implies that there exists a subsequence of (ϕ_n), which we again denote by (ϕ_n), such that $\phi_n \to \phi$ for some $\phi \in C(\Gamma)$ and $\phi = -\mathbf{A}_f\phi$. Since (8.39) is uniquely solvable, this implies that $\phi = 0$, which contradicts the fact that ϕ is a limit of functions of norm one. This completes the proof.

As we have already mentioned the continuity of the mapping $f \to \phi_f$ implies the continuity of the mapping $f \to F$ where F is the far-field pattern. This fact, and the compactness of U, now imply the following theorem (cf. Theorem 8.3).

Theorem 8.7. The functional C_g takes its absolute minimum on the set U.

Let $\Phi^*(g)$ be the set of all functions $f^* \in U$ such that $C_g(f^*)$ is equal to the minimum value of C_g over U. Then we have the following result on the continuous dependence of f on the far-field pattern F (cf. Theorem 8.4).

Theorem 8.8. The set $\Phi^*(g)$ is graph compact.

9

OPTIMAL CONTROL PROBLEMS IN RADIATION AND SCATTERING THEORY

In this final chapter we shall consider two optimal control problems associated with the radiation or scattering by an infinite cylinder, basing our presentation on the results of Angell and Kleinman [1] and Kirsch [1]. In both of these problems we are given a bounded connected domain and our aim is to "control" the boundary data such that the far-field power flux through a given angle is maximized. The difference between the two problems lies in the nature of the boundary data. In the first case the boundary data are assumed to be of Dirichlet type and hence the analysis is considerably simplified due to the linear nature of the problem. From a physical point of view this problem can be viewed as a radiation or an antenna synthesis problem. The second problem we shall consider is a scattering problem with an impedance-type boundary condition, that is, we wish to choose the impedance such that the far-field power flux due to the scattering of a given incident field is maximized. We note that this problem is nonlinear and, except for a change in the functional to be minimized (the "cost" functional), is essentially the same mathematical problem as the inverse scattering Problem A2 of Chapter 6. Indeed it was for this reason that we did not discuss this inverse scattering problem in Chapter 8, preferring to present the mathematical details in the context of an optimal control problem rather than an inverse scattering problem. As in the case of the inverse scattering problem, the concept of compactness is seen to play an essential role. However, from the point of view of optimal control the idea of weak compactness is more appropriate than strong compactness, and hence we shall begin this chapter with a brief review of the basic concepts of weak compactness in Hilbert space.

Although the analysis of this chapter is restricted to two dimensions, the analysis is easily extendable to the three-dimensional case for both acoustic and electromagnetic waves.

9.1 WEAK COMPACTNESS IN HILBERT SPACE

Let H be a complex Hilbert space with inner product $(\cdot, \cdot)$. A sequence (x_n) in H is said to be weakly convergent to an element $x \in H$ if $\lim_{n \to \infty}(x_n, y) = (x, y)$ for every $y \in H$. It is easily verified that the element x is unique.

Definition 9.1. A set K in H is said to be weakly compact if every sequence in K has a weakly convergent subsequence whose limit lies in K.

Theorem 9.2. Let H be a separable Hilbert space. Then every bounded subset of H has a weakly convergent subsequence.

Proof. Since H is separable there exists a countable dense subset $D = \{y_n\}$. Let K denote the given bounded subset of H and (x_n) a sequence in K. Then since $((x_n, y_1))$ is bounded, there exists a subsequence $(x_{n(1)})$ of (x_n) such that $((x_{n(1)}, y_1))$ converges. Similarly, for every integer $j \geqslant 2$ there exists a subsequence $(x_{n(j)})$ of $(x_{n(j-1)})$ such that $((x_{n(j)}, y_k))$ is convergent for $1 \leqslant k \leqslant j$. Hence $(x_{n(n)})$ is a subsequence of (x_n) for which $((x_{n(n)}, y_k))$ convergences for every $k \geqslant 1$.

We now want to show that this subsequence is weakly convergent. For notational convenience we relabel $x_{n(n)}$ by x_n. Let span$\{D\}$ be the subspace of H formed by taking all linear combinations of elements in D. Then for $y \in \text{span}\{D\}$ the linear functional f defined by $f(y) = \lim_{n \to \infty}(x_n, y)$ is well defined and continuous. Since D is dense in H, f has an extension to all of H, and by the Riesz representation theorem there exists an $x \in H$ such that $f(y) = (x, y)$ for $y \in H$. Now let $y \in H$ and $\varepsilon > 0$, $z \in D$, such that $\|y - z\| < \varepsilon$. Then

$$|(x_n - x, y)| \leqslant |(x_n, y - z)| + |(x, y - z)| + |(x_n - x, z)|$$
$$\leqslant \varepsilon\|x_n\| + \varepsilon\|x\| + |(x_n - x, z)|.$$

Hence since (x_n) is bounded and $(x_n, z) \to (x, z)$ for $z \in D$ we have (since ε is arbitrary) that $(x_n, y) \to (x, y)$ for $y \in H$, that is, x is the weak limit of (x_n).

Definition 9.3. A set K in H is said to be weakly closed if the limit of every weakly convergent sequence from K is contained in K.

Note that since strong convergence (i.e., convergence in norm) implies weak convergence, a weakly closed set is necessarily closed.

Definition 9.4. A set K in H is said to be convex if for $x, y \in K$ and $0 \leqslant \lambda \leqslant 1$, we have that $\lambda x + (1 - \lambda) y \in K$.

Theorem 9.5. Every closed, convex set of a Hilbert space is weakly closed.

Proof. Let K be a closed convex set of H and x an element of H that is not in K. Then since every nonempty closed convex subset of a Hilbert space has an element of minimal norm, there is an $x_0 \in K$ such that $\min_{y \in K}\|y - x\| = \|x_0 - x\|$. Without loss of generality we can assume (by translation) that $x = -x_0$. Since $(x, x_0) < 0$, the theorem will be proved if we can show that $\mathrm{Re}(z, x_0) \geqslant 0$ for all $z \in K$. This follows from the fact that if (z_n), $z_n \in K$, is weakly convergent to an element x not in K, then $\mathrm{Re}(z_n - x, x_0) \geqslant -(x, x_0) > 0$, contradicting the fact that $|(z_n - x, x_0)|$ tends to zero. To show that $\mathrm{Re}(z, x_0) \geqslant 0$ for all $z \in K$ we consider the function ϕ: $[0,1] \to \mathbb{R}$ defined by

$$\phi(t) = \|(1-t)x_0 + tz - x\|^2$$

and note that since K is convex, ϕ has its minimum at $t = 0$. Therefore $\phi'(0+)$ is nonnegative, that is,

$$\mathrm{Re}(x_0, z - x_0) \geqslant 0.$$

But since $x = -x_0$ this implies that $\mathrm{Re}(z, x_0) \geqslant \|x_0\|^2 > 0$.

Corollary 9.6. Every closed, bounded, convex subset of a separable Hilbert space is weakly compact.

Proof. This follows immediately from Theorems 9.2 and 9.5.

Definition 9.7. Let K be a subset of a Hilbert space H. A function F: $K \to \mathbb{R}$ is said to be weakly continuous at $x \in K$ if for every sequence (x_n) in K that is weakly convergent to $x \in K$ we have $F(x) = \lim_{n \to \infty} F(x_n)$.

For our purposes the importance of weak compactness is that weakly continuous functions assume their maximum and minimum on weakly compact sets.

Theorem 9.8. Let K be a weakly compact subset of a Hilbert space H and let F: $K \to \mathbb{R}$ be weakly continuous at every point of K. Then

1. F is bounded on K.
2. F achieves its maximum and minimum on K.

Proof. Suppose F is not bounded from above on K. Then there exists a sequence (x_n) in K such that $F(x_n) \geqslant n$. Since K is weakly compact there is a subsequence $(x_{n(k)})$ of (x_n) such that $(x_{n(k)})$ is weakly convergent to an element $x \in K$. Since F is weakly continuous we have $\lim_{k \to \infty} F(x_{n(k)}) = F(x) < \infty$, which is a contradiction. Hence F is bounded above and a similar proof shows that F is bounded below.

Now let $M=\sup_{x\in K}F(x)$ and (x_n) a sequence in K for which $M=\lim_{n\to\infty}F(x_n)$. Then (x_n) has a weakly convergent subsequence $(x_{n(k)})$ such that $(x_{n(k)})$ converges weakly to an element $x\in K$. By the weak continuity of F we now have $M=\lim_{k\to\infty}F(x_{n(k)})=F(x)$ and hence F achieves its maximum on K. A similar proof shows that F also achieves its minimum on K.

In concluding this section we wish to emphasize that the results presented here for the case of a Hilbert space have a straightforward and natural extension to reflexive Banach spaces and, more generally, to reflexive dual systems, for example, $\langle L^\infty(\Omega), L^1(\Omega)\rangle$ for some given set Ω. The reader should have little difficulty in modifying the proofs just given for Hilbert spaces to these more general settings. Such extensions are not presented here since our aim is only to present the basic concepts of weak compactness in their simplest setting (i.e., a Hilbert space) and no attempt is made toward a definitive treatment. However, having given the basic results on weak compactness in a Hilbert space we shall not hesitate to make use, where appropriate, of their natural generalizations to the above-mentioned broader settings, and with the background given in this section we trust that this will cause no difficulty for the reader.

9.2 OPTIMAL CONTROL FOR A RADIATION PROBLEM

In this section we shall consider the problem of choosing the surface current on a bounded obstacle D such as to maximize the far-field power flux through a given angle. In order for a solution to exist to this problem (for both mathematical and physical reasons!) it is necessary to restrict the class of admissible surface currents, and this shall be done by requiring the currents to be in a weakly compact subset of $L^2(\partial D)$. We shall then show that the optimal solution can be constructed by standard procedures. We then consider a class of control sets such that the optimal solution is the limit of a sequence of bang-bang controls. Our analysis is based on the work of Angell and Kleinman [1].

Let D be a bounded connected domain in the plane with C^2 boundary ∂D and unit outward normal ν. Then if $\phi\in L^2(\partial D)$ denotes the surface current on ∂D, we can express the radiating field u^s arising from this current in the form (cf. Section 4.1).

$$u^s(x)=\frac{i}{4}\int_{\partial D}H_0^{(1)}(k|x-y|)\phi(y)\,ds(y);\qquad x\in\mathbb{R}^2\setminus\bar{D},$$

where k is the wave number (assumed positive) and $H_0^{(1)}$ denotes a Hankel function of the first kind of order zero. Then from the asymptotic behavior of Hankel's function we see that at infinity u^s has the asymptotic behavior

$$u^s(x)=\frac{1}{4}e^{i(kr+\pi/4)}\sqrt{\frac{2}{\pi kr}}\,F(\theta;k)+O\left(\frac{1}{r^{3/2}}\right)$$

where (r, θ) are polar coordinates and the far-field pattern F is defined by

$$F(\theta; k) = \int_{\partial D} \phi(y) \exp[-ikr' \cos(\theta - \theta')] \, ds(y) \tag{9.1}$$

with (r', θ') denoting polar coordinates of $y \in \partial D$ (we assume that the origin is contained in D). Note that (9.1) defines a compact operator $\mathbf{F}$: $L^2(\partial D) \to L^2(0, 2\pi)$ and henceforth we shall write (9.1) simply as

$$F = \mathbf{F}\phi. \tag{9.2}$$

In what follows, we shall denote the inner product and norm in $L^2(0, 2\pi)$ by $(\cdot, \cdot)$ and $\|\cdot\|$, respectively, and those in $L^2(\partial D)$ by $(\cdot, \cdot)_{\partial D}$ and $\|\cdot\|_{\partial D}$.

Now let $Q(r; \phi)$ denote the power flux through a circle of radius r due to the surface current ϕ on ∂D. Let C_r denote this circle and assume that D is contained inside the disk bounded by C_r. Then

$$Q(r; \phi) = \int_{C_r} |u^s|^2 \, ds = \gamma \int_0^{2\pi} |F(\theta)|^2 \, d\theta + O\left(\frac{1}{r}\right)$$

where $\gamma = \gamma(k)$ is a positive constant. If we define the far-field radiated power by

$$\begin{aligned} Q(\phi) &= \lim_{r \to \infty} Q(r; \phi) \\ &= \gamma \int_0^{2\pi} |F(\theta)|^2 \, d\theta \end{aligned}$$

then we have $Q(\phi) = \gamma \|\mathbf{F}\phi\|^2$. Now let α be a measurable subset of $[0, 2\pi]$ and $\alpha(\theta)$ its characteristic function. Then, ignoring the unimportant constant γ, we define the (normalized) far-field power flux through α by

$$\begin{aligned} Q_\alpha(\phi) &:= \int_0^{2\pi} \alpha(\theta) |F(\theta)|^2 \, d\theta \\ &= \int_0^{2\pi} \alpha(\theta) |\mathbf{F}\phi|^2 \, d\theta. \end{aligned} \tag{9.3}$$

We can now formulate our optimal control problem as follows: For a given closed, bounded, convex subset U of $L^2(\partial D)$ (called the class of admissible controls) find a $\phi_0 \in U$ (an optimal control) such that $Q_\alpha(\phi_0)$ is the maximum value of Q_α over U.

We shall now establish the existence of an optimal control ϕ_0 as defined above.

Lemma 9.9. The functional Q_α is weakly continuous on $L^2(\partial D)$.

Proof. Let (ϕ_n) be weakly convergent to ϕ in $L^2(\partial D)$. Then since $\mathbf{F}$ is compact it follows from the uniform boundedness principle that $\mathbf{F}\phi_n \to \mathbf{F}\phi$ strongly in $L^2(0,2\pi)$. But

$$|Q_\alpha(\phi_n)-Q_\alpha(\phi)| = \left|\int_0^{2\pi}\alpha(\theta)\left[|(\mathbf{F}\phi_n)(\theta)|^2 - |(\mathbf{F}\phi)(\theta)|^2\right]d\theta\right|$$

$$\leqslant \int_0^{2\pi}\left||(\mathbf{F}\phi_n)(\theta)|^2 - |(\mathbf{F}\phi)(\theta)|^2\right|d\theta$$

$$\leqslant \|(|\mathbf{F}\phi_n| + |\mathbf{F}\phi|)\|^2\|(|\mathbf{F}\phi_n| - |\mathbf{F}\phi|)\|^2$$

$$\leqslant (\|\mathbf{F}\phi_n\| + \|\mathbf{F}\phi\|)^2\|\mathbf{F}\phi_n - \mathbf{F}\phi\|^2.$$

Since $\mathbf{F}\phi_n$ tends to $\mathbf{F}\phi$ strongly in $L^2(0,2\pi)$ we now have that

$$\lim_{n\to\infty}|Q_\alpha(\phi_n)-Q_\alpha(\phi)| = 0,$$

that is, Q_α is weakly continuous.

Theorem 9.10. Let U be a closed, bounded, convex subset of $L^2(\partial D)$. Then there exists a $\phi_0 \in U$ such that

$$Q_\alpha(\phi_0) = \sup_{\phi\in U} Q_\alpha(\phi).$$

Proof. Since $L^2(\partial D)$ is a separable Hilbert space we have from Corollary 9.6 that U is weakly compact. The theorem now follows from Theorem 9.8.

Theorem 9.10 can be strengthened to state that Q_α assumes its optimal value at a point on the boundary of U.

Theorem 9.11. Let U be a closed, bounded, convex subset of $L^2(\partial D)$. Then the functional Q_α takes its optimal value at a point of the boundary of U.

Proof. Let $\phi_1, \phi_2 \in U$. Then since U is convex, $\phi = \lambda\phi_1 + (1-\lambda)\phi_2 \in U$ for $0 \leqslant \lambda \leqslant 1$, and a short calculation shows that

$$Q_\alpha(\phi) = (\lambda^2 - \lambda)Q_\alpha(\phi_1 - \phi_2) + \lambda Q_\alpha(\phi_1) + (1-\lambda)Q_\alpha(\phi_2).$$

Therefore, since $\lambda^2 - \lambda \leqslant 0$ and $Q_\alpha(\phi_1 - \phi_2) \geqslant 0$, we have

$$Q_\alpha(\phi) \leqslant \lambda Q_\alpha(\phi_1) + (1-\lambda)Q_\alpha(\phi_2),$$

that is, Q_α is a convex functional. Now suppose that ϕ_0 is an optimal control. Then, since U is closed, bounded, and convex, by Corollary 9.6 and the Krein–Milman theorem (cf. Royden [1]) we can state that U is equal to the convex hull of its boundary points. Hence we can write ϕ_0 in the form

$$\phi_0 = \sum_{i=1}^{n} \lambda_i \phi_i$$

where ϕ_i, $i=1,\ldots,n$, are on the boundary of U, $\lambda_i \geqslant 0$ for $i=1,2,\ldots,n$, and

$$\sum_{i=1}^{n} \lambda_i = 1.$$

The convexity of Q_α now implies that

$$\begin{aligned} Q_\alpha(\phi_0) &= Q_\alpha\left(\sum_{i=1}^{n} \lambda_i \phi_i\right) \\ &\leqslant \sum_{i=1}^{n} \lambda_i Q_\alpha(\phi_i) \\ &\leqslant \max_{1 \leqslant i \leqslant n} \{Q_\alpha(\phi_i)\} \sum_{i=1}^{n} \lambda_i \\ &= \max_{1 \leqslant i \leqslant n} \{Q_\alpha(\phi)\} \end{aligned}$$

and since Q_α assumes its maximum value at ϕ_0, there exists a ϕ on the boundary of U such that $Q_\alpha(\phi_0) = Q_\alpha(\phi)$.

By using more sophisticated techniques it is possible to show that Q_α takes its optimal value at an extreme point of U, that is, a point $\phi \in U$ that is not properly contained in any set of the form $\lambda\phi_1 + (1-\lambda)\phi_2$, $\lambda \in (0,1)$, $\phi_1, \phi_2 \in U$ (Kirsch [4]).

We shall now show that in the case when U is the unit ball in $L^2(\partial D)$, Theorem 9.11 leads to a constructive procedure for finding an approximation to the optimal control ϕ_0 and the maximum power flux $Q_\alpha(\phi_0)$. To show this we let B denote the closed unit ball in $L^2(\partial D)$, and note that B is closed, bounded, and convex. If we now rewrite the functional Q_α in the form

$$Q_\alpha(\phi) = (\alpha \mathbf{F}\phi, \mathbf{F}\phi) = (\mathbf{F}^*\alpha\mathbf{F}\phi, \phi)_{\partial D}$$

where $\mathbf{F}^*$: $L^2(0,2\pi) \to L^2(\partial D)$ is the adjoint operator to $\mathbf{F}$ defined by

$$(\mathbf{F}^*\psi)(y) := \int_0^{2\pi} \psi(\theta) \exp[ikr'\cos(\theta-\theta')]\,d\theta, \quad y=(r',\theta') \qquad (9.4)$$

for $\psi \in L^2(0,2\pi)$, and define the operator $\mathbf{R}$: $L^2(\partial D) \to L^2(\partial D)$ by

$$\mathbf{R}\phi := \mathbf{F}^*\alpha\mathbf{F}\phi, \qquad (9.5)$$

then choosing $U = B$ we can reformulate our optimal control problem as that of determining ϕ_0, $\|\phi_0\|_{\partial D} = 1$, such that

$$(\mathbf{R}\phi, \phi)_{\partial D} \leqslant (\mathbf{R}\phi_0, \phi_0)_{\partial D}$$

for all $\phi \in L^2(\partial D)$, $\|\phi\|_{\partial D} = 1$.

Theorem 9.12. The operator **R** is compact and self-adjoint. Let λ_0 be the largest eigenvalue of **R**. Then the optimal control ϕ_0 is an eigenfunction of **R** corresponding to the eigenvalue λ_0.

Proof. Since **F** is compact so is $\mathbf{R} = \mathbf{F}^*\alpha\mathbf{F}$ and since α is real valued, **R** is self-adjoint. Hence the spectrum of **R** is discrete and real with zero as its only accumulation point. Furthermore,

$$\sup_{\|\phi\|_{\partial D} = 1} (\mathbf{R}\phi, \phi)_{\partial D} = |\lambda_0|$$

where λ_0 is the eigenvalue of **R** having the largest absolute value (cf. Stakgold [2]). But since α is a characteristic function, $\alpha = \alpha^2$, and therefore

$$(\mathbf{R}\phi, \phi)_{\partial D} = (\mathbf{F}^*\alpha\mathbf{F}\phi, \phi)_{\partial D} = (\alpha\mathbf{F}\phi, \alpha\mathbf{F}\phi) \geqslant 0$$

which implies that **R** is positive and its spectrum nonnegative. It now follows that

$$\sup_{\|\phi\|_{\partial D} = 1} (\mathbf{R}\phi, \phi)_{\partial D} = \lambda_0$$

and if ϕ_0 is a (not necessarily unique) normalized eigenfunction corresponding to the eigenvalue λ_0, then

$$(\mathbf{R}\phi_0, \phi_0)_{\partial D} = (\lambda_0\phi_0, \phi_0)_{\partial D} = \lambda_0.$$

The Galerkin procedure for constructing an approximation to the eigenvalue λ_0 and the eigenfunction ϕ_0 is well known and can be found in any number of textbooks (cf. Stakgold [2]).

The above discussion shows that by choosing U to be the closed unit ball in $L^2(\partial D)$, Theorem 9.11 reduces the candidates for optimal solutions and leads to a constructive procedure for their calculation. We shall now consider a situation in which the control set U has empty interior and hence Theorem 9.11 becomes vacuous. We shall show that in this case there exists a maximizing sequence of extreme points of the control set and that furthermore we can characterize these extreme points as bang-bang controls. To be more specific we consider the control set

$$G = \{\phi \in L^2(\partial D) | \psi_0(y) \leqslant \phi(y) \leqslant \psi_1(y) \text{ a.e. on } \partial D\} \tag{9.6}$$

where ψ_0, ψ_1 are real-valued continuous functions defined on ∂D. It is easily verified that G is closed, bounded, and convex; however, every point of G is a boundary point (we leave this last observation as an exercise for the reader). Since in this case Theorem 9.11 gives no new information, we shall consider, instead of the boundary of G, the set of extreme points of G.

Theorem 9.13. Let G be the class of admissible controls defined in (9.6) and ϕ_0 the optimal control that maximizes Q_α over G. Then there exists a sequence of extreme points (ϕ_n^e) of G such that

$$\lim_{n\to\infty} Q_\alpha(\phi_n^e) = Q_\alpha(\phi_0).$$

Proof. By Corollary 9.6, G is weakly compact. Let G^{ext} be the set of extreme points of G. Then by the Krein–Milman theorem G is the closed, convex hull of G^{ext} and hence there exists a sequence (ϕ_n), where each ϕ_n lies in the convex hull of G^{ext}, such that ϕ_n tends (strongly) to ϕ_0. Then since Q_α is weakly continuous on G

$$\lim_{n\to\infty} Q_\alpha(\phi_n) = Q_\alpha(\phi_0). \tag{9.7}$$

But for every n, $n = 1,2,\ldots,$ we can write

$$\phi_n = \sum_{l=1}^{m_n} \lambda_l^n \phi_l^n$$

where

$$\sum_{l=1}^{m_n} \lambda_l^n = 1$$

and each ϕ_l^n is an extreme point of G. Reasoning now as in Theorem 9.11 we have that for each n there exists an integer l_n, $1 \leqslant l_n \leqslant m_n$, such that

$$Q_\alpha(\phi_n) \leqslant Q_\alpha(\phi_{l_n}^n) \leqslant Q_\alpha(\phi_0) \tag{9.8}$$

where $\phi_{l_n}^n \in G^{\text{ext}}$, and hence from (9.7), (9.8)

$$\lim_{n\to\infty} Q_\alpha(\phi_{l_n}^n) = Q_\alpha(\phi_0).$$

We shall now show that the extreme points of G are "bang-bang" controls, that is, the functions in G^{ext} take on only the extreme values $\psi_0(y)$ or $\psi_1(y)$ at almost all points $y \in \partial D$. In the proof of the following theorem we shall let χ_M denote the characteristic function of a measurable set M and $\mu(M)$ its measure.

Theorem 9.14. Let G^{ext} denote the set of extreme points of G. Then $\phi \in G^{\text{ext}}$ if and only if

$$\phi = \chi_{E_0}\psi_0 + \chi_{E_1}\psi_1$$

almost everywhere where $E_0 \cap E_1 = \varnothing$, $\mu(E_0 \cup E_1) = \mu(\partial D)$.

Proof. Suppose ϕ is not of the prescribed form. Then there exists an $\varepsilon > 0$ and a measurable set $E_2 \subset \partial D$, $\mu(E_2) > 0$, such that for $y \in E_2$

$$\psi_0(y) < \phi(y) - \varepsilon < \phi(y) < \phi(y) + \varepsilon < \psi_1(y).$$

Define the functions ϕ_1 and ϕ_2 by

$$\phi_1(y) := \begin{cases} \phi(y) \quad , & y \in \partial D \setminus E_2 \\ \phi(y) - \varepsilon, & y \in E_2 \end{cases}$$

$$\phi_2(y) := \begin{cases} \phi(y) \quad , & y \in \partial D \setminus E_2 \\ \phi(y) + \varepsilon, & y \in E_2. \end{cases}$$

Then $\phi_1, \phi_2 \in G$ and

$$\phi(y) = \tfrac{1}{2}(\phi_1(y) + \phi_2(y))$$

almost everywhere on ∂D, that is, ϕ is not an extreme point of G. Hence every extreme point must have the form

$$\phi = \chi_{E_0}\psi_0 + \chi_{E_1}\psi_1 \tag{9.9}$$

almost everywhere for a suitable choice of E_0 and E_1.

Now suppose conversely that (9.9) is valid and that

$$\phi = \lambda\phi_1 + (1-\lambda)\phi_2 \tag{9.10}$$

for $0 < \lambda < 1$ and $\phi_1, \phi_2 \in G$. We shall show that for $i = 0, 1$, $\phi_1 = \phi_2 = \psi_i$ on E_i and hence ϕ cannot be expressed as a proper convex combination of other points of G, that is, $\phi \in G^{\text{ext}}$. To see this, first assume that $y \in E_0$. Then from (9.9) we have that $\phi(y) = \psi_0(y)$. Hence from (9.10) we have

$$\lambda(\phi_1(y) - \psi_0(y)) + (1-\lambda)(\phi_2(y) - \psi_0(y)) = 0$$

almost everywhere for $y \in E_0$. Since $0 < \lambda < 1$ and $\phi_i \geqslant \psi_0$ for almost all $y \in E_0$ we must have $\phi_1 = \psi_0$ and $\phi_2 = \psi_0$. The corresponding argument for $y \in E_1$ now completes the proof.

Corollary 9.15. Let G be the class of admissible controls defined in (9.6) and ϕ_0 the optimal control that maximizes Q_α over G. Then ϕ_0 is the limit of a sequence of bang-bang controls.

For further results on the existence of bang-bang controls the reader is referred to Kirsch [2].

9.3 OPTIMAL CONTROL FOR A SCATTERING PROBLEM

We shall now consider the problem of the optimal choice of the impedance of an obstacle in order to maximize the far-field power flux through a given angle. We shall view this as a scattering problem in which the incident field is given, that is, we assume the total field u and the impedance λ satisfy the set of equations

$$u = u^i + u^s \tag{9.11a}$$

$$\Delta u + k^2 u = 0 \quad \text{in} \quad \mathbb{R}^2 \backslash \bar{D} \tag{9.11b}$$

$$\frac{\partial u}{\partial \nu} + \lambda u = 0 \quad \text{on} \quad \partial D \tag{9.11c}$$

$$\left(\operatorname{grad} u^s(x), \frac{x}{|x|}\right) - iku^s(x) = o\left(\frac{1}{|x|^{1/2}}\right) \tag{9.11d}$$

where u^i and u^s denote the incident and scattered waves, respectively, with u^i being an entire solution of (9.11b), the wave number k is positive and fixed, the impedance λ satisfies $\operatorname{Im} \lambda \geqslant 0$, and the scattering obstacle D is bounded, connected, and has C^2 boundary ∂D with unit outward normal ν. Our aim is to choose λ, subject to certain a priori restrictions, such that the far-field power flux through the angle α

$$\tilde{Q}_\alpha(u^s) = \lim_{r \to \infty} \int_0^{2\pi} \alpha(\theta) |u^s(r, \theta)|^2 \, d\theta \tag{9.12}$$

is maximized, where α is the characteristic function of a measurable subset α of $[0, 2\pi]$. Our approach will follow that of Kirsch [1], except that for simplicity we shall assume that the wave number k is less than the first eigenvalue of the interior Dirichlet problem for (9.11b) in D and denote this eigenvalue by k_0. By a suitable modification of the integral equations involved in our analysis (cf. Section 3.7) this restriction can easily be avoided.

For an alternate approach to the above optimal control problem the reader is referred to Angell and Kleinman [2].

We begin our analysis by reformulating (9.11a)–(9.11d) as an integral equation. This is accomplished by looking for a solution u^s to (9.11) in the form

$$u^s(x) = \frac{i}{4} \int_{\partial D} H_0^{(1)}(k|x - y|) \phi(y) \, ds(y)$$

where $H_0^{(1)}$ denotes a Hankel function of the first kind and ϕ is a continuous density to be determined. In a manner that by now is well known one is led to an integral equation for ϕ of the form

$$\phi(x)-\frac{i}{2}\int_{\partial D}\frac{\partial}{\partial\nu(x)}H_0^{(1)}(k|x-y|)\phi(y)\,ds(y)$$

$$-\lambda(x)\frac{i}{2}\int_{\partial D}H_0^{(1)}(k|x-y|)\phi(y)\,ds(y)=-2\left(\frac{\partial u^i}{\partial\nu}(x)+\lambda(x)u^i(x)\right)$$

or

$$\phi-\mathbf{K}'\phi-\lambda\mathbf{S}\phi=g_1+\lambda g_2 \tag{9.13}$$

where

$$(\mathbf{K}'\phi)(x):=\frac{i}{2}\int_{\partial D}\frac{\partial}{\partial\nu(x)}H_0^{(1)}(k|x-y|)\phi(y)\,ds(y)$$

$$(\mathbf{S}\phi)(x):=\int_{\partial D}H_0^{(1)}(k|x-y|)\phi(y)\,ds(y),$$

and $g_1:=-2(\partial u^i/\partial\nu)$, $g_2:=-2u^i$. From the results of Section 3.7 we have that the integral equation (9.13) is uniquely solvable provided $\operatorname{Im}\lambda\geqslant 0$ and $k<k_0$, the first eigenvalue of the interior Dirichlet problem. Furthermore, from the previous section of this chapter we have that u^s has the asymptotic behavior

$$u^s(x)=\frac{1}{4}e^{i(kr+\pi/4)}\sqrt{\frac{2}{\pi kr}}\,F(\theta;k)+O\left(\frac{1}{r^{3/2}}\right)$$

where F is defined by (9.1), (9.2), that is,

$$F=\mathbf{F}\phi \tag{9.14}$$

where $\mathbf{F}$: $C(\partial D)\to C(0,2\pi)$ is compact. Hence we can define the (normalized) power flux through the angle α by

$$Q_\alpha(\phi):=\int_0^{2\pi}\alpha(\theta)|\mathbf{F}\phi(\theta)|^2\,d\theta, \tag{9.15}$$

and define our optimization problem as that of choosing λ such that Q_α is maximized where Q_α is related to λ through the integral equation (9.13). In this context it is advantageous from the point of view of existence results to look for generalized solutions of the optimization problem in the sense that we allow λ to be an element of $L^\infty(\partial D)$ and look for a solution of the integral equation (9.13) in $L^2(\partial D)$, that is, we assume ϕ, g_1, and g_2 are in $L^2(\partial D)$. All the operators are well defined in this case and since the spectrum is unchanged, the integral equation (9.13) is uniquely solvable in $L^2(\partial D)$ for $k < k_0$.

Before stating our generalized optimization problem in a precise fashion we introduce some convenient notational shorthand. In particular let $H = L^2(\partial D)$, $\hat{H} = L^\infty(\partial D)$, and, defining $\tilde{H}$ to be $L^1(\partial D)$, we have that $\langle \hat{H}, \tilde{H} \rangle$ is a reflexive dual system. Let U be a weakly compact subset of $H^+ = \{\lambda \in H \mid \operatorname{Im} \lambda \geqslant 0\}$ with respect to $\langle \hat{H}, \tilde{H} \rangle$. Then our generalized optimization problem, that we shall call problem P, is to determine $(\lambda^*, \phi^*) \in U \times H$ such that $Q_\alpha \colon H \to \mathbb{R}$ is maximized subject to

1. $\phi - \mathbf{K}'\phi - \lambda \mathbf{S}\phi = g_1 + \lambda g_2$.
2. $\lambda \in U$.

Note that from Lemma 9.9 we have that Q_α is weakly continuous on $L^2(\partial D)$.

Theorem 9.16. The mapping $\lambda \to \phi$ is a weakly continuous mapping from H^+ into H, that is, if (λ_n) tends weakly to λ in H^+ then (ϕ_n) is weakly convergent to the image of λ in H.

Proof. We shall first show that the mapping is bounded. Assume that this is not true. Then there exists a sequence (λ_n), $\lambda_n \in H^+$, such that the λ_n are bounded but $\|\phi_n\| \to \infty$, where ϕ_n is the solution of (9.13) with λ set equal to λ_n. Let $\psi_n = \phi_n / \|\phi_n\|$. Then ψ_n satisfies the integral equation

$$\psi_n - \mathbf{K}'\psi_n - \lambda_n \mathbf{S}\psi_n = \frac{1}{\|\phi_n\|}(g_1 + \lambda_n g_2). \tag{9.16}$$

Since the λ_n and ψ_n are bounded, there exist subsequences, which we shall again denote by (λ_n) and (ψ_n), such that (λ_n) and (ψ_n) are weakly convergent to $\lambda \in H^+$, $\psi \in H$, respectively. (The weak convergence of (λ_n) is with respect to the dual system $\langle \hat{H}, \tilde{H} \rangle$.) Then since

$$\lambda_n \mathbf{S}\psi_n - \lambda \mathbf{S}\psi = \lambda_n \mathbf{S}(\psi_n - \psi) + (\lambda_n - \lambda)\mathbf{S}\psi \tag{9.17}$$

and the first term on the right-hand side converges strongly to zero (by the compactness of $\mathbf{S}$), whereas the second term is weakly convergent to zero, we have from (9.16) that

$$\psi - \mathbf{K}'\psi - \lambda \mathbf{S}\psi = 0.$$

From the fact that (9.13) is uniquely solvable we can now conclude that $\psi = 0$, and from (9.16) and (9.17) we have that ψ_n converges strongly to zero. But this is impossible since $\|\psi_n\| = 1$. Hence the mapping $\lambda \to \phi$ is bounded.

Now let (λ_n) be weakly convergent to $\lambda \in H^+$ and ϕ_n the solution of (9.13) with λ replaced by λ_n. Then (ϕ_n) is a bounded sequence and hence from the above argument there exists a subsequence, which we shall again denote by (ϕ_n), such that (ϕ_n) converges weakly to a solution ϕ of (9.13). But since this integral equation is uniquely solvable, the whole sequence (ϕ_n) is weakly convergent to ϕ, that is, the mapping $\lambda \to \phi$ is weakly continuous.

We can now establish the existence of a solution to problem P.

Theorem 9.17. There exists a solution $(\lambda^*, \phi^*) \in U \times H$ of problem P.

Proof. From the weak continuity of Q_α and Theorem 9.16 we have that Q_α achieves its maximum for $\lambda \in U$, $\phi \in H$, where ϕ is related to λ through (9.13). The ordered pair (λ^*, ϕ^*) for which Q_α achieves this maximum is the solution of problem P.

We now want to present a method for obtaining approximations to the solution of problem P. Recall that for the linear optimal control problem discussed in the previous section we were able to obtain an approximation to the optimal control by applying a standard Galerkin procedure for finding the largest eigenvalue of a compact self-adjoint operator, the details of which can be found in any number of textbooks, for example, Stakgold [2]. In the present case, due to the nonlinearity of the mapping from the impedance λ to the far-field pattern F, the situation is considerably more complicated. Nevertheless we shall show in the sequel that approximations to the optimal control can again be found via a Galerkin procedure. Our analysis is again based on Kirsch [1].

Let H_n, $H_n \subset H_{n+1} \subset H$, $n = 1,2,\ldots$, be closed subspaces such that $\cup_{n=1}^{\infty} H_n$ is dense in H and let $\mathbf{P}_n\colon H \to H_n$ be the projection operator of H onto H_n. Then we can formulate a sequence of approximation problems to problem P, denoted by problem P_n, as the problem of determining $(\lambda_n^*, \phi_n^*) \in U \times H_n$ such that $Q_\alpha\colon H_n \to \mathbb{R}$ is maximized subject to

1. $\phi - (\mathbf{P}_n \mathbf{K}')\phi - (\mathbf{P}_n \lambda \mathbf{S})\phi = \mathbf{P}_n(g_1 + \lambda g_2)$.
2. $\lambda \in U$.

Theorem 9.18. There exists an integer n_0 such that for $n \geqslant n_0$, $\lambda \in U$, $g \in H_n$, the equation

$$\phi - (\mathbf{P}_n \mathbf{K}')\phi - (\mathbf{P}_n \lambda \mathbf{S})\phi = g \tag{9.18}$$

is uniquely solvable in H_n.

Proof. Since the integral equation (9.18) is of Fredholm type, by the Fredholm alternative it suffices to show that the only solution of the homogeneous integral equation is the trivial solution, that is, ϕ identically zero. Suppose the theorem were not true. Then there exist integers $n(m)$ and functions $\phi_m \in H_{n(m)}$, $\lambda_m \in U$, $m = 1,2,3,\ldots$ with $\|\phi_m\| = 1$ such that

$$\phi_m - \left(\mathbf{P}_{n(m)} \mathbf{K}'\right)\phi_m - \left(\mathbf{P}_{n(m)} \lambda_m \mathbf{S}\right)\phi_m = 0.$$

There exist weakly convergent subsequences of (ϕ_m) and (λ_m), that we again denote by (ϕ_m) and (λ_m) such that (ϕ_m) is weakly convergent to $\phi \in H$ and (λ_m) is weakly convergent to $\lambda \in U$. By following the arguments of Theorem 9.16 it is easily seen that $(\mathbf{P}_{n(m)}\mathbf{K}')\phi_m$ is strongly convergent to $\mathbf{K}'\phi$ and $(\mathbf{P}_{n(m)}\lambda_m\mathbf{S})\phi_m$ is weakly convergent to $\lambda\mathbf{S}\phi$. Hence

$$\phi - \mathbf{K}'\phi - \lambda\mathbf{S}\phi = 0$$

and we can conclude by the unique solvability of this integral equation that $\phi = 0$, that is, (ϕ_m) is weakly convergent to zero. Since $\mathbf{K}'$ and $\mathbf{S}$ are compact we can conclude now, as in Theorem 9.16, that (ϕ_m) is strongly convergent to zero. But this is a contradiction since $\|\phi_m\| = 1$.

The following existence theorem now follows exactly as the case of Theorem 9.17.

Theorem 9.19. Let $n \geqslant n_0$, where n_0 is defined in Theorem 9.18. Then there exists a solution $(\lambda_n^*, \phi_n^*) \in U \times H_n$ of problem P_n.

Now let (λ^*, ϕ^*) be a solution to problem P and (λ_n^*, ϕ_n^*) a solution to problem P_n, where we assume $n \geqslant n_0$. We would like to show that (λ_n^*, ϕ_n^*) tends to (λ^*, ϕ^*) in some sense as n tends to infinity. In order to establish such a result we first need the following lemma (compare this to Theorem 9.16).

Lemma 9.20. For a given $\lambda \in H^+$ let $\phi_n \in H_n$ be the solution of (9.18) for $g = \mathbf{P}_n(g_1 + \lambda g_2)$, $n \geqslant 0$. Then the mapping $\lambda \to \phi_n$ is a weakly continuous mapping from H^+ into H_n and is uniformly bounded with respect to n.

Proof. The weak continuity of the mapping follows exactly as in Theorem 9.16. In order to show the uniform boundedness of the mapping with respect to n, assume the contrary. Then there exist a subsequence $n(m)$ and a sequence $(\lambda_m, \phi_m) \in U \times H_{n(m)}$ such that $\|\lambda_m\| \leqslant C$ for some positive constant C, $\|\phi_m\| \to \infty$, and ϕ_m is a solution of the integral equation

$$\phi_m - (\mathbf{P}_{n(m)}\mathbf{K}')\phi_m - (\mathbf{P}_{n(m)}\lambda_m\mathbf{S})\phi_m = g_1 + \lambda_m g_2.$$

Setting $\psi_m = \phi_m / \|\phi_m\|$ we have that ψ_m is a solution of

$$\psi_m - (\mathbf{P}_{n(m)}\mathbf{K}')\psi_m - (\mathbf{P}_{n(m)}\lambda_m\mathbf{S})\psi_m = \frac{1}{\|\phi_m\|}(g_1 + \lambda_m g_2).$$

Then since (λ_m) and (ψ_m) are bounded, there exist subsequences, which we shall again denote by (λ_m) and (ψ_m), such that (λ_m) and (ψ_m) are weakly convergent to $\lambda \in H^+$, $\psi \in H_n$. Hence we can conclude that $(\mathbf{P}_{n(m)}\mathbf{K}')\psi_m$ is strongly convergent to $\mathbf{K}'\psi$ and $(\lambda_m\mathbf{S})\psi_m$ is weakly convergent to $(\lambda\mathbf{S})\psi$. From the uniform boundedness principle we have that there exists a positive constant C such that $\|\mathbf{P}_n\| \leqslant C$ for all integers n. Therefore $(\mathbf{P}_{n(m)}\lambda_m\mathbf{S})\psi_m$ is bounded and there exists a subsequence again denoted by $(\mathbf{P}_{n(m)}\lambda_m\mathbf{S})\psi_m$, such that

$(\mathbf{P}_{n(m)}\lambda_m\mathbf{S})\psi_m$ is weakly convergent to an element $\gamma \in H$. But for any positive integer j we have that $(\mathbf{P}_j\mathbf{P}_{n(m)}\lambda_m\mathbf{S})\psi_m$ is weakly convergent to $(\mathbf{P}_j\lambda\mathbf{S})\psi$ and hence $\mathbf{P}_j\gamma = (\mathbf{P}_j\lambda\mathbf{S})\psi$ for every integer j, that is, $\gamma = \lambda\mathbf{S}\psi$. Therefore $(\mathbf{P}_{n(m)}\lambda_m\mathbf{S})\psi_m$ is weakly convergent to $\lambda\mathbf{S}\psi$. The unique solvability of (9.13) now implies that $\psi = 0$ and arguing as in Theorem 9.16, we can now conclude that (ψ_n) converges strongly to zero. But this is impossible since $\|\psi_n\| = 1$. Thus the mapping $\lambda \to \phi_n$ is uniformly bounded with respect to n.

Theorem 9.21. Let Q^* and Q_n^* be the optimal values of problems P and P_n, respectively, where $n \geqslant n_0$, and let Φ and Φ_n denote the sets of solutions to these problems. Then

1. $\lim_{n\to\infty} Q_n^* = Q^*$.
2. Every sequence $(\lambda_n^*, \phi_n^*) \in \Phi_n$ contains a weakly convergent subsequence and every weak limit point lies in Φ.

Proof. Lemma 9.20 implies that there exists a weakly convergent subsequence of (λ_n^*, ϕ_n^*) that we again denote by (λ_n^*, ϕ_n^*). Let (λ^*, ϕ^*) be any weak limit point of (λ_n^*, ϕ_n^*). By our previous arguments we can conclude that $(\lambda^*, \phi^*) \in U \times H$. Now let $(\hat{\lambda}, \hat{\phi}) \in U \times H$ be any solution of problem P. Let $\hat{\phi}_n$ be a solution of (9.18) with $\lambda = \hat{\lambda}$ and $g = \mathbf{P}_n(g_1 + \hat{\lambda} g_2)$. Then by our previous arguments we can conclude that $\hat{\phi}_n$ is strongly convergent to $\hat{\phi}$ and

$$
\begin{aligned}
Q^* \geqslant Q_\alpha(\phi^*) &= \lim_{n\to\infty} Q_\alpha(\phi_n^*) \\
&= \lim_{n\to\infty} Q_n^* \\
&\geqslant \lim_{n\to\infty} Q_\alpha(\hat{\phi}_n) \\
&= Q_\alpha(\hat{\phi}) = Q^*.
\end{aligned}
$$

Therefore $Q^* = Q_\alpha(\phi^*) = \lim_{n\to\infty} Q_n^*$ and the theorem is proved.

In closing we shall briefly make a few remarks on the existence of bang-bang controls for the optimization problem considered in this section. Due to the nonlinear nature of this problem, results are considerably harder to obtain than in the case of the linear optimization problem considered in Section 9.2. To state one result that is known (Kirsch [2]), let R be a compact and convex subset of the closed upper half plane with nonempty interior and let the class of admissible impedances be given by

$$U = \{\lambda \in L^\infty(\partial D) | \lambda(\mathbf{x}) \in R \text{ almost everywhere on } \partial D\}.$$

Let (λ^*, ϕ^*) be a solution of problem P and u^* the corresponding total field. Define w to be the unique solution of the following "adjoint" scattering problem:

$$w = w^i + w^s \tag{9.19a}$$

$$\Delta w + k^2 w = 0 \quad \text{in} \quad \mathbb{R}^2 \setminus \bar{D} \tag{9.19b}$$

$$\frac{\partial w}{\partial \nu} + \lambda^* w = 0 \quad \text{on} \quad \partial D \tag{9.19c}$$

$$\left(\operatorname{grad} w^s(x), \frac{x}{|x|}\right) - ikw^s(x) = o\left(\frac{1}{|x|^{1/2}}\right) \tag{9.19d}$$

where

$$w^i(x) = \int_0^{2\pi} \alpha(\theta) \overline{\mathbf{F}\phi^*(\theta)} e^{-ik(\xi, x)}\, d\theta \tag{9.20}$$

with $\xi = (\cos\theta, \sin\theta)$. Then it has been shown by Kirsch [2] that

$$\lambda^*(x) \in \partial R \quad \text{for all} \quad x \notin N(w) \cap N(u^*)$$

where $N(w)$ and $N(u^*)$ denote the set of zeros of the functions w and u^*, respectively, restricted to ∂D. Although the set $N(w) \cup N(u^*)$ can be very wild in general, it has been shown by Kirsch [2] that it does not contain any disk, that is, a set that is the intersection of a ball with ∂D.

REFERENCES

J. F. Ahner

1. The exterior Dirichlet problem for the Helmholtz equation. *J. Math. Anal. Appl.* **52**, 415–429 (1975).
2. The exterior Robin problem for the Helmholtz equation. *J. Math. Anal. Appl.* **66**, 37–54 (1978).

J. F. Ahner and R. E. Kleinman

1. The exterior Neumann problem for the Helmholtz equation. *Arch. Rational Mech. Anal.* **52**, 26–43 (1973).

T. S. Angell, D. Colton, and A. Kirsch

1. The three dimensional inverse scattering problem for acoustic waves. *J. Diff. Eq.*, **46**, 46–58 (1982).

T. S. Angell and R. E. Kleinman

1. Generalized exterior boundary-value problems and optimization for the Helmholtz equation. *J. Optimization Theory and Applications* **37**, 469–497 (1982).
2. Scattering control by impedance loading. In *International U.R.S.I. Symposium on Electromagnetic Waves*, Munich, 223B/1–223B/4 (1980).

T. S. Angell and M. Z. Nashed

1. Operator-theoretic and computational aspects of ill-posed problems in antenna theory. In *Proc. Int. Symp. Math. Theory of Networks and Systems*, Delft University of Technology, The Netherlands, 449–511 (1979).

F. V. Atkinson

1. On Sommerfeld's "Radiation Condition." *Philos. Mag.* **40**, 645–651 (1949).

G. Backus and F. Gilbert

1. Uniqueness in the inversion of inaccurate gross earth data. *Phil. Trans. Royal Soc. London* **266**, 123–197 (1970).

B. Baker and E. T. Copson

1. *The Mathematical Theory of Huygens' Principle.* Clarendon Press, Oxford (1950).

M. Bertero, C. De Mol, and G. A. Viano

1. The stability of inverse problems. In *Inverse Scattering Problems in Optics*, H. P. Baltes, ed., Springer-Verlag, Berlin, 161–214 (1979).

R. P. Boas

1. *Entire Functions.* Academic Press, New York (1954).

R. P. Boas and R. C. Buck

1. *Polynomial Expansions of Analytic Functions.* Springer-Verlag, Berlin (1964).

J. C. Bolomey and W. Tabbara

1. Numerical aspects of coupling between complementary boundary value problems. *IEEE Trans. Ant. and Prop.* **AP-21**, 356–363 (1973).

H. Brakhage and P. Werner

1. Über das Dirichletsche Aussenraumproblem für die Helmholtzsche Schwingungsgleichung. *Arch. Math.* **16**, 325–329 (1965).

R. Burridge

1. *Some Mathematical Topics in Seismology*. Lecture notes, Courant Institute of Mathematical Sciences, New York University, New York (1976).

A. J. Burton and G. F. Miller

1. The application of integral equation methods to the numerical solution of some exterior boundary-value problems. *Proc. Royal Soc. London* **A323**, 201–220 (1971).

D. Colton

1. On the inverse scattering problem for axially symmetric solutions of the Helmholtz equation. *Q. J. Math*. **22**, 125–130 (1971).
2. A reflection principle for solutions to the Helmholtz equation and an application to the inverse scattering problem. *Glasgow Math. J*. **18**, 125–130 (1977).
3. The inverse scattering problem for a cylinder. *Proc. Royal Soc. Edinburgh* **84A**, 135–143 (1979).
4. *Analytic Theory of Partial Differential Equations*. Pitman Publishing, London (1980).
5. The inverse electromagnetic scattering problem for a perfectly conducting cylinder. *IEEE Trans. Ant. and Prop*. **AP-29**, 364–368 (1981).
6. Stable methods for determining the surface impedance of an obstacle from low frequency far field data. *Applicable Analysis*, **14**, 61–70 (1982).
7. The inverse scattering problem for acoustic waves. In *Proc. 1982 Dundee Conference on Ordinary and Partial Differential Equations*, Springer-Verlag Lecture Notes in Mathematics, Vol. 964, 143–161 (1982).

D. Colton and A. Kirsch

1. The determination of the surface impedance of an obstacle from measurements of the far field pattern. *SIAM J. Appl. Math*. **41**, 8–15 (1981).
2. Stable methods for solving the inverse scattering problem for a cylinder. *Proc. Royal Soc. Edinburgh* **89A**, 181–188 (1981).
3. Dense sets and far field patterns in acoustic wave propagation, *SIAM J. Math. Anal.* **15**, 996–1006 (1984).

D. Colton and R. E. Kleinman

1. The direct and inverse scattering problems for an arbitrary cylinder: Dirichlet boundary conditions. *Proc. Royal Soc. Edinburgh* **86A**, 29–42 (1980).

D. Colton and R. Kress

1. Iterative methods for solving the exterior Dirichlet problem for the Helmholtz equation with applications to the inverse scattering problem for low frequency acoustic waves. *J. Math. Anal. Appl.* **77**, 60–72 (1980).
2. The impedance boundary value problem for the time harmonic Maxwell equations. *Math. Meth. in the Appl. Sci.* **3**, 475–487 (1981).
3. The unique solvability of the null field equations of acoustics. *Q. J. Mech. Appl. Math.*, *36*, 87–95 (1983).

R. Courant and D. Hilbert

1. *Methods of Mathematical Physics, Vol. 1.* Interscience, New York (1965).

C. L. Dolph

1. The integral equation method in scattering theory. In *Problems in Analysis*, R. C. Gunning, ed., Princeton University Press, Princeton, N.J. 201–227 (1970).

H. Engl and R. Kress

1. A singular perturbation problem for linear operators with an application to electrostatic and magnetostatic boundary and transmission problems. *Math. Meth. in the Appl. Sci.* **3**, 249–274 (1981).

A. Erdélyi

1. Singularities of generalized axially symmetric potentials. *Comm. Pure Appl. Math.* **9**, 403–414 (1956).

A. Erdélyi, W. Magnus, F. Oberhettinger, and F. G. Tricomi

1. *Higher Transcendental Functions*, *Vol. 2*. McGraw-Hill, New York (1953).

I. Fredholm

1. Sur une classe d'équations functionelles. *Acta Math*. **27**, 365–390 (1903).

P. R. Garabedian

1. An integral equation governing electromagnetic waves. *Q. Appl. Math*. **12**, 428–433 (1955).
2. *Partial Differential Equations*. Wiley, New York (1964).

R. P. Gilbert

1. *Function Theoretic Methods in Partial Differential Equations*. Academic Press, New York (1969).
2. *Constructive Methods for Elliptic Equations*. Springer-Verlag Lecture Notes in Mathematics, Vol. 365, Springer-Verlag, Berlin (1974).

J. Giroire

1. Integral equations for exterior problems for the Helmholtz equation. Ecole Polytechnique, Palaisau, France, Rapport Interne No. 40 (1978).

J. Giroire and J. C. Nedelec

1. Numerical solution of an exterior Neumann problem using a double layer potential. *Math. Comp*. **32**, 973–990 (1978).

G. Gray

1. Low frequency iterative solution of integral equations in electromagnetic scattering theory. University of Delaware, Appl. Math. Inst. Tech. Report 35A (1978).

D. Greenspan and P. Werner

1. A numerical method for the exterior Dirichlet problem for the reduced wave equation. *Arch. Rational Mech. Anal*. **23**, 288–316 (1966).

V. Gülzow

1. Über das Grenzverhalten von stationären elektromagnetischen Feldern bei kleinen Frequenzen unter magnetischen Randbedingungen. *Diplomarbeit*, Göttingen (1981).

N. M. Günter

1. *Die Potentialtheorie und ihre Anwendungen auf Grundaufgaben der mathematischen Physik*. Teubner-Verlag, Leipzig (1957).

S. I. Hariharan

1. Inverse scattering for an exterior Dirichlet problem. *Q. Appl. Math*., **40**, 273–286 (1982).

P. Hartman and C. Wilcox

1. On solutions of the Helmholtz equation in exterior domains *Math. Zeit*. **75**, 228–255 (1961).

P. Henrici

1. On the domain of regularity of generalized axially symmetric potentials. *Proc. Amer. Math. Soc*. **8**, 29–31 (1957).

G. C. Hsiao and W. Wendland

1. A finite element method for some integral equations of the first kind. *J. Math. Anal. Appl*. **58**, 449–481 (1977).

D. S. Jones

1. Integral equations for the exterior acoustic problem. *Q. J. Mech. Appl. Math*. **27**, 129–142 (1974).
2. *Methods in Electromagnetic Wave Propagation*. Clarendon Press, Oxford (1979).

K. Jörgens

1. *Lineare Integraloperatoren*. Teubner-Verlag, Stuttgart (1970).

A. Kirsch

1. Optimal control of an exterior Robin problem. *J. Math. Anal. Appl*. **82**, 144–151 (1981).
2. A weak bang-bang principle for the control of an exterior Robin problem. *Applicable Analysis* **13**, 65–75 (1982).
3. Private communication (1979).
4. Private communication (1981).

5. A numerical method for an inverse scattering problem. University of Delaware, Appl. Math. Inst. Tech. Report, 127A, 1982.
6. The Robin problem for the Helmholtz equation as a singular perturbation problem, *Num. Funct. Anal. and Optimization* **8**, 1–20 (1985).

R. Kittappa and R. E. Kleinman

1. Acoustic scattering by penetrable homogeneous objects. *J. Math. Phys.* **16**, 421–432 (1975).

A. Klein

1. Über das Verhalten der Lösungen von Randwertaufgaben aus der Theorie akustischer und electromagnetischer Schwingungen für kleine Frequenzen. *Diplomarbeit*, Göttingen (1980).

R. E. Kleinman

1. Iterative solutions of boundary value problems. In Springer-Verlag Lecture Notes in Mathematics, Vol. 561, *Function Theoretic Methods for Partial Differential Equations*, Springer-Verlag, Berlin, 298–313 (1976).

R. E. Kleinman and G. F. Roach

1. Boundary integral equations for the three-dimensional Helmholtz equation. *SIAM Review* **16**, 214–236 (1974).
2. On modified Green's functions in exterior problems for the Helmholtz equation, *Proc. Royal Soc. London* **A383**, 313–332 (1982).

R. E. Kleinman and W. Wendland

1. On Neumann's method for the exterior Neumann problem for the Helmholtz equation. *J. Math. Anal. Appl.* **57**, 170–202 (1977).

W. Knauff

1. Ein numerisches Verfahren zur Lösung eines Aussenraumproblems für die vektorielle Helmholtzgleichung. *NAM-Bericht* Nr. 28, Göttingen (1981).

W. Knauff and R. Kress

1. On the exterior boundary-value problem for the time-harmonic Maxwell equations. *J. Math. Anal. Appl.* **72**, 215–235 (1979).
2. A modified integral equation method for the electric boundary-value problem for the vector Helmholtz equation. *ISNM* **53**, 157–170 (1980).

R. Kress

1. Über die Integralgleichung des Pragerschen Problems. *Arch. Rational Mech. Anal.* **30**, 381–400 (1968).
2. Ein Iterationsverfahren für eine Klasse von Funktionalgleichungen zweiter Art. *J. Reine und Angew. Math.* **238**, 207–216 (1969).
3. Die Behandlung zweier Randwertprobleme für die vektorielle Poissongleichung nach einer Integralgleichungsmethode. *Arch. Rational Mech. Anal.* **39**, 206–226 (1970).
4. On the limiting behaviour of solutions to boundary integral equations associated with time harmonic wave equations for small frequencies. *Math. Meth. in the Appl. Sci.* **1**, 89–100 (1979).
5. On boundary integral equation methods in stationary electromagnetic reflection. In Springer-Verlag Lecture Notes in Mathematics, Vol. 846, *Ordinary and Partial Differential Equations*, Springer-Verlag, Berlin, 210–226 (1980).
6. On the existence of a solution to a singular integral equation in electromagnetic reflection. *J. Math. Anal. Appl.* **77**, 555–566 (1980).
7. A singular perturbation problem for linear operators with an application to the limiting behaviour of stationary electromagnetic wave fields for small frequencies. *Meth. Verf. Math. Phys.* **21**, 5–30 (1981).

R. Kress and G. F. Roach

1. On the convergence of successive approximations for an integral equation in a Green's function approach to the Dirichlet problem, *J. Math. Anal. Appl.* **55**, 102–111 (1976).
2. Transmission problems for the Helmholtz equation. *J. Math. Phys.* **19**, 1433–1437 (1978).

R. Kress and W. Spassov

1. On the condition number of boundary integral operators for the exterior Dirichlet problem for the Helmholtz equation, *Numer. Math.* **42**, 77–95 (1983).

W. D. Kupradse

1. Existence and uniqueness theorems in diffraction theory. *Doklady. Akad. Nauk. USSR* **5**, 1–5 (1934).
2. Integral equations for electromagnetic waves. *Doklady. Akad. Nauk USSR* **4**, 1–5 (1934).
3. *Randwertaufgaben der Schwingungstheorie und Integralgleichungen*. Deutscher Verlag der Wissenschaften, Berlin (1956).

R. Kussmaul

1. Ein numerische Verfahren zur Lösung des Neumannschen Aussenraumproblems für die Helmholtzsche Schwingungsgleichung. *Computing* **4**, 246–273 (1969).

P. D. Lax and R. S. Phillips

1. *Scattering Theory*. Academic Press, New York (1967).

R. Leis

1. Über das Neumannsche Randwertproblem für die Helmholtzsche Schwingungsgleichung. *Arch. Rational Mech. Anal.* **2**, 101–113 (1958).
2. Zur Dirichletschen Randwertaufgabe des Aussenraums der Schwingungsgleichung. *Math. Zeit.* **90**, 205–211 (1965).
3. *Vorlesungen über partielle Differentialgleichungen zweiter Ordnung*. Bibliographisches Institut, Mannheim (1967).

B. Ja. Levin

1. *Distribution of Zeros of Entire Functions*. American Mathematical Society, Providence (1964).

R. C. MacCamy

1. Low frequency acoustic oscillations. *Q. Appl. Math.* **23**, 247–256 (1965).

A. Majda

1. High-frequency asymptotics for the scattering matrix and the inverse problem of acoustic scattering. *Comm. Pure Appl. Math.* **30**, 165–194 (1977).

E. Martensen

1. *Potentialtheorie*. Teubner-Verlag, Stuttgart (1968).

P. A. Martin

1. On the null field equations for the exterior problems of acoustics. *Q. J. Mech. Appl. Math.* **33**, 385–396 (1980).

A. W. Maue

1. Über die Formulierung eines allgemeinen Beugungsproblems durch eine Integralgleichung. *Zeit. Physik* **126**, 601–618 (1949).

J. R. Mautz and R. F. Harrington

1. *H*-field, *E*-field, and combined-field solution for conducting bodies of revolution. *Archiv. f. Elektronik u. Übertragungstechnik* **32**, 159–164 (1978).
2. A combined-source solution for radiating and scattering from a perfectly conduction body. *IEEE Trans. Ant. and Prop.* **AP-27**, 445–454 (1979).

W. L. Meyer, W. A. Bell, B. T. Zinn, and M. P. Stallybrass

1. Boundary integral solutions of three dimensional acoustic radiation problems. *J. of Sound and Vibration* **59**, 245–262 (1978).
2. Prediction of the sound field radiated from axisymmetric surfaces. *J. Acoustic Soc. America* **65**, 631–638 (1979).

S. G. Mikhlin

1. *Mathematical Physics, an Advanced Course*. North-Holland, Amsterdam (1970).

C. Müller

1. Über die Beugung elektromagnetischer Schwingungen an endlichen homogenen Körpern. *Math. Ann.* **123**, 345–378 (1951).

2. Zur Methode der Strahlungskapazität von H. Weyl. *Math. Zeit.* **56**, 80–83 (1952).
3. Randwertprobleme der Theorie elektromagnetischer Schwingungen. *Math. Zeit.* **56**, 261–270 (1952).
4. Radiation patterns and radiation fields. *J. Rat. Mech. Anal.* **4**, 235–246 (1955).
5. *Grundprobleme der mathematischen Theorie elektromagnetischer Schwingungen.* Springer-Verlag, Berlin (1957).

C. Müller and H. Niemeyer

1. Greensche Tensoren und asymptotische Gesetze der elektromagnetischen Hohlraumschwingungen. *Arch. Rational Mech. Anal.* **1**, 305–358 (1961).

M. Z. Nashed

1. Operator-theoretic and computational approaches to ill-posed problems with applications to antenna theory. *IEEE Trans. Ant. and Prop.* **AP-29**, 220–231 (1981).

J. C. Nedelec

1. Curved finite element methods for the solution of singular integral equations on surfaces in $\mathbb{R}^3$. *Comp. Math. Appl. Mech. Engin.* **8**, 61–80 (1976).

Z. Nehari

1. *Conformal Mapping.* McGraw-Hill, New York (1952).

C. Neumann

1. Zur Theorie des logarithmischen und des Newtonschen Potentials. *Berichte über die Verhandlungen der Königlich Sächsichen Gesellschaft der Wissenschaften zu Leipzig* **22**, 45–56, 264–321 (1870).

O. I. Panich

1. On the question of the solvability of the exterior boundary-value problems for the wave equation and Maxwell's equations. *Russian Math. Surveys* **20**, 221–226 (1965).

J. Plemelj

1. *Potentialtheoretische Untersuchungen.* Preisschriften der Fürstlich Jablonowskischen Gesellschaft zu Leipzig, Teubner-Verlag, Leipzig (1911).

C. Pommerenke

1. *Univalent Functions.* Vandenhoeck and Ruprecht, Göttingen (1975).

F. Rellich

1. Über das asymptotische Verhalten der Lösungen von $\Delta u + \lambda u = 0$ in unendlichen Gebieten. *Jber. Deutsch. Math. Verein.* **53**, 57–65 (1943).

F. Riesz

1. Über lineare Funktionalgleichungen. *Acta Math.* **41**, 71–98 (1918).

A. Roger

1. Newton–Kantorovich algorithm applied to an electromagnetic inverse problem. *IEEE Trans. Ant. and Prop.* **AP-29**, 232–238 (1981).

H. L. Royden

1. *Real Analysis.* Macmillan, New York (1963).

C. Ruland

1. Ein Verfahren zur Lösung von $(\Delta + k^2)u = 0$ in Aussengebeiten mit Ecken. *Applicable Analysis* **7**, 69–79 (1978).

J. Schauder

1. Über lineare, vollstetige Funktionaloperationen. *Studia Math.* **2**, 183–196 (1930).

S. Silver

1. *Microwave Antenna Theory and Design.* M.I.T. Radiation Laboratory Series Vol. 12, McGraw-Hill, New York (1949).

B. D. Sleeman

1. The three-dimensional inverse scattering problem for the Helmholtz equation. *Proc. Camb. Phil. Soc.* **73**, 477–488 (1978).
2. Two-dimensional inverse scattering and conformal mappings. *IMA J. Appl. Math.* **27**, 19–31 (1981).

3. The inverse problem of acoustic scattering. University of Delaware, Appl. Math. Inst. Tech. Report 114A (1981).

R. T. Smith

1. A class of inverse scattering problems in acoustics. Ph.D. Thesis, University of Delaware (1982).

A. Sommerfeld

1. Die Greensche Funktion der Schwingungsgleichung. *Jber. Deutsch. Math. Verein.* **21**, 309–353 (1912).

I. Stakgold

1. *Boundary Value Problems of Mathematical Physics, Vol. 2.* Macmillan, New York (1968).
2. *Green's Functions and Boundary Value Problems.* Wiley, New York (1979).

J. A. Stratton and L. J. Chu

1. Diffraction theory of electromagnetic waves. *Phys. Rev.* **56**, 99–107 (1939).

A. N. Tikhonov and V. Y. Arsenin

1. *Solutions of Ill-Posed Problems.* Winston, Washington, D.C. (1977).

F. Ursell

1. On the exterior problems of acoustics. *Proc. Cambridge Philos. Soc.* **74**, 117–125 (1973).
2. On the exterior problems of acoustics II. *Proc. Cambridge Philos. Soc.* **84**, 545–548 (1978).

I. N. Vekua

1. Metaharmonic functions. *Trudy Tbilisskogo matematichesgo Instituta* **12**, 105–174 (1943).
2. *New Methods for Solving Elliptic Equations.* North-Holland, Amsterdam (1967).

P. C. Waterman

1. Matrix formulation of electromagnetic scattering. *Proc. IEEE* **53**, 805–812 (1965).
2. New formulation of acoustic scattering. *J. Acoustic Soc. America* **45**, 1417–1429 (1969).

W. Wendland

1. Die Fredholmsche Alternative für Operatoren, die bezüglich eines bilinearen Funktionals adjungiert sind. *Math. Zeit.* **101**, 61–64 (1967).
2. Bemerkungen über die Fredholmschen Sätze. *Meth. Verf. Math. Phys.* **3**, 141–176 (1970).
3. Integral equation methods for boundary value problems: applications and their solution, to appear.

P. Werner

1. On the exterior boundary value problem of perfect reflection for stationary electromagnetic waves. *J. Math. Anal. Appl.* **7**, 348–396 (1963).
2. On an integral equation in electromagnetic diffraction theory. *J. Math. Anal. Appl.* **14**, 445–462 (1966).
3. On the behaviour of stationary electromagnetic wave fields for small frequencies. *J. Math. Anal. Appl.* **15**, 447–496 (1966).
4. Über das Verhalten elektromagnetischer Felder für kleine Frequenzen in mehrfach zusammenhängenden Gebieten. *J. f. Reine v. Angew. Math.* 1, **278/279**, 365–397 (1975), 2, **280**, 98–121 (1976).

H. Weyl

1. Kapazität von Strahlungsfeldern. *Math. Zeit.* **55**, 187–198 (1952).
2. Die natürlichen Randwertaufgaben im Aussenraum für Strahlungsfelder beliebiger Dimensionen und beliebigen Ranges. *Math. Zeit.* **56**, 105–119 (1952).

C. H. Wilcox

1. A generalization of theorems of Rellich and Atkinson. *Proc. Amer. Math. Soc.* **7**, 271–276 (1956).
2. An expansion theorem for electromagnetic fields. *Comm. Pure Appl. Math.* **9**, 115–134 (1956).

P. Wilde

1. Über das Grenzverhalten von stationären akustischen und elektromagnetischen Feldern in der Umgebung von Eigenfrequenzen. *Diplomarbeit*, Göttingen (1981).

Index